Guilt

Guilt

A Force of Cultural Transformation

Edited by

KATHARINA VON KELLENBACH AND
MATTHIAS BUSCHMEIER

OXFORD
UNIVERSITY PRESS

OXFORD
UNIVERSITY PRESS

Oxford University Press is a department of the University of Oxford. It furthers the University's objective of excellence in research, scholarship, and education by publishing worldwide. Oxford is a registered trade mark of Oxford University Press in the UK and certain other countries.

Published in the United States of America by Oxford University Press
198 Madison Avenue, New York, NY 10016, United States of America.

Library of Congress Cataloging-in-Publication Data
Names: Kellenbach, Katharina von, 1960– editor. | Buschmeier, Matthias, editor.
Title: Guilt : a force of cultural transformation /
[edited by] Katharina von Kellenbach and Matthias Buschmeier.
Description: New York, NY : Oxford University Press, [2022] |
Includes bibliographical references.
Identifiers: LCCN 2021029405 (print) | LCCN 2021029406 (ebook) |
ISBN 9780197557440 (paperback) | ISBN 9780197557433 (hardback) |
ISBN 9780197557471 (online) | ISBN 9780197557457 (UPDF) |
ISBN 9780197557464 (epub)
Subjects: LCSH: Guilt. | Moral motivation. | Social justice.
Classification: LCC BF575.G8 G839 2022 (print) | LCC BF575.G8 (ebook) |
DDC 152.4/4—dc23
LC record available at https://lccn.loc.gov/2021029405
LC ebook record available at https://lccn.loc.gov/2021029406

DOI: 10.1093/oso/9780197557433.001.0001

Contents

Acknowledgments

This book has grown out of the discussions of an interdisciplinary research group that convened for a ten-month residential fellowship in 2018–2019 at the ZiF (Zentrum für interdisziplinäre Forschung, or Center for Interdisciplinary Research), Bielefeld University's Institute for Advanced Study in Germany. This collaborative interdisciplinary research project received generous support and funding from the ZiF and Bielefeld University, as well as the German Research Council (DFG) and the Fritz-Thyssen-Foundation over a three year period.

For the residential portion of the project, the administration and staff of the ZiF created extraordinary working conditions, which made life easy and comfortable for fellows and conference participants alike, who traveled far and wide to the ZiF for short- and long-term stays. We want to thank all of them here without listing individual names, as well as the fellows and invited guests for their engagement in vigorous debate and openness to entertain new ideas and perspectives.

Without our editor, Laura M. Rosell, we would not have been able to assemble this multilingual and interdisciplinary book as quickly and efficiently. The chapters by Tanja Penter, Meinolf Schumacher, and Klaus Günther were translated by Katharina von Kellenbach before being sent through Laura M. Rosell's capable editorial hands. The Bielefeld University student assistants Johanna Leefken, Katharina Krause, and Florian Stühlmeyer helped to prepare the manuscript for print. We want to thank our home institutions, St. Mary's College of Maryland and Bielefeld University, for releasing us from teaching duties during the residential period at the ZiF. Our partners, Gudrun Buschmeier and Ulrich Ettler, have kept us sane and functional, and supported us through the completion of this volume.

Contributors

John Borneman is Professor of Anthropology and Director of the Program in Contemporary European Politics and Society at Princeton. He has worked ethnographically in Germany since 1982 and in Syria since 1999, with publications on, for instance, the influence of political division on the practices of kinship and the changing nature of political authority, questions of fieldwork methods, and the analysis of emotional transference in fieldwork (2011, 2009, 2007, 1997, 1992). His current project focuses on the *Incorporation* of Syrian refugees in Germany.

Matthias Buschmeier is Associate Professor (Akademischer Oberrat) for German and Comparative Literature at Bielefeld University. He is the editor of a special issue on Guilt and Culture of the *Journal for Cultural Poetics (KulturPoetik)*. He has published extensively on German and European Literature as well as on questions of cultural theory. Currently, he is researching "European Literatures of Military Occupations 1938–53." Matthias Buschmeier was co-convenor of the international research group *Felix Culpa? Guilt as Culturally Productive Force* at the Center for Interdisciplinary Research in Bielefeld 2018–2020.

Susan Derwin is Director of UC Santa Barbara's Interdisciplinary Humanities Center and Professor of German and Comparative Literature at the University of California, Santa Barbara. Her teaching and writing have been devoted to the issues of social reintegration confronting military veterans and to publicly-engaged humanities. She is the founding director of the University of California Veterans Summer Writing Workshop and of Foundations in the Humanities, a correspondence program for incarcerated individuals operating in multiple California prisons.

Nelly van Doorn-Harder teaches Religious Studies at Wake Forest University (USA) and at the Center for Islamic Theology (CIT) at the Vrije Universiteit Amsterdam. She is the editor of the journal *Interreligious Studies and Intercultural Theology* (ISIT). Her research concerns Islam in Indonesia and Coptic Christianity in Egypt, focusing on issues of leadership, gender, human rights, and interfaith relations. Currently, her research focuses on Indonesian National Commission for Women's Rights (Komnas Perempuan) and their activities to combat violence against women, including domestic violence and child marriage.

Saskia Fischer is a research associate (wissenschaftliche Mitarbeiterin) in the German Department at Leibnitz University Hannover. She published her dissertation on *Ritual und Ritualität im Drama nach 1945* in 2019. From 2017 to 2019 she was the academic coordinator of the research group *Felix Culpa? Guilt as Culturally Productive Force* at the Center for Interdisciplinary Research, Bielefeld. With Maria-Sibylla Lotter, she is the co-editor of *Guilt, Forgiveness, and Moral Repair* forthcoming with Palgrave Macmillan. Her main research areas are German literature in the 20th century, specifically on the Shoah and camp literature, as well as on literature after 1945, and theatre as ritual.

Parvis Ghassem-Fachandi is Associate Professor in the Department of Anthropology at Rutgers University and received his PhD in Anthropology from Cornell University in 2006. He is the author of *Pogrom in Gujarat: Hindu Nationalism and Anti-Muslim Violence in India* (Princeton University Press 2012), and edited *Violence: Ethnographic Encounters* (New York: Berg Publishers 2009). In 2006–2007 he was post-doctoral fellow at the *Center for Religion and Media* at New York University, United States. He was a fellow at the *Institut d'Études Avancée de Nantes* in France in 2012 and at the *Indian Institute of Advanced Studies* in Shimla, India, in 2016.

Klaus Günther is Professor of Legal Theory, Criminal Law and Criminal Procedure at Goethe-University of Frankfurt, and Prof. h.c. from the Universidad del Rosario (Bogotá/Columbia); He leads the Cluster of Excellence "EXC 243 Formation of Normative Orders" at Goethe University since 2007; serves at the Board of Directors of the Institute of Social Research (IfS) Frankfurt since 2001; He is the author of *Schuld und kommunikative Freiheit. Studien zur individuellen Zurechnung strafbaren Unrechts im demokratischen Rechtsstaat* (Guilt and Communicative Freedom) (2005).

Dominik Hofmann is completing his PhD dissertation at Bielefeld Graduate School in History and Sociology. He studied political and social sciences in Würzburg, Salamanca, and Bielefeld. His main areas of research are sociological theory, especially sociological systems theory, and sociology of law. His current research focuses on the phenomenon of impunity with a special attention to Latin America.

Victor Igreja is Senior Lecturer and teaches Social Justice at the University of Southern Queensland, Australia. His research focuses on war memories, aesthetics of healing, silence, and reconciliation in Mozambique. He studied violence in Australia and Timor Leste and has published his findings in the *British Journal of Sociology, British Journal of Psychiatry, Journal of Traumatic Stress*, Social Science and Medicine, Current Anthropology, *International Journal of Transitional Justice*, Comparative Studies in Society and History, Anthropological Quarterly, *Journal of the Royal Anthropological Institute*, and Encyclopedia of Transitional Justice, and International Encyclopedia of Anthropology.

Katharina von Kellenbach is Professor Emerita of Religious Studies at St. Mary's College of Maryland and project coordinator at the Evangelische Akademie zu Berlin. She co-convened the ZiF research group *Felix Culpa? Guilt as Culturally Productive Force* in Bielefeld 2018–2020. Her areas of expertise include Jewish-Christian relations, Holocaust Studies, feminist theology, and interreligious dialogue. She published *Anti-Judaism in Feminist Religious Writings* (Oxford University Press 1994), *The Mark of Cain: Guilt and Denial in the Lives of Nazi Perpetrators* (Oxford University Press 2013). She is currently working on *Composting Guilt: The Purification of Memory after Atrocity* (Oxford University Press 2022).

Maria-Sibylla Lotter is Professor for Moral Philosophy at Ruhr University of Bochum. She published *Scham, Schuld, Verantwortung* (Frankfurt/M.: Suhrkamp 2012) where she develops a general description of guilt which is applicable to different cultural contexts. It is based on a conceptual dissolution of the western amalgam of guilt and *blame*worthiness. Her research crosses the legal, psychological, theological discourses on guilt and reclaims it as a fundamental category of defining the self in interaction with social contexts. With Saskia Fischer, she is the co-editor of *Guilt, Forgiveness, and Moral Repair* forthcoming with Palgrave Macmillan.

Ethel Matala de Mazza is Professor of German and European Literature at Humboldt University Berlin. She is interested in the intellectual and literary history of the political imaginary, and the interrelations between law and literature, genre theory, and the aesthetics of modern mass culture. She has worked intensively on the importance of narratives for the enactment of the political and made first steps toward a cultural theory to grasp the logic of enduring violent conflicts (2011).

Tanja Penter is Professor for Eastern European History at the University of Heidelberg. She has published extensively on Soviet postwar trials against German war criminals and Soviet collaborators. She participated in a comparative international research project on compensation programs for forced laborers and focused on the practice and effects of compensation programs for Nazi victims in Russia and Belarus. Her second book examines the experiences of violence in the Donbass region under Nazi and Stalinist rule (2010). Currently, she is involved in a trilateral research project on "Civilian Victims at the Eastern Front in WWII" with Russian and Ukrainian researchers.

Meinolf Schumacher is Professor of Medieval German Literature at Bielefeld University. His current research focuses on historical semantics and the literary history of compassion and consolation. He published his book *Sündenschmutz und Herzensreinheit* (Munich: Fink 1996) on impurity and purification as metaphors in Medieval Latin and German literature.

Lisa B. Spanierman is Professor and Head of the Faculty of Counseling and Counseling Psychology in the College of Integrative Sciences and Arts at Arizona State University. She is Fellow of the American Psychological Association (APA) and from 2012 to 2015 served as Vice President for Scientific Affairs for APA's Society of Counseling Psychology. Her research focuses on racial microaggressions and white individuals' racial affect (e.g., white guilt, white empathy, and white fear).

Valerij Zisman is a PhD student in the Philosophy Department of Bielefeld University and the recipient of a scholarship from the *Studienstiftung des deutschen Volkes*. His area of competence is in the field of normative ethics, moral psychology, and political philosophy. He is currently completing his dissertation on the ethics of state punishment.

Introduction

Guilt as a Force of Cultural Transformation

Matthias Buschmeier and Katharina von Kellenbach

Guilt and its adjudication have become a highly visible and political topic in recent times. In international diplomacy, debates over race and gender, the litigation of corporate wrongdoing and calls for accountability, apology (Barkan 2000) and reparations are all proliferating. Wilfried McClay states "that one could not begin to understand the workings of world politics today without taking into account a whole range of morally charged questions of guilt" (McClay 2017, 48). Issues of national identity and memory, transitional justice, post-colonialism, ecology and climate change, and the rise of populism are discussed through the prism of guilt. There is also a growing backlash against this global rise of morality in politics that seemingly compels everyone to live in an "age of apology" (Gibney et al. 2008; Berger 2012, 8), which can be manipulated for political gain and monetary benefit (Torpey 2009). How do we, and how *should* we, deal with this "strange persistence of guilt" (McClay 2017) and its effects on individuals and collectives?

An interdisciplinary research group convened for a ten-month residential fellowship at the Center for Interdisciplinary Research (the Zentrum für Interdisziplinäre Forschung, or ZiF), the Institute for Advanced Study of Bielefeld University in Germany, to discuss the role of guilt in cultural transformation. This book is the result of our deliberations across our diverse disciplines and cultural contexts. The contributions assembled here approach guilt as an enduring phenomenon that must be addressed via interpersonal relations, legal processes, artistic productions, and political discourses, if a society or culture wants guilt to become generative or lead to lasting change. The case studies presented here examine primarily guilt relations in the aftermath of collective violence rather than individual crime or personal wrongdoing. While guilt can be an emotion or an individual affect involving feelings of pain, fear, rage, and shame, as recently argued by psychologists of emotions, we have been more interested in the culturally constructed (Wong and Tsai 2007) and complex *supra*-individual, culturally mediated language of law and ethics, religion and politics, as well as literature and psychology.

Matthias Buschmeier and Katharina von Kellenbach, *Introduction* In: *Guilt*. Edited by: Katharina von Kellenbach and Matthias Buschmeier, Oxford University Press.
DOI: 10.1093/oso/9780197557433.003.0001

Guilt is often understood primarily as an emotion—subjective by nature—and relevant only in the personal and private domain. But psychology, including psychoanalysis, insists on the interrelatedness and social inscription of guilt feelings. Furthermore, empirical research in social psychology has shown that guilt is a prosocial emotion that serves to re-establish injured relationships and balances unequal power arrangements between individuals (Baumeister, Stillwell, and Heatherton 1994; Dekel et al. 2016) and groups (Branscombe and Doosje 1998, 2004; Allpress et al. 2010; Allpress and Brown 2013). While we speak of guilt's "productive" potential, we are aware of the experience of emotional distress and even psycho-emotional pathology. Nevertheless, guilt exceeds the emotional realm of individuals and plays an important role in regulating social relations.

Guilt becomes transformative (the authors of this volume submit) in cultural expressions, artistic representation, legal deliberation, ritual negotiation, historical documentation, and religious supplication. The painful dynamics of guilt, shame, remorse, regret, and contrition require such cultural techniques so that individuals and communities can transform these social-emotional experiences into a productive, rather than destructive, force. The chapters assembled in this book examine the social, spiritual, psychological, and cultural pressures that drive guilt into representation, sometimes after considerable time has passed, and over several generations. Without resorting to metaphysics, it seems false to assume that people will get away with atrocity, even if they succeed in hiding the evidence and suppressing the voices of victims. Moral reminders "haunt" individuals and communities, sometimes over an excruciatingly long period of time, and external observers often exert pressure to confront historical wrongdoing, thereby demanding a reckoning even after generations. Such calls for accountability are increasingly a reality.

Guilt is meditated by socio-cultural structures, such as ideas, values, norms, narratives, symbols, rituals, and aesthetics. Early on, the anthropologist Ruth Benedict, in *The Chrysanthemum and the Sword* (1946), differentiated between guilt and shame as fundamentally distinct cultural reactions to wrongdoing, ascribing the former to the "West" and the latter to Japanese and, by extension, East Asian cultures. This dichotomy is far too simple (for an overview of this discussion, see Creighton 1990). Obviously, responses to wrongdoing, especially on the collective level, are culturally specific. But unlike the core argument of Benedict's book, these differences are not easily mapped out in terms of guilt cultures on the one side and shame cultures on the other. These differences do not lead to an Occident-Orient reification of the two terms. Rather, these differences consist of the preponderant weight that one assigns to an experience (a weight that may be informed by time period and institutions), and of the cultural variability in how shame is—or is not—transformed into guilt. At certain historical moments, cultural practices change as new guilt semantics and practices shift to address and come to terms with particular events. Therefore, this book interweaves specific case studies with broader theoretical reflections on the conditions necessary for transformative guilt.

Guilt is often primarily understood through the "Western" canon of biblical religions, ancient Greek literature, Roman law, and Enlightenment philosophy. Between Freud's theory of the unconscious and Nietzsche's famous rejection of guilt morality as a form of Christian repression in *On the Genealogy of Morality* ([1887] 2006), the ZiF research group debated the cultural and religious limitations of guilt conceptualizations. For instance, in what ways is "guilt" a distinctly "Western" concept with no equivalent in other cultural or religious traditions? Hypothetically, if a religion, such as Buddhism, emphasizes the acceptance of transience and the universality of suffering over memory and righteousness, can we speak of guilt consciousness there (Katchadourian 2010)? Would it be legitimate to interpret Buddhist meditative rituals of purification (Rösch 2010, 2012) or Cambodian practices to commemorate the victims of Khmer Rouge mass violence through the language of guilt (Kidron 2012)? There are also good reasons to be suspicious of the rise of global guilt discourses in the form of human rights and transitional justice campaigns. Not much comparative research currently exists on how guilt is conceptualized and manifested around the world. This book provides preliminary answers to some of these questions and affirms the global need for cultural strategies to represent guilt relations. It is a contribution toward a more nuanced understanding of the world's—and history's—diversity of guilt concepts and the practices that have emerged from different religious, cultural, and local traditions.

The notion of "transformation" is also profoundly ambivalent. It should not be understood as a normative term, but only as a descriptive one. To say that guilt is transformative does not necessarily mean that guilt serves the greater good. Guilt becomes transformative in retribution and retaliation as much as in reconciliation and the repair of relationships. Practices of guilt attribution and denial, feelings of remorse and requests for release, as well as negotiations over repair, redress, and transformation, can all inspire renewed violence just as they can prevent violence. Without minimizing the inherent ambivalence of guilt discourses, the contributors of this volume seek to understand the particular symbolic, legal, ritual, political, and aesthetic representations that create the conditions for prosocial resolutions to antecedent harm.

We approach guilt not as a static condition but as a controversial negotiation that is mediated in the social sphere. Guilt is more than an emotion. It is a construct that plays critical functions in societies, while both its functions and those societies change over time. There is a temporality to guilt that extends laterally (i.e., how much guilt is acceptable within a given timeframe) as well as longitudinally (i.e., how guilt for past atrocities is transmitted across generations). Guilt has both a synchronic and a diachronic dimension.

Synchronically, wrongdoing occurs in the present and affects contemporary life in social communities, although circumstances may prevent naming, describing, or even perceiving certain wrongdoings as crimes. As Victor Igreja's chapter on Timor-Leste shows, even when a truth commission documents atrocities, it may be prudent to avoid guilt discourses in order to avoid laying blame, foster conciliatory relations, and risk stoking resentment.

Meanwhile, diachronically, past wrongdoing has long-term effects that are not apparent in the synchronic dimension. One can also *feel guilty* even though one has done no wrong. The origins of guilt feelings are often obscure, sometimes buried deeply in the unconscious, and at times the result of diffuse intergenerational transmission. People can feel guilty when they have witnessed wrongdoing, no matter that they were incapable of resisting or even of naming the events in question *as* acts of wrongdoing. One can also feel guilty for past wrongdoing by virtue of one's status as the recipient of a legacy with intergenerational effects. For any number of reasons such as these, guilt feelings can arise even though one may not be blameworthy or at fault. Sometimes, for instance, even direct victims of traumatic violence feel guilty; such guilt is known as "survivor guilt," and it can be a symptom of PTSD (post-traumatic stress disorder), which counts guilt feelings as one of its various possible symptoms (Leys 2007). Guilt feelings resulting from trauma require different therapeutic strategies and interventions. The suffering attendant to this type of guilt is real and cannot be dismissed; psychological research has shown that the emotional experience of "actual" guilt for culpable wrongdoing is structurally similar to "existential" guilt (Hoffman 1976), and the psyche applies the same coping mechanisms to avoid either type of experience (Montada et al. 1986). "Feeling bad" for what others have done testifies to the existence of a supra-individual guilt that affects people intergenerationally across time and collectively as members of ethnic and other identity groups. Guilt is therefore not only a direct personal consequence of wrongdoing but also a multilevel, communicative process across time and space. In its historic dimension, it is transmitted through cultural remembrance as well as in the unconscious. In both cases, it shapes people's sense of self and of their very right to exist.

Psychology uses different theoretical models to explain the origins and social functions of guilt feelings. Roy Baumeister has posited that guilt feelings are a "mechanism of human relatedness" and that guilt is, by nature, a "fundamentally social" emotional phenomenon (Baumeister, Stillwell, and Heatherton 1994, 243). As a social emotion, guilt initiates interpersonal practices for coping with conflict and stress. Psychologists and psychiatrists know from clinical experience that people who have not personally participated in wrongdoing can and do feel guilty over the deeds of their communities, past and present, and are often willing to engage in reparative action—sometimes more so than the actual perpetrators.

To conclude: guilt connects communities to the past and generates new social orders in the future.

The Cultural Analysis of Guilt

The diverse understandings of culture that shape this volume depend on our disciplines. Some focus primarily on the history of ideas and textual traditions, while others principally study social practices. Some are mainly concerned with norms and values, while others analyze how certain concepts function concretely within social orders. This diversity of approaches reflects, in turn, the diversity that characterizes the scholarly literature on culture in general, and on guilt in particular. Distinguishing between these approaches is not to say that they do not overlap; indeed, they are presented here as Weberian "ideal types" rather than sharply distinguished dimensions. They are meant as heuristic constructs that help to understand and differentiate the various attempts to circumscribe the complex phenomenon of guilt.

Nominalists focus on the linguistic definition and semantic meaning of certain terms. For nominalists, the context and use of words determines their meaning. The history of ideas and issues of translatability become central to this approach. For instance, it has been argued that guilt was a fundamentally different concept in pagan antiquity than in the Christian world (Gagné 2013). In fact, Eric R. Dodds (1951) extrapolated and applied Benedict's (1946) geographic distinction of Asian and Western cultures to the temporal dimension of ancient Greek shame-honor cultures prior to Plato on the one side and Jewish and Christian guilt cultures on the other. A nominalist might argue that "guilt," as a term, is essentially wrapped up in the biblical—and especially Christian—religious traditions, and that it therefore must remain alien and inappropriate to "non-Western" and non-monotheistic cultures. This is because, by the logic of nominalism, a term carries a cultural- and context-specific meaning that cannot be separated from its original usage, no matter where else in time or space an experience might appear. The authors of this volume have debated the applicability of "Western" guilt discourses to various global conflicts and peace efforts and have asked whether, in the absence of explicit guilt discourse, communities still "feel" something akin to guilt in the wake of violation and wrongdoing.

Normative approaches consider guilt primarily in relation to moral and legal frameworks of individual violation and transgression. Blame and fault are assigned to offenders, who are morally discredited and punished. The law prescribes punishment in proportion to an offense, when violations are grave enough to require the intercession of court systems. Granting a third party (such as the court system) the authority to punish instead of placing that authority in the hands of the victim(s) is supposed to end or prevent retaliatory violence while re-establishing the balance of power in the relationship between victims and perpetrators. Advocates for a narrower normative conception of guilt give the legal system primary responsibility for adequately addressing guilt by punishing perpetrators and vindicating victims' rights. That said, it is mainly individuals, not collectives, who are charged and convicted in national and international courtrooms; whether and under what conditions collectives can be charged or held legally accountable is subject to intense debate in legal and moral philosophy.

Phenomenological approaches—a term we use here not in its strictly philosophical sense, but rather its anthropological one—focus on the observation and description of social and cultural practices. Phenomenologists do not take the absence of certain terms to mean the absence of certain experiences and practices. To the phenomenologist, culture is more than linguistics. All communities and cultures respond to transgression, and the *phenomenon* of guilt attribution and exculpation persists even in the absence of explicit guilt discourses. Obviously, phenomenologists agree with nominalists that cultural differences arising from religion, history, and geography are important and intriguing. But it is the embodied *reality* of various responses to wrongdoing that is given preference by phenomenologists, as opposed to nominalists' concern with textual traditions and the history of ideas.

Functionalist approaches ask how certain cultural concepts, be they defined primarily linguistically or phenomenologically, serve to stabilize social order by mediating the competing demands of interconnected individuals who must, in fact, cooperate (Malinowski 1944). Guilt, in this respect, can be considered a response to threats and stresses in the social fabric. For instance, evolutionary psychologists (e.g., Breggin 2015), describe guilt and shame as integral emotions in the formation and maintenance of social groups. Recent studies have shown that people are more inclined to trust individuals who demonstrate a sense of guilt proneness (Levine et al. 2018). Guilt stabilizes the validity of social expectations. The consciousness and expression of guilt seem to contribute to trust-building in interpersonal relationships. Ultimately, a functionalist might argue that guilt protects individuals against the threat of (temporary or permanent) exclusion from the group; since ostracism was life-threatening for much of human history, the capacity to feel or anticipate guilt served to regulate maladaptive desires for immediate satisfaction. Freud's conceptualization of culture as guilt-driven belongs to the functional, non-moralistic approach.

These categories overlap and appear in different forms in the chapters of this book.

The Language of Guilt

While the nominalist position insists on linguistic and conceptual clarification, guilt remains an essentially "contested concept" (Collier, Hidalgo, and Maciuceanu 2006; Connolly [1974] 2008) and a polysemic term that means wildly different things to different people, depending on their cultural, religious, political, linguistic, and disciplinary usage and contexts. Instead of lamenting the term's notorious haziness, however, the contributions in this book embrace its multiple meanings and accept the absence of a singular definition (Moos and Engert 2016). Accordingly, different terms for guilt and its remedies are used throughout this book, including guilt and guilt feelings, culpable wrongdoing, fault, blame, responsibility, contrition, and accountability.

Nonetheless, it is important to orient the reader around our (broad) operating concept of guilt: guilt refers to acts and omissions that violate a normative framework and are committed with intention, malice, and forethought. In this normative sense, guilt is always individual: someone is charged in courts of law, where evidence is presented and witnesses are heard, as well as circumstances and motivations weighed, before a verdict of guilt or innocence is pronounced. We want to distinguish here between the transgressive *act* and the *process* by which such acts are turned into guilt discourses. As Klaus Günther argues in chapter 5, a transgressive act compels a community to discharge its *terminus ad quem* responsibility and to find and convict the individual offender for his or her *ex post facto* responsibility. Criminal law is a system of legal and social discourses that constructs norms and sends messages about responsibility for their violation. Its procedural rules form a communicative process by which *ex post facto* responsibility situated in the past turns into future-oriented *terminus ad quem* responsibility.

Guilt, in the sense of accountability, responsibility, and liability, is always individual (Lotter 2012, 246–8). Hannah Arendt famously distinguished guilt from responsibility in her critique of Karl Jaspers's essay *The Question of German Guilt* ([1946] 2000). She feared that expanding the term "guilt" beyond the legal sphere and the realm of individual accountability would enable the exculpation of perpetrators. In her essay "Collective Responsibility," Arendt wrote:

> Guilt, unlike responsibility, always singles out; it is strictly personal. It refers to an act, not to intentions or potentialities. It is only in a metaphorical sense that we can say we feel guilty for the sins of our fathers or our people or mankind, in

> short, for deeds we have not done, although the course of events may well make us pay for them. (Arendt [1968] 1987, 43)

Arendt preferred the term "collective responsibility" to preserve a strong concept of individual guilt as grounds for juridical liability. In her essay "Organized Guilt and Universal Responsibility," she insisted that guilt presupposes culpable wrongdoing for it to be prosecuted and sanctioned. Drawing on Hegel's famous "right for punishment for the perpetrator," she argued:

> So long as punishment is the right of the criminal—and this paradigm has for more than two thousand years been the basis of the sense of justice and right of Occidental man—guilt implies the consciousness of guilt, and punishment evidence that the criminal is a responsible person. (Arendt [1945] 2005, 127)

Arendt's concept of guilt relies on a strong notion of the individual as free and self-aware of the consequences of his or her behavior, an idea that is rooted in Aristotelian philosophy and parsed by the Jewish and Christian traditions (Siedentop 2015).

Collective and Individual Guilt

The primacy of the individual in guilt discourses is increasingly being challenged, especially in the aftermath of the Shoah. There is now a rich body of scholarly literature on *collective responsibility* (May and Hoffmann 1991; French and Wettstein 2006; Isaacs 2011). Following the early forays of Karl Jaspers, who wrote in the shadow of the Nuremberg International Military Tribunal (which applied the Nuremberg Principles for the first time in history), guilt is increasingly approached as a supra-individual and collective phenomenon. While national and international criminal courts continue to prosecute mainly individuals, the world is struggling to frame issues of complicity in response to atrocity (Card 2002), mass violence, collective evil (Vetlesen 2005), and state crimes that implicate individuals at various levels of agency. While Jaspers denied the possibility of collective guilt as a matter of legal principle, he proposed three more categories of "moral," "political," and "metaphysical" guilt for thinking through the complexities of complicity (Jaspers [1946] 2000).

Individuals act within chains of command and webs of complicity. Intriguingly, the ancient tribal strategies of conflict resolution that adjudicated the guilt of ancestors, families, and tribes, as well as related collectives, are making a comeback (Schmidt 1999; Anderson 2009; Dietrich 2010; Gagné 2013). The German law professor and novelist Bernhard Schlink speaks of becoming "guilty by solidarity" and of being entangled in "webs of guilt" (Schlink 2010, 16). Individuals who fail to break with the historical wrongdoings of their in-group make moral and political choices in the present; failure to dissociate from perpetrators incurs moral guilt.

Tracy Isaacs calls this failure a "personal moral guilt" (Isaacs 2011, 78). Members of collectives consider themselves somehow accountable even if they cannot be held legally responsible (Young 2011). Isaacs calls this "membership guilt feelings" (Isaacs 2011, 75), which she distinguishes from collective guilt. In her view, the terminology of collective guilt should be limited to collective agents, such as states, corporations, political parties, churches, and any other corporate entity that is organized to pursue particular ends. As individual agents within such corporate systems agree to work toward shared goals and participate with intention, they thereby become implicated in collective guilt. Isaacs defines collective guilt as "blameworthy moral responsibility at the collective level. Collectives with the capacity for intentional action are guilty whenever they are blameworthy for their actions" (71). Individuals incur "collective guilt" when they participate in the pursuit of corporate wrongdoing, but they experience "membership guilt," instead, when they are descendants of historical injustice, inheriting not only profits, privileges, and benefits but also liabilities and blame for the harms done by their (literal or figurative) ancestors.

Inherited "membership" guilt is fiercely fought over and negotiated in the cultural arena of identity politics, where debates over white guilt, colonialism, and racism rage (Bruckner 2018; Steele 2007). Terms such as "white guilt" and "colonial guilt" presuppose the existence of guilt feelings, whether that guilt is ascribed as being of the "membership" or "collective" variety (Branscombe and Doosje 2004). The use of such identity categories is often criticized, as we will show, for falsely blaming entire collectives for past moral failures and for turning guilt into a merely "metaphorical sense" (Arendt [1968] 1987, 43).

The Politics of Guilt

The current heated debate over guilt in the politics of the antiracism, women's, human rights, and ecology movements is, in itself, a testament to the transformative potential of guilt. Across Europe and the United States, a growing chorus of conservative commentators has begun to criticize guilt attributions as a weapon of progressive and identity politics. They denounce the discourse of privilege, which implicates white people, heterosexual men, Europeans, etc. (Bruckner 2018; Steele 2007) in histories of oppression and exploitation as derogatory and defamatory.

As early as the 1970s, the Jewish Iraqi-British historian Elie Kedourie turned against what he called the "Chatham House Version" of history (Kedourie 1970); in his view, a left-liberal historiography creates "the canker of imaginary guilt" (Kedourie 2000, 220), which paralyzes great powers like the British Empire and leads to a shamefaced self-image. According to Kedourie, however, it was not the colonization of the Middle East that led to violence and ethnic cleansing; rather, it was the collapse of colonialism. That is to say, it was not the British colonial masters who were to blame, but rather the decolonizers who had adopted the concept of aggressive nation-states from the "West" in order to emancipate themselves from imperialism. To tell the colonization of the Middle East as the guilt story of the "West" only elides the brutal tendencies of "Third World" nationalists, who are far worse than the colonial powers they replaced. (So went Kedourie's view.)

At the end of the 1990s, the conservative African American literary scholar Shelby Steele transformed this idea. Steele identifies guilt as the mechanism by which the left-liberal establishment constantly seeks to prove its moral superiority over all other groups. Today Steele regards the Obama presidency, for example, as the expression and culmination of *White Guilt* (Steele 2007), meaning that white Americans elected him only to defend themselves against decades of liberal elites' accusations of being racist, which they had internalized irrationally. Affirmative action, racial justice, and environmental protection policies are all products of white guilt (Steele 2008). By contrast, former U.S. president Donald Trump embodied a politics that claimed to be free of liberal guilt manipulations, despite the fact that Trump himself made frequent guilt accusations. Steele spoke on behalf of those Americans who were uneasy with what they considered a blame culture of political correctness and cancel culture, rooted in arbitrary notions of historical obligation.

In Britain, the Oxford theologian Nigel Biggar defended the ethics of colonialism in an op-ed in *The Times* titled "Don't Feel Guilty about Our Colonial History" (November 30, 2017). Coming to the defense of the American political scientist Bruce Gilley (2017), whose controversial essay "The Case for Colonialism" in the *Third World Quarterly* resulted in the resignation of the entire editorial board, Biggar argued that "Western" colonialism was mainly beneficial and its use of power legitimate. Biggar not only denied the legitimacy of arguments advanced by post-colonial scholars and movements but further maintained that any admission of "guilt will make us vulnerable to willful manipulation" (Biggar 2017) and undermine national pride. Only denial of guilt makes it possible to act freely and confidently in the world.

In France, the philosopher Pascal Bruckner considers the entire "West" arrested in a "guilt complex," as his book *La tyrannie de la penitence: Essai sur le masochisme occidental* (2006) was translated into German under the title of

Schuldkomplex (2008). Such a masochistic "tyranny of guilt" undermines the "West's" ability to conduct interventionist policies to protect ethnic minorities and nonwhite peoples against corrupt and autocratic leaders, because European and white American leaders are paralyzed by the fear of repeating colonial guilt. In his latest book, *An Imaginary Racism: Islamophobia and Guilt* (2018), Bruckner extends this argument to the "Western" attitude toward Islam. Starting with Edward Said's *Orientalism* (1978), he argues that the "West" falsely accepted charges of guilt for its perspective and interactions with Islam. Today, he fears that this self-criticism has morphed into an unwillingness to criticize Islam intellectually (unlike Christianity, Hinduism, Buddhism, and Judaism) out of fear of being labeled Islamophobic or racist.

In France, but also in Germany, conflicts with Islam are regularly explained as a feature of guilt processes. In his essay, whose title translates to "Take My Guilt Upon You: On the Political Rift between the Orient and the Occident" (our translation), which appeared in the German weekly *DIE ZEIT* (no. 4, 2003), the Deputy chief of mission at the Israeli embassy in Berlin, Mordechay Lewy, argues that Islam is rooted in shame cultures that lack concepts of original sin, individual freedom, personal responsibility, and therefore historical collective guilt. Since the religions of the Occident—Judaism and Christianity—practice confession and atonement for guilt, its adherents are steeped in self-criticism and self-reflexivity. For Lewy, the guilt culture of the "West" is, on the one hand, morally superior, and yet simultaneously disadvantaged and obstructed in a "clash of civilizations."

As critical as these authors are, they acknowledge guilt as a transformative political force, albeit in the form of pathological and paralyzing feelings. Interestingly, they transmute guilt from the concrete realm of history into the semantic field of unsubstantiated, illusory, and imaginary guilt feelings. Bruckner applies basic Freudian concepts of guilt-and-shame relations that result in masochism (cf. the descriptive summary by Piers 1953). The "trick" here is to slip the reality of legal and political guilt into the uneasiness of burdensome emotions in need of treatment. Therapy, rather than justice, becomes the remedy.

Kedourie, Steele, Gilley, Biggar, and Bruckner are established scholars in their fields. But they provide ammunition for white supremacists, whose nationalist-chauvinist agenda spilled into public view with the ascendancy of President Trump (Hage 2012). Populist books, such as *The Myth of White Guilt* by Reed Benson (2018), an American professor of religion, make the denial of guilt central to this agenda. Systematically, Benson rejects guilt in chapters on slavery, the genocide of native peoples, and the Holocaust, concluding with a chapter titled "Eco-Guilt." Benson's guilt denial rests on three strategies: first, he claims that today's ethical and political standards cannot be applied in judging history (e.g., the slave trade); second, he repudiates "cherry picking history" for its selective emphasis on white crimes; and third, he charges those who stand to gain from such critical historiography with outward deception and falsification:

> This double standard is maintained in the mind of White people through guilt. They have been told so often how oppressive Whites are, with such vehement indignation and carefully selected stories, while others that show a broader perspective are ignored. These "cherry picked" narratives included White lynchings of Blacks, White atrocities against American Indians, White abuse of Chinese immigrants, White disdain of Mexicans, and so forth. (Benson 2018, 11)

The denial of guilt serves to establish self-righteous innocence and victim status. As victims of a culture war, waged by Black and brown or Muslim people, along with their advocates among the "cultural elites," white America and Europeans see themselves humiliated and disempowered on moral grounds. The "new right" claims victimization and the prerogative to defend itself against expectations of political change, as Benson indicates in his subtitle: *Defending our History and Culture*.

Guilt denial resists calls for cultural and political change. The critique of guilt as a term in political thought tends to reject the analysis of guilt relationships, and aims to reinterpret the past in order to avoid any indebtedness for the present. Such a rejection of responsibility dismisses obligations to vindicate and empower victims (and their descendants) and releases perpetrators from accountability and reparation. What is at stake in these conflicts over the narration of history are the power imbalances that have accrued over time. As guilt percolates through the search for resolution to historical legacies of injustice, its temporality and productivity become manifest.

But even on the political Left, as Lisa Spanierman reveals in her chapter in this volume, guilt discourse is suspected to be potentially harmful because it indicts groups for historic and collective crimes they have not personally committed. Accusations of guilt are loaded speech acts that serve to empower victim groups and aim to shift power imbalances. Inculpating someone may mean grasping for the moral high ground, and it may entail the abdication of personal and political responsibility. A victim is innocent, passive, powerless. Victimhood releases one from burdensome responsibilities, which are shifted and attributed to allegedly powerful perpetrators; accordingly, victim groups, while disadvantaged, disempowered, and degraded, can still try to claim moral superiority. The invocation and negotiation of guilt claims aims to change the attitudes and behavior of perpetrator(s)—which triggers both old and new resentments.

The Temporality of Guilt

How much guilt can be accepted and digested by a particular community at any given time, and how much is transmitted across generations, is context-specific. Guilt unfolds as a temporal process (Goldie 2012, 62), with particular timelines that fluctuate as communities' normative expectations shift in the wake of transgressive events. This "process nature" of guilt is apparent in the concept's language: guilt must be experienced, accused, confessed, defended against, judged, punished, regretted, repented, atoned for, apologized for, pardoned, remitted, and forgotten. All of these verbs indicate that guilt builds upon important cultural and social techniques (evolving from communities' need to establish a social order) that both respond to and reduce uncertainty and disorder, as Dominik Hofmann explains in reference to Niklas Luhmann's theory in chapter 7.

Historically, this becomes evident in the fact that certain offenses are not considered violations at the time of their commission. Slavery was considered legally permissible and morally acceptable. The recognition that the capture, trafficking, sale, and perpetual enslavement of human beings violates human rights and constitutes a crime against humanity evolved over time and as a result of profound social, political, legal, moral, and sometimes even literal military battles. Hence, there are social and temporal dimensions to the perception of guilt.

As noted above, synchronically, communities experience and commit atrocities without fully perceiving or naming certain forms of suffering as crimes. But there is also a diachronic dimension by which past guilt is projected into the future, becomes symptomatic in unpredictable ways, and thus connects different times and generations in experiencing such different manifestations of guilt. In guilt claims, communities negotiate normative expectations while searching for language and symbolic representations that ground relationships of respect and redress. Social orders are built and rebuilt in dynamic temporal processes of negotiation such as these. While slavery, for instance, was officially abolished in the decades between 1846 (Denmark), 1865 (United States), and 1888 (Brazil), and finally by the League of Nations in 1926, the ability of formerly enslaved peoples to level charges of guilt or to demand reparations has remained severely restricted to this day. Atrocities on the scale of the Atlantic slave trade are generally denied by the beneficiaries and victors, whose version of history is upheld by the systemic suppression of victims' voices. History is written by the victors, as Walter Benjamin rightly noted, and shaped by their perspectives. In guilt discourses, survivors and their descendants take the power to remember and to write new historical accounts. Such narratives emerge, sometimes decades after the events, accompanied by protest and civil unrest. Charges of guilt are loaded speech acts that command a response. They might trigger resentment, violence, and attempts at suppression, but also creative initiatives that seek to articulate certain truths in literary and artistic productions, rituals of reconciliation, and the development of memorial projects. Yet even where violence results, guilt's highly effective potential for repairing social relations is no less manifest.

Ultimately, the experience and feeling of guilt prompts articulation, even in totalitarian situations. Suppressed histories do not vanish. Memories and recorded testimonies, archived and hidden documents, and fictional accounts all circulate well beyond any system of control and eventually surface to challenge the official, carefully fabricated doctrine of history. The Armenian Genocide is one prominent example (Akçam 2012). Even a century after the events, the virulence of guilt and its passionate denial continue unabated. Guilt on this scale cannot be forgotten, contained, or repressed, and those who argue the benefits of collective forgetting (Meier 2010) neglect the physiological and psychological reality of trauma. While collective amnesia may be in the interest of immediate political pacification and conflict resolution, it does not serve deeper claims to justice or peaceable relations. It takes intense cultural effort to articulate and transform the guilt of the past before relations can be re-established on the basis of shared memory. Such a process also takes time and creative energy—and once the idea of a shared history is challenged by revisionism, new spaces of potential conflict open up.

Guilt's temporal dimension gives it an enduring presence that entails more than the commemoration of history, which has been the focus of literary scholars (Assmann and Conrad 2010; Erll 2008). Guilt binds generations together across time and establishes identities within landscapes of memory that give rise to rights and responsibilities, as well as debts and obligations. In the case of atrocities, special *acts of symbolic expression* are required in order to find the language capable of translating the unbearable and unspeakable into guilt stories. Guilt so grave is impossible to fully address in the courtroom. At the same time, guilt's symbolic transmutation into legal proceedings, monetary restitution, rituals of apology, and theater performances make it bearable and forgivable. In the words of the cultural theorist Jan Assmann, "Any guilt culture creates forms for dealing with the lasting and accumulating character of guilt. [. . .] Such forms of nondetermined guilt are a problem for which guilt cultures need to cultivate solutions" (Assmann 1999, 107; our translation). Money, law, rites and rituals (including both profane and religious), dramatic performances, fictional narratives, and movies are only *some* of the "forms" that provide a means by which guilt can be addressed (Buschmeier 2018). Their shape and outlook differ enormously across space and time.

The productivity of guilt is not inherently positive or necessarily prosocial. Indeed, even in economics, productivity can be "good" or "bad," "high" or "low." In general, the productivity of guilt should be approached with a value-neutral perspective, through the lens of change. "Productive" refers to processes that generate new (but not unequivocally better) outcomes and characterizes the force of guilt as *generative* and/or *transformative*. A *generative* force produces something new from various components that were available but unused beforehand. A *transformative* force reshapes the outlook of something that was already there. Evaluating the outcomes that result as positive or negative is a secondary step. Something new is not necessarily better than the old. Transformation can also lead to total destruction. Each "productive" process entails very individual circumstances. All of this is true of guilt, in that, depending on the context and the circumstances, it can be positive or negative, constructive or destructive.

This book examines four areas of transformative guilt: first, the domain of interpersonal social relations; second, the sphere of the law and the juridical management of guilt; third, the field of the arts, which creates words, performances, spaces, and images for the unspeakable and the unsayable; and fourth, the public realm of politics, in which guilt is articulated in ritual apologies and political negotiations.

Guilt as a Prosocial Force in Interpersonal Relations

The first section of this book examines the different conditions under which guilt serves prosocial ends. The authors approach this task from the disciplines of medieval studies, social psychology, religious studies, and anthropology. They examine very different contexts: the social benefits of penance in Christian Europe, the role of guilt feelings in the resistance to police brutality and racial injustice in United States, the feminist appeal to guilt to protect girls from early marriage in Indonesia, and the layered interactions of guilt feelings surrounding the welcome and integration of Syrian refugees in Germany. What these essays have in common is the conviction that guilt serves prosocial ends, whether in the form of "good works," which lubricated welfare systems in the Middle Ages without taxation; in the form of white sensitivity and empathy with victims of racism in the United States; or in the establishment of Germany's modern-day "welcome culture."

The medievalist Meinolf Schumacher retraces the European Christian debate over the sacrament of penance during the Reformation, which wrought enormous social and political changes. Roman Catholic penitential teachings obligated people to perform external acts of charity that provided social services to others. The social welfare system, developed over centuries, was grounded in the penitential demands of *satisfactio operis* as a precondition for reconciliation. Once indulgences, questioned by the Reformation, were abolished in Protestant territories, this system collapsed. Secular taxation followed, as well as the rise of a "bad conscience" without possibility for relief, which created the heavy burden of guilt in modernity.

In the case of the United States, writes Lisa Spanierman, white Americans who accept responsibility for the historical legacy of enslavement and white privilege are more willing to support affirmative action and to engage in solidarity with Black and brown Americans. Social psychology has conducted empirical studies and developed tests to confirm the prosocial functions of guilt in the creation and maintenance of relationships, despite guilt's sometimes paralyzing and deleterious effects.

In Indonesia, feminist Muslim activists use guilt and shame strategies to challenge and reform the widespread practice of child marriage. The battle to criminalize child marriage in the law, to change cultural practices, and to reframe the interpretation of the Qur'an and the Hadith involves negotiations with powerful men, especially the male guardians, who are responsible for the welfare of young girls and can therefore be held accountable for high divorce rates and the lifelong impoverishment of women and their children. Guardians can be shamed for their failure to protect women and children. Such guilt strategies aim to change the discourse about women's rights and misogynist practices, among others, by appealing to the need for education, which is in the long-term national interest.

The anthropologist John Borneman analyzes the confluence and interaction between Syrian refugees' survivor guilt and Germany's historical guilt in the *Willkommenskultur* following the arrival of one million Syrian refugees in Germany in 2015. Often unconscious, this negotiation of guilt has both enhanced and obstructed the complex processes of social incorporation of refugees into German society. The author argues that bringing guilt from the unconscious into conscious articulation turns guilt into a productive tool to enhance relationships between newcomers and German host communities.

While guilt, in each of these chapters, is highly ambivalent, the authors locate its beneficial prosocial functions within the obligation to care for people—and for values—that deserve protection and repair.

Transforming Guilt into (Restorative) Justice

The second section examines guilt's symbolic transmutation into the language of law and legal proceedings. The law is the primary institution that is supposed to deliver justice by declaring guilt and innocence; in law, the destructive dynamics of violence and violation are transformed into orderly court procedures and amenable penalties that mollify and pacify. But the law cannot (and must not) consider *all* circumstances and consequences of guilt. Jan Assmann observes:

> If guilt is determined, then punishment leads to expiation. That is what social institutions of jurisdiction and enforcement are there for. However, this only applies to a small part of the guilt to be accrued in a guilt culture. The more complex and confusing a society becomes, the more opportunities there are to transgress and violate the norms on which respect it is founded. (Assmann 1999, 107–8; our translation)

Speaking from the disciplines of law, philosophy, sociology, and history, these essays reflect on law as a creative construct that affirms social order in the wake of disruption and destruction. Guilt finds expression, which is an ongoing process.

Klaus Günther argues in his chapter that the productivity of criminal law becomes obvious when one considers not only its individual attribution of guilt, but also the communicative processes that precede individual verdicts, and addresses the community as a whole. He takes up Nikolas Rose's concept of "responsibilization." Both Günther and Rose rely on Foucault's analysis of the post-welfare state and the emerging neoliberal control society to argue against punishment and harsh treatment as inevitable and natural outcome of criminal convictions (Rose 2000). Misconduct and the violation of norms implicate the community in *terminus ad quem* responsibility, which is discharged in convicting an individual by attributing *ex post facto* responsibility for an offense. Guilt must be circumscribed and delimited in communicative processes that involve not only perpetrator and victim but, in fact, the entire society. But the productive potential of guilt is obstructed by harsh treatment and punishment that follows conviction, because it incentivizes perpetrators to deny responsibility. Punishment renders convicts passive rather activating and recruiting their willingness to take responsibility in acts of repair and reparation that benefit victims and society.

The philosopher Valerij Zisman also questions the ethics of punishment and defends restorative justice as a viable alternative to the infliction of harm on perpetrators in the context of criminal law. He argues that acknowledgment of one's guilt and feelings of remorse and/or empathy serve a productive function in restorative justice processes because they help repair relationships between victims and perpetrators, as well as with the community at large, and they compel changes to offenders' behavior. Criminal law should embrace restorative justice in order to avoid morally problematic practices of inflicting harsh treatment on convicts; instead, the criminal justice process should endorse prosocial and relationship-enhancing strategies for addressing guilt.

Outright impunity, on the other hand, argues sociologist Dominik Hofmann, fails to address and transform guilt. The term "impunity" can mean a wide variety of phenomena, from arbitrary presidential pardons to broadly written amnesty laws, and from corruption and the rampant abuse of power to selective prosecutions. In all of these instances, Hofmann posits, the law is experienced as an "absent presence," unable or unwilling to transmute guilt into sentences and remedies. Such failures on the part of the law turn guilt into a nebulous and amorphous presence that percolates through and pollutes communities. Hofmann locates its force not in metaphysics but in the world of "discourse and shared expectations" about justice and equality. Left unaddressed, ultimately, guilt turns destructive. Its containment by legal institutions is required to transform it into something generative, and the resulting benefits restore order and vindicate victims.

The historian Tanja Penter compares the criminal trials of German and Soviet participants in the Nazis' mass murder of patients in Ukraine and asks which roles these criminal trials (and the guilt therein negotiated) played in the transformative process leading from war to peace. There were stark differences in the investigation, trial processes, and punishment of German perpetrators and Soviet collaborators under their respective post-war court systems. Penter describes these differences and asks whether they ultimately resulted in different attitudes toward the memory of National Socialism: while Soviet authorities used the court as a stage for cracking down on a perceived lack of loyalty to Soviet rule in formerly German-occupied territories, the West German courts tried to minimize disruption and avoided pronouncing most individuals guilty.

Guilt as Creative Irritation

Especially in the wake of atrocities, special *acts of symbolic expression* are required to translate the unbearable and unspeakable into guilt stories that can be incorporated into collective identity. The articulation of guilt in creative works enables communities to confront the implications of their histories and their current social and political lives. Hence, film, theater, ritual, and literature become critically important vehicles for audiences to absorb and digest their complicity in collective evils. Film and theater, respectively, became spaces for cultural negotiation and social creativity in Joshua Oppenheimer's docudrama *The Act of Killing* and Peter Weiss's play *The Investigation*, both of which confronted audiences with their guilt and complicity decades after the events portrayed. Oppenheimer restaged the killings of Indonesian communists and turned contrition into performance, while Peter Weiss brought a rendition of the first Frankfurt Auschwitz Trial to the stage in the 1960s in West Germany. Works such as these create narratives of guilt and landscapes of memories, which bind generations together and transmit histories of guilt and trauma alike. In this way, artists, writers, and other creatives can help to drive guilt's transformation, by finding aesthetic and narrative expression for subjects that are often culturally silenced and rendered taboo.

The religious studies scholar Katharina von Kellenbach explores, through Joshua Oppenheimer's docudrama *The Act of Killing* on the anti-communist violence in Indonesia, the inscrutability of contrition in the process of repentance. The film occasioned heated debates over the depth and veracity of the repentance of main protagonist Anwar Congo's contrition. Kellenbach argues that contrition should not be reduced to internal matters of the heart, but rather should be understood as performative spectacle and symbolic speech act, both of which impact political conditions on the ground. While the depth and veracity of one's contrition can never be proven, its external performance creates spaces for individual and political reckoning with wrongdoing.

The literary scholar Saskia Fischer describes Peter Weiss's play *The Investigation* and its impact on post-war Germany's cultural, intellectual, and political landscape. Weiss was keenly aware that Aristotle's idea of theater as catharsis—i.e., theater that was simply meant to purify and transport its audience—could no longer be the purpose and function of theater after Auschwitz, as guilt on the level of Auschwitz cannot be cleansed, forgiven, or truly resolved. What, then, was the purpose of theater in a nation that had embraced humanist education and Enlightenment values while unleashing total war on Jews? In the face of Auschwitz and the complete moral breakdown of civilization, writers and stage directors no longer found it possible to trust the classic aesthetic forms and functions of drama. Fischer asks whether staged dramas and other artistic performances can ultimately contribute to a society's critical recognition of its own historical guilt.

The anthropologist Parvis Ghassem-Fachandi examines a Hindu farmer who becomes the spirit medium of a Muslim saint and builds a Muslim shrine in a village that wrestles with the implications of anti-Muslim pogroms in Gujarat, India. While never articulated by the villagers, this ritual functions as connective tissue between communities in the absence of guilt claims. The lack of conversation about responsibility and guilt that Ghassem-Fachandi encounters more broadly in his fieldwork in Gujarat raises important questions about the repetitive cycles of violence that have plagued the region.

The literary scholar Susan Derwin examines the concept of "moral injury" through the literature of the United States' returning veterans. Plagued by depression and high suicide rates, veterans struggle to come to terms with guilt feelings that take root in the fissures between warrior culture and civil culture.

In the absence of the country's willingness to take responsibility for its multiple wars, warriors must come to terms—largely alone—with their actions that have transgressed civilian norms. Unless a community acknowledges its own political accountability for bloodshed in war, veterans are forced to bear sole responsibility for violence. For veterans, finding language to express their experiences breaks their isolation and forces their community to face the realities of contemporary warfare. Guilt thus becomes generative of change once it can be shared in language and symbolic expression, developed in literary projects. It allows veterans to rejoin civil society and forces their community to confront the violence and guilt committed in its name.

The Politics of Guilt Negotiations

Guilt confessions communicate a recognition of victim groups' human worth and dignity after these individuals were rendered mere objects by a transgression. When perpetrators acknowledge guilt, they affirm a shared system of mores, acknowledge the harm that was done to the victims, express their regret about the injurious actions, pledge to behave in nontransgressive ways in the future, and declare their desire to return to (or establish) a harmonious relationship. This is why reconciliation based on forgetting, while seemingly facilitating a fresh start, like Germany's *Stunde Null* (Hour Zero) in 1945, cannot suffice to rebuild relations between victim and perpetrator communities.

In this last section, a philosopher, a social justice scholar, and two literary scholars reflect on public statements and political apologies as an enduring quality of guilt in national and international politics. It is in the realm of politics that guilt is proclaimed and contested, negotiated and performed.

The philosopher Maria-Sibylla Lotter discusses the function of apologies for historical wrongdoing, whether they are volunteered by representatives of perpetrator communities or demanded by victim groups. Are such apologies productive or, instead, manipulative with respect to changing power arrangements? While apologies can repair moral order by signaling a shared understanding of values, they can also become ritualized in political conflicts. When guilt claims are expressed at a remove from blameworthy wrongdoing, Lotter argues, they sometimes in fact *discourage* the sort of engagement in structural change that might otherwise remedy social injustices. In such cases, apologies are not effective tools of redress; instead, claims of guilt-feeling can become attempts at disempowerment and degradation.

The social justice scholar Victor Igreja critiques the establishment of global paradigms for transitional justice, which he calls "reproductive guilt" in a nod to Paul Ricoeur. Instead, Igreja points to the multiple layers of political violence that the former Portuguese colony of Timor-Leste endured as an unintended consequence of the decolonization process that began in 1974 and was attended by waves of bloodshed, both internal and external, in the form of an Indonesian military invasion. Guilt and innocence, Igreja argues, cease to be clear categories, and this ambiguity militates against truth and reconciliation commissions that aim to distinguish the guilty from the innocent. Instead, he proposes that we view transformative guilt in the very endurance of complex guilt relations, which must be endured as an "unresolved" and ambiguous lack of criminal accountability. Further, he questions principles of transitional justice that aim to channel guilt into the legal domain of national and international courts, because these approaches dismiss the local complexity of power and agency. Guilt finds a variety of outlets in social and symbolic spaces beyond the courtroom.

The scholar of German literature Ethel Matala de Mazza explores the aftermath of World War I, when the Versailles Treaty declared Germany guilty of starting the war and imposed enormous reparation debts. This colossal defeat became mythologized in the German imagination as the stab-in-the-back legend and created an aggressive political climate in the Weimar Republic. When the politician Kurt Eisner declared his readiness to issue a public confession of guilt to re-establish international relations, he was assassinated for it. The history of political apology has generally ignored this precedent. The chapter considers the idea of the political transformativity of guilt in the 1920s. In response, many intellectuals desperately tried to shift the question of guilt from one of political responsibility to, rather, a structural analysis of the German defeat, which created new cultural typologies (old vs. new economies) that identified areas for cultural change.

The literary scholar Matthias Buschmeier analyzes speeches delivered by Germany's federal presidents on commemorative occasions. The chapter argues that by emphasizing shame rather than guilt in the country's response to the Shoah, German officials, to the present day, have intentionally misrepresented the early guilt campaigns of the Allied Forces in 1945. Fearing legal prosecution and financial penalties, the German populace understood collective guilt as a legal concept, and thus fiercely rejected it. Meanwhile, German officials expressed shame, thereby acknowledging some responsibility. Shame, however, shifts the burden back to victim groups, who are then expected to *lift* the burden of shame from their erstwhile oppressors as a goodwill gesture in facilitating reconciliation. In their commemorative speeches, in this way, West German presidents missed the chance to foster a productive understanding of guilt that would depend on open acknowledgment and the courage to face juridical and financial consequences. Their emphasis instead on feelings of shame instigated aggressive attitudes toward victim groups, who have come to be viewed as the cause and the wellspring of the nation's enduring experience of shame.

References

Akçam, Taner. 2012. *The Young Turks' Crime against Humanity: The Armenian Genocide and Ethnic Cleansing in the Ottoman Empire*. Princeton, NJ: Princeton University Press.

Allpress, Jesse A., Fiona Kate Barlow, Rupert Brown, and Winnifred R. Louis. 2010. "Atoning for Colonial Injustices: Group-Based Shame and Guilt Motivate Support for Reparation." *International Journal of Conflict and Violence* 4, no. 1: 75–88.

Allpress, Jesse A., and Rupert Brown. 2013. "Nie Wieder: Group-Based Emotions for In-Group Wrongdoing Affect Attitudes toward Unrelated Minorities." *Political Psychology* 34, no. 3: 387–407.

Anderson, Gary A. 2009. *Sin: A History*. New Haven, CT: Yale University Press.

Arendt, Hannah. (1945) 2005. "Organized Guilt and Universal Responsibility." In *Essays in Understanding, 1930–1954: Formation, Exile, and Totalitarianism*, edited by Jerome Kohn, 121–31. New York: Harcourt.

Arendt, Hannah. (1968) 1987. "Collective Responsibility." In *Amor Mundi: Explorations in the Faith and Thought of Hannah Arendt*, edited by James W. Bernauer, 43–50. Dordrecht: Springer.

Assmann, Aleida, and Sebastian Conrad, eds. 2010. *Memory in a Global Age: Discourses, Practices and Trajectories*. Basingstoke, UK: Palgrave Macmillan.

Assmann, Jan. 1999. "Das Herz auf der Waage: Schuld und Sünde im Alten Ägypten." In *Schuld*, edited by Tilo Schabert and Detlev Clemens, 99–148. Munich: Wilhelm Fink.

Barkan, Elazar. 2000. *The Guilt of Nations: Restitution and Negotiating Historical Injustices*. New York: W. W. Norton.

Baumeister, Roy F., Arlene Stillwell, and Todd F. Heatherton. 1994. "Guilt: An Interpersonal Approach." *Psychological Bulletin* 115, no. 2: 243–67.

Benedict, Ruth. 1946. *The Chrysanthemum and the Sword: Patterns of Japanese Culture*. New York: Houghton Mifflin Harcourt.

Benson, Reed Amussen. 2018. *The Myth of White Guilt: Defending Our History and Culture from Liberalists, Socialists, and other Neo-Marxist Liars*. Schell City, MO: Watchman Outreach Ministries.

Berger, Thomas U. 2012. *War, Guilt, and World Politics after World War II*. Cambridge: Cambridge University Press.

Biggar,Nigel2017."Don'tFeelGuiltyaboutOurColonialHistory."*Times*(London),November 30, 2017. https://www.thetimes.co.uk/article/don-t-feel-guilty-about-our-colonial-history-ghvstdhmj.

Branscombe, Nyla R., and Bertjan Doosje. 1998. "Guilty by Association: When One's Group Has a Negative History." *Journal of Personality and Social Psychology* 75, no. 4: 872–86.

Branscombe, Nyla R., and Bertjan Doosje, eds. 2004. *Collective Guilt: International Perspectives*. Cambridge: Cambridge University Press.

Breggin, Peter R. 2015. "The Biological Evolution of Guilt, Shame and Anxiety: A New Theory of Negative Legacy Emotions." *Medical Hypotheses* 85, no. 1: 17–24.

Bruckner, Pascal. 2006. *La tyrannie de la penitence: Essai sur le masochisme occidental*. Paris: Librairie générale française.

Bruckner, Pascal. 2008. *Der Schuldkomplex: Vom Nutzen und Nachteil der Geschichte für Europa*. Translated by Michael Bayer. Munich: Pantheon.

Bruckner, Pascal. 2018. *An Imaginary Racism: Islamophobia and Guilt*. Translated by Steven Rendall and Lisa Neal. Cambridge: Polity Press.

Buschmeier, Matthias. 2018. "Felix Culpa?—Zur kulturellen Produktivkraft der Schuld." *Communio* 47: 38–50.

Card, Claudia. 2002. *The Atrocity Paradigm*. New York: Oxford University Press.

Collier, David, Fernando Daniel Hidalgo, and Andra Olivia Maciuceanu. 2006. "Essentially Contested Concepts: Debates and Applications." *Journal of Political Ideologies* 11, no. 3: 211–46.

Connolly, William E. (1974) 2008: "Essentially Contested Concepts." In *Democracy, Pluralism and Political Theory*, edited by Samuel Allen Chambers and Terrell Carver, 257–79. London: Routledge.

Creighton, Millie R. 1990. "Revisting Shame and Guilt Cultures: A Forty-Year Pilgrimage." *Ethos* 18, no. 3: 279–307.

Dekel, Sharon, Daria Mamon, Zahava Salomon, Olivia Lanman, and Gabriella Dishy. 2016. "Can Guilt Lead to Psychological Growth Following Trauma Exposure?" *Psychiatry Research* 236: 196–98.

Dietrich, Jan. 2010. *Kollektive Schuld und Haftung: Religions- und rechtsgeschichtliche Studien zum Sündenkuhritus des Deuteronomiums und zu verwandten Texten*. Tübingen: Mohr Siebeck.

Dodds, Eric R. 1951. *The Greeks and the Irrational*. Berkeley: University of California Press.

Erll, Astrid, ed. 2008. *Cultural Memory Studies: An International and Interdisciplinary Handbook*. Berlin: De Gruyter.

French, Peter A., and Howard K. Wettstein, eds. 2006. *Shared Intentions and Collective Responsibility*. Boston: Blackwell.

Gagné, Renaud. 2013. *Ancestral Fault in Ancient Greece*. Cambridge: Cambridge University Press.

Gibney, Mark, Rhoda E. Howard-Hassmann, Jean-Marc Coicaud, and Niklaus Steiner, eds. 2008. *The Age of Apology: Facing Up to the Past*. Philadelphia: University of Pennsylvania Press.

Gilley, Bruce. 2017. "The Case for Colonialism." *Third World Quarterly* 95: 1–17.

Goldie, Peter. 2012. *The Mess Inside: Narrative, Emotion, and the Mind*. Oxford: Oxford University Press.

Hage, Ghassan. 2012. *White Nation: Fantasies of White Supremacy in a Multicultural Society*. Hoboken, NJ: Taylor and Francis.

Hoffman, Martin. L. 1976. "Empathy, Role-Taking, Guilt, and Development of Altruistic Motives." In *Moral Development and Behaviour*, edited by Thomas Lickona, 124–43. New York: Holt, Rinehart and Winston.

Isaacs, Tracy. 2011. *Moral Responsibility in Collective Contexts*. Oxford: Oxford University Press.

Jaspers, Karl. (1946) 2000. *The Question of German Guilt*. Translated by E. B. Ashton and with a new introduction by Joseph W. Koterski. New York: Fordham University Press.

Katchadourian, Herant A. 2010. *Guilt: The Bite of Conscience*. Stanford, CA: Stanford University Press.

Kedourie, Elie. 1970. *The Chatham House Version and Other Middle-Eastern Studies*. London: Weidenfeld & Nicolson.

Kedourie, Elie. 2000. *In the Anglo-Arab Labyrinth: The McMahon-Husayn Correspondence and Its Interpretations 1914–1939*. London: Frank Cass.

Kidron, Carol A. 2012. "Alterity and the Particular Limits of Universalism: Comparing Jewish-Israeli Holocaust and Canadian-Cambodian Genocide Legacies." *Current Anthropology* 53, no. 6: 723–54.

Levine, Emma E., T. Bradford Bitterly, Taya R. Cohen, and Maurice E. Schweitzer. 2018. "Who Is Trustworthy? Predicting Trustworthy Intentions and Behavior." *Journal of Personality and Social Psychology* 115, no. 3: 468–94.

Lewy, Mordechay. 2003. "Der Islam und der Westen: Nimm meine Schuld auf dich. Wie tief ist der politische Riss zwischen Orient und Okzident? Die Religionsgeschichte gibt Antworten." *DIE ZEITE*, January 16, 2003. https://www.zeit.de/2003/04/Schuld_im_Islam.

Leys, Ruth. 2007. *From Guilt to Shame: Auschwitz and After.* Princeton, NJ: Princeton University Press.

Lotter, Maria-Sibylla. 2012. *Scham, Schuld, Verantwortung: Über die kulturellen Grundlagen der Moral.* Berlin: Suhrkamp.

Malinowski, Bronislaw. 1944. *A Scientific Theory of Culture and Other Essays.* Chapel Hill: University of North Carolina Press.

May, Larry, and Stacey Hoffman, eds. 1991. *Collective Responsibility: Five Decades of Debate in Theoretical and Applied Ethics.* Lanham, MD: Rowman & Littlefield.

McClay, Wilfred M. 2017. "The Strange Persistence of Guilt." *The Hedgehog Review* 19, no. 1: 40–55.

Meier, Christian. 2010. *Das Gebot zu vergessen und die Unabweisbarkeit des Erinnerns: Vom öffentlichen Umgang mit schlimmer Vergangenheit.* Munich: Siedler.

Montada, Leo, C. Dalbert, B. Reichle, and M. Schmitt. 1986. "Urteile über Gerechtigkeit, 'existentielle Schuld' und Strategien der Schuldabwehr." In *Moralische Zugänge zum Menschen—Zugänge zum moralischen Menschen: Beiträge zur Entstehung moralischer Identität*, edited by Fritz Oser, W. Althof, and D. Garz, 205–25. Munich: Kindt.

Moos, Thorsten, and Stefan Engert, eds. 2016. *Vom Umgang mit Schuld: Eine multidisziplinäre Annäherung.* Frankfurt am Main: Campus.

Nietzsche, Friedrich. (1887) 2006. *On the Genealogy of Morality.* Edited by Keith Ansell-Pearson. Translated by Carol Diethe. Cambridge: Cambridge University Press.

Piers, Gerhart. 1953. "Shame and Guilt: A Psychoanalytic Study." In *Shame and Guilt: A Psychoanalytic and a Cultural Study*, edited by Gerhart Piers and Milton B. Singer, 15–58. New York: Norton.

Rösch, Petra H. 2010. "Vergehen reinigen und Verdienste erwerben: Die Familie Chen stiftet eine Buddhistische Stele." In *Bild und Ritual-Visuelle Kulturen in historischer Perspektive*, edited by Claus Ambos, Petra Rösch, and Stefan Weinfurter, 29–42. Darmstadt: Wissenschaftliche Buchgesellschaft.

Rösch, Petra H. 2012. "The Purifying Power of the Buddhas." In *How Purity Is Made*, edited by Petra H. Rösch and Udo Simon, 217–44. Wiesbaden: Harrassowitz.

Schlink, Bernhard. 2010. "Collective Guilt?" In *Guilt about the Past*, edited by Bernhard Schlink, 5–22. London: House of Anansi Press.

Schmidt, Konrad. 1999. "Kollektivschuld? Der Gedanke übergreifender Schuldzusammenhänge im Alten Testament und im Alten Orient." *Zeitschrift für altorientalische und biblische Rechtsgeschichte* 5: 193–222.

Siedentop, Larry. 2015. *Inventing the Individual: The Origins of Western Liberalism.* London: Penguin.

Steele, Shelby. 2007. *White Guilt: How Blacks and Whites Together Destroyed the Promise of the Civil Rights Era.* New York: Harper Perennial.

Steele, Shelby. 2008. "Obama's Post-Racial Promise." *Los Angeles Times*, November 5, 2008. https://www.latimes.com/archives/la-xpm-2008-nov-05-oe-steele5-story.html.

Torpey, John. 2009. "An Avalanche of History: The 'Collapse of the Future' and the Rise of Reparations Politics." In *Historical Justice in International Perspective: How Societies Are Trying to Right the Wrongs of the Past*, edited by Manfred Berg and Bernd Schäfer, 21–39. Cambridge: Cambridge University Press.

Vetlesen, Arne Johan. 2005. *Evil and Human Agency*. Cambridge: Cambridge University Press.

Wong, Yin, and Jeanne Tsai. 2007. "Cultural Modes of Shame and Guilt." In *The Self-Conscious Emotions: Theory and Research*, edited by Jessica L. Tracy, Richard W. Robins, and June Price Tangney, 209–23. New York: Guilford.

Young, Iris Marion. 2011. *Responsibility for Justice*. New York: Oxford University Press.

PART I

GUILT AS A PROSOCIAL FORCE IN INTERPERSONAL RELATIONS

1

Guilt as a Positive Motivation for Action?

On Vicarious Penance in the History of Christianity

Meinolf Schumacher

One way to speak of the cultural productivity of guilt is to consider the meritorious actions that result from or are mandated by guilt. Can guilt become productive by causing compensatory action that would otherwise not happen? This question involves two different angles: on the one hand, there is the question of justice and the related obligation to vindicate the victims of serious crimes (as far as this is possible at all). On the other hand, there is the potential of guilt to inspire meaningful and socially valuable actions that go beyond the individual offense and beyond benefitting individual victims. In this latter, more "open" sense, the productivity of guilt consists of the ways in which guilt can be instrumentalized to motivate morally good action in general.

However, the world—particularly the Western world—exhibits a profound aversion to the instrumentalization of guilt, even if done with the best of intentions. For example, it is a disparagement and an insult to accuse individuals of attempting to buy their way out of guilt. And when charitable or political organizations use guilt to mobilize action (e.g., in the campaign against climate change, when carbon offsets and climate certificates are offered to reduce travelers' guilt feelings and ecological footprints), they are accused of practicing *Ablasshandel* (trade in indulgences), in a nod to the Church's pre-Reformation guilt negotiations. From whence does this aversion to instrumentalizing guilt originate, and what might its alternatives be? A brief look into the history of Christian repentance can help us to identify, understand, and contextualize the answers.

Christian Penance in the Middle Ages and the Effect of Good Actions

Christianity's first sacrament to facilitate the forgiveness of sins is baptism, which can be received only once. However, early Christians' hope that the grace of baptism would prevent all sinning henceforth proved unrealistic and required the

Meinolf Schumacher, *Guilt as a Positive Motivation for Action?* In: *Guilt*. Edited by: Katharina von Kellenbach and Matthias Buschmeier, Oxford University Press. © Oxford University Press 2022. DOI: 10.1093/oso/9780197557433.003.0002 development of a second sacrament for the forgiveness of sins. The Church Fathers imaginatively referred to this second sacrament as *tabula post naufragium*—that is, a board to cling to after a shipwreck—the idea being that the sinner could use this "board" (or, in this case, sacrament) to stay afloat long enough to swim safely to heaven's shores (Rahner 1964, 432–72). The sacrament of penance was created to restore the baptismal grace that had been lost.

There are three important conditions for this sacrament to be persuasive: first, the conviction that God lets no sin go unpunished (*nullum peccatum impunitum*). This divine legal principle (Angenendt 1994) prohibits impunity (cf. Dominik Hoffmann in this volume) and grounds the church's practices surrounding repentance. Not only is *contritio cordis* (contrition of the heart) a precondition for forgiveness (Konstan 2010, 2018), but it must be followed by *confessio oris* (oral confession) as well as *satisfactio operis* (penance in the narrow sense), in order to expiate guilt that would allow the sinner to be accepted back into a state of divine grace (Enxing and Gautier 2019). Interestingly, while in the English language the terms "penance" and "repentance" derive from Latin *poena* and are therefore connected with punishment and suffering (cf. Katharina von Kellenbach in this volume), the German term *Buße* is related to the verb *bessern* (to make better, to improve). Literally, in the German language, *Buße* is understood as the process of "making good" or better (Weisweiler 1930; Stutz 1986). The acts prescribed by the third step of satisfaction consist of prayer, charity, and alms, as well as service and good works of all kinds. Satisfaction requires sacrifice, and these ritual acts and substitutionary gestures are rooted in antecedent sacrificial practices. It is not only the rich and powerful who can make sacrifices, offer compensatory payments, and perform penance in order to reconcile with God; the poor also offer "spiritual alms"—in the form of prayer. (And this is true regardless of whether one's poverty is the result of involuntary circumstances *or* a voluntary religious vow.)

For every trespass (i.e., sin), a penalty must be endured in order to balance the scales of justice. It does not matter—and this is the second condition—whether this happens in this life or in purgatory, an imaginary realm within the afterlife. Meanwhile, a person's sufferings in the current life are credited toward the penalties in the hereafter. Every sin is punished only once, for God, as a just judge, follows the Roman principle of *ne bis in idem*. This prohibition of double jeopardy, which we still have in criminal law today (Landau 1970), explains why people punish themselves in acts of repentance; repentance, in part, forestalls divine punishment—and it is preferable to forestall, because earthly punishment is finite, whereas divine punishment might be endless. In other words, whatever is suffered now will not have to be endured in the hereafter.

Third, it is not obligatory that this punishment be incurred exclusively by the sinner; penitential compensation is mandatory, but a substitute penitent is acceptable. Although every sin must be punished, someone can serve as a surrogate and participate in another's recompense. Here, the idea of the scapegoat and of substitutionary animal sacrifices, with a long history across different religions, comes into play (Girard 1986; Schwager 1987); it is in this tradition that Christ serves as the sacrificial lamb that paid the ransom for the redemption of humankind (Schaede 2004). Today, however, this approach to justice strikes us as odd: for instance, if someone had been sentenced to a five-year prison term and four friends donated one year of prison each, that person would be released after the first year. We no longer accept such a concept of shared justice in which penance is incumbent upon the entire group. This is, however, exactly the notion that underlies the concept of substitutionary atonement.

There are several reasons why a person might complete penitential obligations for another. One might be motivated by the practice of general virtues of empathy and compassion or by the Christian commandments of mercy and charity. Furthermore, by atoning for another, mutual relationships of obligation and solidarity are established (Angenendt 1997, 305). Before the Reformation, repentance was not an individual affair between the sinner and God but rather a communal effort; the idea of substitutionary repentance created a world of human solidarity before a divine judge. We can and must stand up for one another, at least as long as we live—and even after death, once the sinner has passed on into purgatory, this mandate remains binding for the living. For instance, prayers, fasts, charitable donations, and physical hardship can be donated to the "poor souls" in purgatory. The performance of such "works of mercy" on behalf of the dead was considered an essential element of Christian piety. Penitential observance was expected not only for family and other close relations but for the entire church community. In that respect, medieval Christian piety created a culture of solidarity among the living that extended as well to the dead.

At the same time, vicarious penance was thought to accumulate merit; it was not entirely unselfish. Medieval authors praised this type of penance as eminently "useful": as a good deed, such an intercession was in itself meritorious and earned dispensation for the petitioner, as well as for the beneficiary of the substitutionary penance (Ohly 1992, 199–201). Honorius Augustodunensis (12th century) falsely attributes this concept to the Bible by (mis)quoting it: "The Scripture says: he who prays for another, liberates himself" (*Scriptura ait: qui pro alio rogat, seipsum liberat*) (Honorius 1854, 830). This principle becomes widely popular and remains so all the way into the 20th century, when Thomas Mann, in turn, adopts this verse in his novel *Doctor Faustus*: "Mark, whoso for other pray / Himself he saves that waye" (Mann 1949, 472). Thereby, the prayer that was intended to speed the redemption of another also helped to alleviate the penance of the generous supplicant.

Donors of charitable funds also secured future penitential prayer by grateful beneficiaries (Borgolte 2018). The generous endowment of poorhouses, leper houses, old age homes, and grants for impoverished students were not just a good deed in the present; they also canceled out potential punishment for one's sins in the future, since such donations created institutions and programs that benefitted generations of recipients who would naturally be inclined to show their gratitude by praying regularly for these generous benefactors and their salvation. Each annual name day and anniversary would be an occasion to remember and pray for the individual's deliverance. The same balancing principle of justice obtains: a service demands its reward, just as a trespass demands its punishment. Therefore, taking advantage of a service and a gift without repaying by gratitude would have been regarded as a kind of theft by medieval legal opinion. Caring for the poor obligated the poor to repay with prayer. Social welfare in the Middle Ages therefore depended on the economics of vicarious penance, which made guilt productive.

Meister Eckhart and Martin Luther against Justification by Works

Today, atheists are not alone in pointing out that this connection between guilt, repentance, and "good" action is problematic. For instance, the German cultural critic Karlheinz Deschner bemoans that "one did not donate to remedy social grievances, to raise the standard of living, to promote art, science, and education, but only to save one's own soul. Hence, one donated to oneself" (Deschner 1990, 448, translation kvk)! Full of contempt, he adds: "And that is precisely what makes the business of Christian charity so disgusting. It rests on nothing more than the principle of *do ut des* [give and take], which is basically Old Testament retributive thinking, a very banal, primitive, but highly effective moral of reward and punishment for the masses" (Deschner 1990, 448).

Indeed, the tradition of penance was criticized long before the 19th and 20th centuries. Already the mystic Meister Eckhart in the Middle Ages criticizes what he suspects was a cost-benefit approach to penance. Ascetic practices, he maintains, are supposed to be a radically spiritual path to God (*vita purgativa*, or "purgative life") rather than a reward and punishment system. He extends the exclusive monastic-mystical practices to the daily piety of lay Christians. And he emphasizes goodwill over external deeds in his definition of the good. It is never about the end but the act itself, which must be performed without ever asking why (*sunder warumbe*) (Sölle 2001, 59–62). This requires the renunciation of all self-referential motivations, including any heavenly or earthly rewards, such as the expiation of guilt. "The just man acts, but he acts without consideration of selfish interest" (Langer 2004, 328, translation kvk). Long before Martin Luther, Eckhart claims that it is never the good deed that sanctifies; rather, we sanctify the deed ("wan diu werk enheiligent uns niht, sunder wir suln diu werk heiligen") (Eckhart 2008, 342).

Eckhart anticipates Martin Luther's doctrine of justification by more than two centuries. In Luther's treatise *Von der Freiheit eines Christenmenschen* (*On the Freedom of a Christian*) (1520, § 21), he argues that it is not the act that constitutes goodness, "which makes a person pious and just before God, but rather acting out of free love and '*umsonst*'" (Luther 1897, 31, translation kvk). Luther's central message finally is

> that faith alone, by pure grace through Christ and His word, makes a person completely just and blessed. And that no work, no commandment is necessary for a Christian to be saved, but that he is free from all commandments and, out of pure freedom, does "*umsonst*" all that he does and nothing with which he seeks his benefit or happiness (§ 23, Luther 1897, 32).

In Luther's use of the German adverb *umsonst*—meaning unpaid, gratuitous, and for free—human action responds to God's action of unconditional grace. From this experience of unearned grace, Luther concludes that the believer "should act freely, joyfully, and gratuitously in response to a Father, who has showered abundant gifts." From such a faith flows the "love and desire for God, which results in a free, willing, and joyful life that is dedicated to the unconditional service to the neighbor" (§ 27, Luther 1897, 35–6). Luther's teachings are directed, above all, against the *do ut des* exchange of give and take, which is deeply engrained in human culture (Mauss 1966; cf. Därmann 2010). Receiving a gift without being able to reciprocate seems to be a shameful experience of powerlessness—but this is exactly what Luther affirms as mankind's position in relation to God. He insists that humanity cannot make a deal with God, humans cannot bribe or manipulate God as they might do with a human being, which Luther would consider a form of idolatry: "Those who want to enter heaven with their own works, draw Christ down to us. But it must be the other way around: Christ must pull us up, or all is lost" (Luther 1915, 230). Martin Luther insists on the fundamental asymmetry in the God-human relationship, which both individualizes the believer and renders the believer passive, thus making penance pointless and vicarious penance impossible.

While Eckhart and Luther agree that good deeds ought to be done and are to be judged by their intention, they cannot be instrumentalized for making amends to atone for guilt. Thus, according to Luther, "in all acts, the intention of the agent should be directed firmly to serve and benefit other people. He should not imagine anything other than what the others need" ("*Freiheit*" § 26; Luther

1897, 35). In other words, believers must, therefore, completely renounce themselves and their interests. Or, to put it more fundamentally: Eckhart and Luther deny that there could be two or more intentions for a morally good action. They obviously also imply that, as soon as self-interest comes into play, self-interest becomes the actual intention. Luther's position is marked by the Christian polemic against the "Pharisees," which is a standard anti-Jewish caricature, and closely linked to the trope of hypocrisy. A hypocrite pretends to act for unselfish reasons but pursues separate and secondary agendas. The charge of hypocrisy, notes Hannah Arendt (1990, 98–109), becomes especially virulent in early modern times, in part because of Luther's logic that anyone who acts not entirely unselfishly must therefore act for hypocritical reasons.

The big difference between Eckhart and Luther, however, is that Eckhart influenced the history of spirituality, while Luther changed the world. The Reformation coincided with the spread of European colonialism and the Enlightenment rise of science and secularism. Hence, Luther's idea that unselfishness is the requisite condition for genuinely good action appears again and again in both the secular and the religious imagination. One can see just how much of Luther's theology permeates secular thought, as in, for instance, the following ironic passage from the novel *Mrs. Dalloway* (1925) by Virginia Woolf, whose protagonist asserts, "she thought there were no Gods; no one was to blame; and so she evolved this atheist's religion of doing good for the sake of goodness" (Woolf 2015, 70).

Social Consequences of the Reformation in Dealing with Guilt

Luther's rejection of "justification by works (*Werkgerechtigkeit*)" (Mühlen 2004) had unanticipated consequences. He expected that people who were no longer obligated to atone for their sins would more readily and joyfully commit to serving their fellow human beings and the common good (Oberman 1986, 219). Instead, the opposite occurred and authorities faced "the collapse of church donations, endowments, and tithing systems due to unwillingness to pay" among new Protestant Christians (Kaufmann 2009, 507, translation kvk). Given the perspective presented in the current chapter, this outcome is not surprising (Wegner 2018, 101), since now these Christians were ideologically deprived of the possibility to make guilt productive through good acts. Without an "incentive structure" (to put it in modern terms), people tend to be less inclined to donate.

No one was as surprised and disappointed about this as Luther himself. On October 31, 1525, he complained to Elector Johann of Saxony about the financial decline of the parishes (Luther 1933, 595). In a late sermon on the topic of the works of mercy (Mt 25: 31–36), he complained bitterly about Protestant countries' failure to donate voluntarily. Under the papacy, people had been ready to give. Now, just after the "rediscovery of the gospel," not even in Luther's own Wittenberg were there enough donations to finance the churches and schools and to support the poor. The only way to maintain these institutions in the wake of this shift was to use church property—that is, what previous generations had donated for their own (and for vicarious) penance. At the same time, the Protestant princes increasingly began to confiscate these very same church properties for their own purposes (Luther 1911). Neither the common man nor the nobility was prepared to act altruistically.

Incidentally, Luther does not seem to have taken this sobering reality as occasion to rethink his doctrine of justification in light of the consequences for social life. Nor did he ultimately have to, since political rulers increasingly began to assume responsibility for social welfare and education (Kaufmann 2009, 714); in place of donations by which people paid their debts to God, they now had to pay taxes to the state. Thus, the Reformation contributed to the "intensification of early modern statehood" (Kaufmann 2009, 504). Although this claim is still debated among historians (Geremek 1994; Jütte 1994), it can be argued that the Reformation was one of the causes of the rise of the tax-financed welfare state (Gabriel and Reuter 2017).

Nevertheless, Protestants developed a penitential piety of social engagement and political responsibility. Many churches became involved in helping the poor, the sick, and victims of war and violence, and such service activities were understood as the fruit of repentance. For instance, in postwar Germany, two organizations were founded with the explicit intention of atoning for Nazi Germany's crimes. Aktion Sühnezeichen, which is known in the English-speaking world as "Action Reconciliation Service for Peace," literally means in German, "Action Sign of Atonement" and was founded in Berlin in 1958 to explicitly confess and address Germany's guilt (Skriver 1962). There is also a women's religious order, called "Evangelical Sisters of Mary," cofounded by Mother Basilea Schlink, which is dedicated to repentance for German Christian crimes committed against the Jews (Faithfull 2014). Such an activist faith approach to guilt and atonement seemingly contradicts what Martin Luther said about faith and grace. At least that is the assessment of Norbert Bolz, who criticizes such supposed "sentimental humanitarianism" as a betrayal of Luther (Bolz 2016).

At least for Germany, it should be remembered that the modern Protestant Diaconia organizations were not founded by the church leaders and instead emerged from "below"—that is, from Pietist movements. Unlike the orthodox Lutherans, the Pietists did not want to leave the responsibility for the community and the well-being of neighbors to the "caring authority" (*fürsorgliche Obrigkeit*). And unlike the Calvinist-influenced Puritans, for example, the Pietists found their own religious validation less in economic success than precisely in "committing an act of love for one's neighbor" (Spener 1676, 112, translation kvk). Philip Gorski captures the essence of Pietism in a formula: "Not private accumulation but public service was the sign of salvation" (Gorski 1993, 293). Presumably, such public service also alleviated feelings of guilt.

Making Guilt Productive Today: Three Examples

After so much contemplation of the evolution of penance throughout the history of Christianity, the question remains: What can we do if we do not want to recommend the medieval Catholic-style practice of penance for people of today? That is, what might provide relief from the burden of guilt—beyond the concept of "restorative justice"? And what can offer exoneration in those cases where there are no (more) direct victims with whom one can be reconciled? We cannot hold out relief by simply letting "bygones be bygones," and we cannot promise release without some kind of accountability. Therefore, the conditional process that prescribed painful insight (*contritio*) and shameful confession (*confessio*)—the hallmarks of the old sacrament of penance—are worthy of retention for the secular world.

Three practices of particular importance in comprehending some of these contemporary guilt negotiations are (1) the concept of court-imposed community service, which evokes the concept of good deeds that atone for guilt; (2) vicarious engagement that mitigates environmental sins, such as climate change; and (3) the use of money, such as charitable donations in the expiation of guilt for past offenses.

Further to the first, many legal systems offer the possibility of performing community service as a substitute for punishment. Criminal justice theorists do not impose community service to benefit the victims but rather to prevent offenders who are too poor to pay bail or financial penalties being sent to prison; alternatively, community service sentences can be intended as a means of facilitating resocialization. Usually, community service hours are meant to benefit the larger community, and the precise community service assignment need not always relate to the nature of the offender's misconduct. Therefore, it is quite possible, for example, to perform community service at an animal shelter for non-animal-related offenses, such as stealing or other minor crimes. Although court-imposed and externally mandated, these "good works" have gained in popularity, and they obviously serve to alleviate the conscience.

As for the second major topic in contemporary guilt negotiations—namely, vicarious engagement—the dilemma is especially interesting because it involves communal wrongs, such as the exacerbation of climate change. In a recent study, social psychologists asked what motivates people to become involved in climate protection. Their results show that individuals are drawn to environmental activism because of their guilty consciences, even if they do not feel personally responsible for particular facets of environmental damage (Rees and Bamberg 2014). As the psychologists note, "moral emotions" motivate prosocial behavior. This relationship between individual guilt and "group-based" guilt evokes the medieval concept of vicarious penance (Ferguson and Branscombe 2010). Environmental activism is a form of "open" penance that is performed on behalf of others (e.g., major polluters, political decision-makers)—and hence a form of productive guilt.

The third topic, money-as-expiation (or, so to speak, the matter of "indulgences," to borrow from the language of previous eras), is often criticized for instrumentalizing a guilty conscience. In the contemporary world, this occurs in the form of donations to "good causes." Aid agencies are often accused of "doing business with a guilty conscience" (e.g., Schoop 2017), which is reminiscent of the Reformation's accusations of indulgence trading as "business with sin" (Laudage 2016). This puts NGOs and charitable organizations in a defensive position. Many fundraisers, therefore, have shifted toward branding their messages in positive emotion, along the lines of "Helping makes you happy!" Such an approach corresponds to the concept of "warm-glow giving," a giving that is done in order to feel better. Andreoni (1990) calls such giving "impure altruism." It is "impure" because it involves selfish motives, such as one's own joy of giving, and is therefore not "pure altruism"—the idea being that the presence of mixed motives fundamentally contaminates and devalues the action itself (cf. Schumacher 1996, 2019). Hence, for instance, making a donation that helps both another and oneself (in the latter case, by making oneself feel warm and fuzzy) is regarded as less morally good. This was not the expectation in the Middle Ages, however, wherein the goal of appeasing one's own conscience and securing one's own salvation was not considered to pose a conflict of interests with the goal of doing good for others. It is only because of the Reformation idea of faith and grace that sincerity has precluded the presence of impure and ambivalent motivations.

Nevertheless, we should accept that an action can be morally good, even if it has more than one intention; that is, even if it is not completely "unselfish." It is certainly bad if one conceals self-interest in one's actions while claiming that one is doing something for the sole sake of doing good; that would be hypocrisy. And it is just as bad to let self-interest be one's dominant motivation; that would constitute selfishness and egotism. But is it blameworthy to do something that serves other individuals or the broader society, even though it will also serve you? What if, for example, I learn through doing good deeds that I am not a very bad person, despite my own guilt? Or if I find some kind of emotional relief in making amends for injustices that members of my family, my nation, my church, etc. have committed? In such cases, the healing and catharsis that I find through helping society would indeed mean that my good deeds ultimately benefited more than just myself.

What contemporary society can learn from the Middle Ages is how to harness the power of guilt for socially beneficial outcomes. Offering opportunities to relieve a guilty conscience through social service, vicarious activism, and charitable donation can be an occasion to reflect on other social areas where guilt can become productive today.

References

Andreoni, James. 1990. "Impure Altruism and Donations to Public Goods: A Theory of Warm-Glow Giving." *The Economic Journal* 100: 464–77. https://doi.org/10.2307/2234133.

Angenendt, Arnold. 1994. "Deus, qui nullum peccatum impunitum dimittit: Ein 'Grundsatz' der mittelalterlichen Bußgeschichte." In *Und dennoch ist von Gott zu reden: Festschrift Herbert Vorgrimler*, edited by Matthias Lutz-Bachmann, 142–56. Freiburg: Herder.

Angenendt, Arnold. 1997. *Geschichte der Religiosität im Mittelalter*. Darmstadt: Wissenschaftliche Buchgesellschaft.

Arendt, Hannah. (1963) 1990. *On Revolution*. London: Penguin Books.

Bolz, Norbert. 2016. *Zurück zu Luther*. Paderborn: Fink.

Borgolte, Michael. 2018. *Weltgeschichte als Stiftungsgeschichte: Von 3000 v. u. Z. bis 1500 u. Z.* Darmstadt: Wissenschaftliche Buchgesellschaft.

Därmann, Iris. 2010. *Theorien der Gabe zur Einführung*. Hamburg: Junius.

Deschner, Karlheinz. 1990. *Kriminalgeschichte des Christentums*. Vol. 3. Reinbek bei Hamburg: Rowohlt.

Eckhart, Meister. 2008. "Die rede der underscheidunge." In *Meister Eckhart: Werke*. Vol. 2, edited by Niklaus Largier, 334–432. Frankfurt am Main: Deutscher Klassiker Verlag.

Enxing, Julia, and Dominik Gautier, eds. 2019. *Satisfactio. Über (Un-)Möglichkeiten von Wiedergutmachung*. Leipzig: Evangelische Verlagsanstalt.

Faithful, George. 2014. *Mothering the Fatherland: A Protestant Sisterhood Repents for the Holocaust*. New York: Oxford University Press.

Ferguson, Mark A., and Nyla R. Branscombe. 2010. "Collective Guilt Mediates the Effect of Beliefs about Global Warming on Willingness to Engage in Mitigation Behavior." *Journal of Environmental Psychology* 30: 135–42. https://doi.org/10.1016/j.jenvp.2009.11.010.

Gabriel, Karl, and Hans-Richard Reuter, eds. 2017. *Religion und Wohlfahrtsstaatlichkeit in Deutschland: Konfessionen—Semantiken—Diskurse*. Tübingen: Mohr Siebeck.

Geremek, Bronisław. 1994. *Poverty: A History*. Translated by Agnieszka Kolakowska. Oxford: Blackwell.

Girard, René. 1986. *The Scapegoat*. Translated by Yvonne Freccero. Baltimore: Johns Hopkins University Press.

Gorski, Philip S. 1993. "The Protestant Ethic Revisited: Disciplinary Revolution and State Formation in Holland and Prussia." *American Journal of Sociology* 99: 265–316. https://doi.org/10.1086/230266.

Honorius Augustodunensis. 1854. "Speculum Ecclesiae." In *Patrologia latina*, vol. 172, edited by Jacques-Paul Migne, 813–1108. Paris: Migne.

Jütte, Robert. 1994. *Poverty and Deviance in Early Modern Europe*. Cambridge: Cambridge University Press.

Kaufmann, Thomas. 2009. *Geschichte der Reformation*. Frankfurt am Main: Verlag der Weltreligionen.

Konstan, David. 2010. *Before Forgiveness: The Origins of a Moral Idea*. Cambridge: Cambridge University Press.

Konstan, David. 2018. "Reue." In *Reallexikon für Antike und Christentum*, vol. 28, 1216–241. Stuttgart: Hiersemann.

Landau, Peter. 1970. "Ursprünge und Entwicklung des Verbotes doppelter Strafverfolgung wegen desselben Verbrechens in der Geschichte des kanonischen Rechts." *Zeitschrift der Savigny-Stiftung für Rechtsgeschichte: Kanonistische Abteilung* 56: 124–56. https://doi.org/10.7767/zrgka.1970.56.1.124.

Langer, Otto. 2004. *Christliche Mystik im Mittelalter: Mystik und Rationalisierung—Stationen eines Konflikts*. Darmstadt: Wissenschaftliche Buchgesellschaft.

Laudage, Christiane. 2016. *Das Geschäft mit der Sünde: Ablass und Ablasswesen im Mittelalter*. Freiburg: Herder.

Luther, Martin. 1897. "Von der Freyheyt eynisz Christen menschen [1520]." In *D. Martin Luthers Werke: Kritische Gesamtausgabe*, vol. 7, 12–38. Weimar: Böhlau.

Luther, Martin. 1911. "Predigt am 26. Sonntag nach Trinitatis [1537]." In *D. Martin Luthers Werke: Kritische Gesamtausgabe*, vol. 45, 324–29. Weimar: Böhlau.

Luther, Martin. 1915. "Hauspostille [1544]." In *D. Martin Luthers Werke: Kritische Gesamtausgabe*, vol. 52, 1–843. Weimar: Böhlau.

Luther, Martin. 1933. "Brief an Kurfürst Johann, 31. Oktober 1525." In *D. Martin Luthers Werke: Kritische Gesamtausgabe. Briefwechsel*, vol. 3, 594–96. Weimar: Böhlau.

Mann, Thomas. 1949. *Doctor Faustus: The Life of the German Composer Adrian Leverkühn, as Told by a Friend*. Translated by Helen Tracy Lowe-Porter. London: Secker & Warburg.

Mauss, Marcel. 1966. *The Gift: Form and Functions of Exchange in Archaic Societies*. Translated by Ian Cunnison. London: Cohen & West.

Mühlen, Karl-Heinz zur. 2004. "Werkgerechtigkeit." In *Historisches Wörterbuch der Philosophie*, vol. 12, 553–56. Basel: Schwabe.

Oberman, Heiko A. 1986. *Luther: Mensch zwischen Gott und Teufel*. Munich: Deutscher Taschenbuch Verlag.

Ohly, Friedrich. 1992. *The Damned and the Elect: Guilt in Western Culture*. Translated by Linda Archibald. Cambridge: Cambridge University Press.

Rahner, Hugo. 1964. *Symbole der Kirche: Die Ekklesiologie der Väter*. Salzburg: Müller.

Rees, Jonas H., and Sebastian Bamberg. 2014. "Climate Protection Needs Societal Change: Determinants of Intention to Participate in Collective Climate Action." *European Journal of Social Psychology* 44: 466–73. https://doi.org/10.1002/ejsp.2032.

Schaede, Stephan. 2004. *Stellvertretung: Begriffsgeschichtliche Studien zur Soteriologie*. Tübingen: Mohr Siebeck.

Schoop, Florian. 2017. "Das Geschäft der Hilfswerke mit dem schlechten Gewissen." *Neue Zürcher Zeitung*, December 19, 2017. https://www.nzz.ch/meinung/das-geschaeft-der-hilfswerke-mit-dem-schlechten-gewissen-ld.1339739.

Schumacher, Meinolf. 1996. *Sündenschmutz und Herzensreinheit: Studien zur Metaphorik der Sünde in lateinischer und deutscher Literatur des Mittelalters*. Munich: Fink.

Schumacher, Meinolf. 2019. "Weeds among the Wheat: The Impurity of the Church between Tolerance, Solace, and Guilt Denial." *CrossCurrents* 69: 252–63. https://doi.org/10.1111/cros.12376.

Schwager, Raymund. 1987. *Must There Be Scapegoats? Violence and Redemption in the Bible*. Translated by Maria L. Assad. San Francisco: Harper & Row.

Skriver, Ansgar. 1962. *Aktion Sühnezeichen: Brücken über Blut und Asche*. Stuttgart: Kreuz.

Sölle, Dorothee. 2001. *The Silent Cry: Mysticism and Resistance*. Translated by Barbara and Martin Rumscheidt. Minneapolis: Fortress Press 2001.

Spener, Philipp Jakob. 1676. *Pia Desideria: Oder Hertzliches Verlangen/ Nach Gottgefälliger Besserung der wahren Evangelischen Kirchen/ sampt einigen dahin einfältig abzweckenden Christlichen Vorschlägen*. Frankfurt am Main: Johann David Zunner. http://www.deutschestextarchiv.de/spener_piadesideria_1676.

Stutz, Elfriede. 1986. "Der 'büßende' Gott." In *Sprache und Recht: Beiträge zur Kulturgeschichte des Mittelalters; Festschrift Ruth Schmidt-Wiegand*, vol. 2, edited by Karl Hauck, Karl Kroeschell, Stefan Sonderegger, Dagmar Hüpper, and Gabriele von Olberg, 944–56. Berlin: de Gruyter.

Wegner, Gerhard. 2018. "Luthers Freiheitsschrift als Ideologie." In *Ambivalenzen der Nächstenliebe: Soziale Folgen der Reformation*, edited by Johannes Eurich, Dieter Kaufmann, Urs Keller, and Gerhard Wegner, 95–121. Leipzig: Evangelische Verlagsanstalt.

Weisweiler, Josef. 1930. *Buße: Bedeutungsgeschichtliche Beiträge zur Kultur- und Geistesgeschichte*. Halle/Saale: Niemeyer.

Woolf, Virginia. 2015. *Mrs. Dalloway*. Edited by Anne E. Fernald. Cambridge: Cambridge University Press.

2

White Guilt in the Summer of Black Lives Matter

Lisa B. Spanierman

In the wake of the brutal killings of George Floyd, Breonna Taylor, Ahmaud Arbery, and so many other Black individuals by police officers and white vigilantes, the United States and the world at large have witnessed a marked shift in racial attitudes among white people. Whereas white support for the Black Lives Matter movement stood at 40% after a police officer killed Michael Brown in 2014, white support increased to 61% in June 2020 (Pew Research Center 2016, 2020). During the months following Mr. Floyd's death in 2020, white Americans protested in support of Black Lives Matter en masse with their Black, Indigenous, and People of Color (BIPOC)[1] counterparts. Already in September 2020, however, even after the widely publicized example of police officers shooting Jacob Blake in the back in Kenosha, Wisconsin, white support for Black Lives Matter began decreasing toward levels prior to Mr. Floyd's death (Marquette University Law School Poll 2020). At the same time, white journalists and elected officials (e.g., Mayor Ted Wheeler of Portland, Oregon) continue to use the language of Black Lives Matter to address anti-Black and systemic racism in their public comments. What drove this striking shift in whites' acknowledgement of systemic racism and their willingness to go into the streets to protest on behalf of racial justice? Was it merely performative?[2] Why did it diminish so rapidly? Will subsequent racist incidents elicit another increase in white support? While answers to these questions are surely multifaceted, I focus in this chapter on the role of white guilt. I explore whether white guilt—considered a self-reflexive emotion in the discipline of psychology[3]—is a factor in such support for racial justice.

White guilt is a controversial and complex topic in the United States that has been criticized by both the conservative right and the progressive left. On all sides of the political spectrum, white guilt feelings have not been understood as productive, but rather as manipulative, futile, impotent, and performative. Conservatives allege that BIPOC and their white allies use guilt as a malevolent tool to maneuver white people and institutions to behave in ways that benefit BIPOC without merit.[4] Critics on the left charge that white people themselves

Lisa B. Spanierman, *White Guilt in the Summer of Black Lives Matter* In: *Guilt.* Edited by: Katharina von Kellenbach and Matthias Buschmeier, Oxford University Press.
DOI: 10.1093/oso/9780197557433.003.0003

use guilt as a defensive device to avoid being called racist and to escape personal and collective responsibility for challenging racial injustice.

Conservative Critique: White Guilt as a Manipulative Tool

The conservative critique of white guilt focuses on how BIPOC and their white allies use guilt as a tool to manipulate white people and institutions. In an article in *The American Conservative*, for instance, the political scientist George Hawley (2017)[5] offered the following to understand the conservative argument:

> One of the most persistent tropes on the racial right is that the major cultural institutions in the United States aggressively push a story of white guilt [. . .]. According to this narrative, white Americans face a constant barrage of derision, persistently hearing about the evils of their white-supremacist ancestors and the unfairness of their current unearned privilege. They are told that their racial sins can never be truly washed away, but they can achieve partial atonement by signing onto various progressive causes, especially generous immigration policies and policies designed to uplift African-Americans.

Also exemplifying this position, the American conservative author and public intellectual Shelby Steele (2006) described white guilt as an unintended consequence of the civil rights era, wherein white Americans were coerced into acknowledging the wrongdoing of slavery and the subsequent oppression of African American people. In Steele's telling, this acknowledgment of racism stigmatized all whites as racist and transferred moral authority from whites to BIPOC. The U.S. government was obliged to enact policies and laws (e.g., affirmative action and the Voting Rights Act of 1965) to dissociate from a racist past and to reclaim moral authority.

Central to the conservative understanding of white guilt is the notion that racism no longer exists except among a fringe element of U.S. society. Conservative critics of white guilt ignore color-blind racism—the dominant racial ideology of the post–civil rights era (Bonilla-Silva 2003; Neville et al. 2013, 455)—and deny structural racism. These critics disregard policies and practices that sustain racial inequities, such as redlining (i.e., racial discrimination in financial lending practices) and the use of IQ tests as a measure of innate intelligence (despite evidence that environmental factors influence IQ scores, the test has been used to designate certain racial groups as biologically inferior; Smith 1995). Conservative critics of white guilt contend that BIPOC need to stop playing the race card as an excuse for their lack of success and should instead pull themselves up by their bootstraps.

Right-wing media frequently features conservative criticism of white guilt that likens the concept to a misguided and controlling religion. For instance, *City Journal* columnist and contributing editor Coleman Hughes (2018) argued that liberal whites have to go to "church" to learn the history of racial oppression in the United States ad nauseam as a mechanism to keep them as guilty as possible. Kyle Smith (2020) in the *National Review* described a "new woke religion" or "white-guilt cult" in which white "wokesters"[6] must chant an antiracist liturgy to seek absolution and get closer to the divine. And on the television show *Tucker Carlson Tonight* in June 2020, the conservative commentator Matt Walsh discussed the "legitimacy of white guilt." He claimed that "America's new religion" is "group humiliation," and avowed education indoctrinates white children to hate themselves because they are racists. Facebook commenters resonated with the Walsh video clip and responded vehemently that they do not feel guilty:

> No guilt here [. . .] I can't help that I was born white [. . .] and, I did nothing wrong! We must stop these liberal professors from indoctrinating our children. (June 12, 2020)

> No guilt no apology. Looking for sympathy, check the dictionary you'll find it between shit and syphilis. (June 12, 2020)

> I dont [*sic*] feel guilty about a damn thing. It wasn't me and I'll never feel personal guilt about anything that happened before I was even born. (June 12, 2020)

These and hundreds of Facebook commenters denied experiencing guilt feelings on the basis that they personally had nothing to do with the enslavement of Africans or any subsequent racial oppression. They are unable to see how they benefit from more than 400 years of structural racism in the United States, and therefore do not *feel* guilt and do not see a need *to engage* in any reparative or restorative behaviors.

Liberal and Progressive Critiques: White Guilt as Paralysis and Performance

Left-wing criticisms of white guilt—liberal and progressive—hinge on the idea that guilt serves no productive function for racial justice. Critics argue that white guilt is too paralyzing to be productive (Kuttner 2018; Wolf 2020). Instead, guilt is viewed as self-flagellation for one's racism and unearned privilege. Thus, the liberal understanding of white guilt involves at least some realization of the horrors of racism and that one benefits unfairly from a system that masquerades as a meritocracy. When white people gain awareness of the enormity of structural racism and white privilege, they may become self-absorbed and focus obsessively on their feelings of white guilt. The person experiencing white guilt may become overly concerned with whether or not they appear racist and thereby inadvertently subvert efforts to understand how structural racism is maintained and reproduced (Leonardo 2004, 140). In these instances, white guilt interferes with critical reflection about racism because whites get mired in feeling individually blameworthy for racism. Interestingly, both the liberal and conservative critiques implicate the avoidance of appearing racist; in the conservative critique, the impetus is external (i.e., manipulation by BIPOC), whereas in the liberal critique, the impetus is internal (i.e., I want to think of myself, or at least appear, as a good person).[7]

The heightened focus on white people's experience of racial guilt in the liberal frame re-centers whiteness and the white person who is experiencing the guilt rather than combatting racial injustice. White people may become so consumed by their own racial guilt and efforts to alleviate it that they are unable to listen to or empathize with the experiences of BIPOC. Thus, white guilt is conceived as an "emotional trap" that keeps many white Americans stuck in a lack of understanding, empathy, and action (Nile and Straton 2003, 2). Progressive critiques of white guilt assert that white people who experience racial guilt tend to "perform" antiracism to assuage that guilt. Moreover, when white guilt becomes the focus, responsibility often is placed on BIPOC to soothe the white person experiencing guilt. "Taking on the alleviation of white guilt as an antiracist project keeps whiteness at the center of antiracism" and does nothing to tackle the problem of systemic racism or disrupt the racial status quo (Thompson 2003, 24). What we see in practice, referred to as performative (but not transformative) antiracism, might entail, for instance, posting a black square on Instagram on "Blackout Tuesday" or marching in a Black Lives Matter protest with no intention to engage further or commit to a racial justice agenda. These examples of "performative allyship" to assuage white guilt tend to work from a place of white American individualism, rather than emphasizing solidarity with BIPOC (Amponsah and Stephen 2020, 10).

In summary, across the political spectrum, critiques of white guilt explicitly or implicitly suggest that under no circumstances is such guilt helpful. While I dismiss the conservative critique of white guilt as BIPOC's manipulative tool—because such a critique relies on the assumption that systemic racism does not exist—I have witnessed numerous examples of white guilt in myself and others as defensive, paralyzing, narcissistic, and performative. I acknowledge the validity of left-wing critiques and at the same time argue that white guilt is complex and nuanced. To this end, I use psychological science to explore the conditions under which white guilt disrupts the racial status quo and seeks to repair relationships between white people and BIPOC in ways that could in fact lead to social change.

Relationship-Enhancing Functions of Guilt Feelings

In contrast to the commonplace critiques of white guilt, scholarship in psychology on personal guilt feelings (not specific to race) elucidates how guilt might be a force for good. In their 1994 review of the literature, social psychologist Roy Baumeister and colleagues argued that guilt, an unpleasant emotional state, is both an internal and social phenomenon that functions to maintain and enhance interpersonal relationships. Although most often related to a particular behavior or wrongdoing, some individuals may feel guilty without having done anything wrong—a type of guilt by association. People may also feel guilty about an inequity in their favor or when they are rewarded unfairly (Baumeister, Wotman, and Stillwell 1993, 383). In other words, specific past or present transgressions are not always involved in guilt.

Baumeister, Stillwell, and Heatherton (1994, 1995) identified several ways that guilt feelings strengthen social bonds. Importantly, guilt motivates a wide range of behaviors that preserve and strengthen relationships. For example, experiencing guilt feelings may motivate individuals to reflect on their actions and subsequently make adjustments to future behavior (Baumeister et al. 2007, 173), facilitate an empathic connection with others (Tangney 1991, 605), and increase cooperation in relationships (De Hooge, Zeelenberg, and Breugelmans 2007, 1032). Additionally, evidence suggests that people's reparative efforts are proportional to their guilt feelings and the strength of the relationship in need of repair. That is, stronger guilt feelings and stronger relationships elicit greater reparative efforts (Baumeister, Stillwell, and Heatherton 1995, 192). Another relationship-enhancing function of guilt is that it can be used by someone with lesser power to influence a person with greater power; however, psychological research indicates that guilt as an interpersonal influence technique is problematic, as the conservative commentators on white guilt suggest, because it may cause resentment (187). Largely, though, since 1994, empirical research in psychology has supported Baumeister and colleagues' initial claims about the relationship-enhancing, prosocial functions of guilt feelings (Ent and Baumeister 2016; Tignor and Colvin 2017). This finding has been evidenced among children as young as three years (Vaish, Carpenter, and Tomasello 2016, 1779).

Collective Guilt: Antecedents and Consequences

Psychological researchers have extended their earlier concern with the individual to focus on the phenomenon of *collective* (or group-based) guilt, defined as "distress that group members experience when they accept that their in-group is responsible for immoral actions that harmed another group" (Branscombe and

Doosje 2004, 3). Similar to individual guilt, collective guilt is a self-focused emotion; in this case however, the "self" pertains to one's group identity as perpetrator of a wrongdoing.

Collective guilt has been linked to acknowledging collective transgressions in the present, such as present-day discrimination toward a particular group (Branscombe, Doosje, and McGarty 2002), and also to recognizing the collective misdeeds that one's group committed in the past (Doosje et al. 1998, 882).[8] Because group-based guilt is an unpleasant emotion, similar to individual guilt, one may engage in actions to alleviate collective guilt feelings. Research to date has used experimental laboratory designs and/or employed a self-report measure of collective guilt (e.g., "I feel guilty about the negative things my ancestors did to other groups," and "I can easily feel guilty for the bad outcomes brought about by members of my group"; Branscombe, Slugoski, and Kappen 2004).

Empirical investigations have focused on antecedents and consequences of collective guilt in a wide array of national contexts. With respect to collective guilt's antecedents (i.e., the factors that engender collective guilt feelings), evidence suggests that the more one identifies with an in-group (the perpetrator or beneficiary group), the stronger one experiences collective guilt feelings (Klandermans, Werner, and Doorn 2008, 346). Not surprisingly then, the more one can distance oneself from the in-group, the less likely one is to experience collective guilt feelings (Peetz, Gunn, and Wilson 2010, 603). Across several studies, in-group advantage framing, in contrast to out-group disadvantage framing, has been linked to higher levels of collective guilt (Greenaway, Fisk, and Branscombe 2017; Harth, Kessler, and Leach 2008; Powell, Branscombe, and Schmitt 2005). For example, among a group of foreign visitors to Nepal from the United States, Europe, and Australia, collective guilt levels were higher when visitors were primed to think about disparities between their nation and Nepal in terms of their in-group privilege in comparison to the disadvantages that Nepalis face (Greenaway et al. 2017, 680). Moreover, messages about disparities or negative historical information about one's group are more powerful when delivered by an in-group member (Doosje et al. 2006, 335; Greenaway, Fisk, and Branscombe 2017, 680).

Empirical investigations also have identified political ideology and justification of disparities as antecedents of collective guilt. Research participants with a liberal ideology tend to score higher on collective guilt than those with a politically conservative ideology (ANES 2016; Klandermans, Werner, and Doorn 2008, 343). Furthermore, when one is able to justify or legitimize group disparities with the belief that these disparities are fair, collective guilt is reduced (Mallett and Swim 2007, 65). On the other hand, collective guilt is increased when one appraises their own group to be responsible for the disparity (Imhoff, Bilewicz, and Erb 2012, 739; Krauth-Gruber and Bonnot 2020, 68; Mallett and Swim 2007, 66; Zimmermann et al. 2011, 835). In short, collective guilt increases when one is unable to justify disparities and instead acknowledges in-group responsibility (Mallett and Swim 2007, 65).

That said, the empirical investigations in which I am most interested here are those that have focused on the *consequences* of collective guilt to understand it as a socially productive force. Across a wide variety of national contexts (e.g., Australia, Canada, Chile, Germany, South Africa, the Netherlands, and the United States), collective guilt shows strong implications for enhancing intergroup relations. In a sample of non-Indigenous Australians for example, researchers found that higher levels of collective guilt predicted support for an official government apology to Aboriginal peoples and Torres Strait Islanders (McGarty et al. 2005, 668). Research across several studies also documented how collective guilt influenced reparation intentions, such as support for material compensation for the target group (Brown et al. 2008; Imhoff, Bilewicz, and Erb 2012; Imhoff, Wohl, and Erb 2013; Klandermans, Werner, and Doorn 2008; Krauth-Gruber and Bonnot 2020; Leach, Iyer, and Pedersen 2006; Zebel et al. 2008). In other studies, researchers have indicated that the type of reparations matter, such that collective guilt tends to predict material compensation but not equal-opportunity policies (Iyer, Leach, and Crosby 2003, 126). Collective guilt also tends to be a stronger predictor of reparation intentions rather than of political action (Iyer, Schmader, and Lickel 2007, 584; Leach, Iyer, and Pedersen 2006, 1243).

Interestingly, collective guilt also serves as a mechanism through which other factors influence reparation intentions. For example, Klandermans and colleagues (2008, 346) observed that the influence of political ideology on support for affirmative action in South Africa was mediated by collective guilt. In other words, a more liberal ideology predicted greater collective guilt, which in turn predicted more supportive attitudes toward affirmative action. In another study, collective guilt mediated the link between empathic perspective-taking and collective action (Mallet et al. 2008, 465). Specifically, when dominant group members (e.g., heterosexual or white) took the perspective of the minority group, they experienced collective guilt and were motivated to take action on behalf of the minority group. Therefore, collective guilt has direct effects on collective action and also helps explain how other variables influence prosocial outcomes.

In what follows, I argue that a particular form of collective guilt—*white guilt*—also has interpersonal, reparative possibilities. Despite staunch criticism from both sides of the political spectrum, white guilt can indeed be a socially productive force.

The Socially Productive Dimensions of White Guilt

Researchers in social and counseling psychology have undertaken the project of investigating white guilt using empirical methods. They define white guilt as characterized by three interrelated properties: a focus of attention on the in-group (i.e., white people), a sense of group responsibility for a transgression or disparity (e.g., contemporary or historical racial injustice), and an unpleasant feeling that people prefer to allay through restitution or avoidance (Grzanka, Frantell, and Fassinger 2020, 49; Iyer, Leach, and Pedersen 2004, 346).

Psychological scientists have developed and used a variety of instruments to investigate white guilt. Here, I discuss findings from studies that primarily employed two measures of white guilt. First, Swim and Miller's (1999, 503) five-item measure of white guilt assesses participants' remorse about past and current disparities, racial privilege, and association with the white race. Sample items include: "I feel guilty about the past and present social inequality of Black Americans (i.e., slavery, poverty)," "I feel guilty about the benefits and privileges that I receive as a White American," and "When I learn about racism, I feel guilt due to my association with the White race." Second, Spanierman and Heppner's (2004) five-item measure of white guilt involves experiencing remorse, shame, and responsibility about one's privileged position in a racialized social system. Items include, "Sometimes I feel guilty about being White" and "Being White makes me feel personally responsible for racism." Evidence across multiple studies has supported the reliability and validity of these measures.[9] Although beyond the scope of the current discussion, Grzanka and colleagues (2020, 51) developed a promising scenario-based measure to assess white guilt proneness.

Generally similar to the empirical findings using variations of Branscombe and colleagues' collective guilt measure, researchers identified antecedents and consequences of white guilt. For example, awareness of white privilege (i.e., unearned benefits granted on the basis of race) and acknowledgement of individual and structural racism are consistent antecedents of white guilt (Leach, Iyer, and Pedersen 2006, 1238); in order to experience white guilt feelings, one must demonstrate at least some awareness of racial privilege and oppression. Additional cross-sectional investigations have identified cultural sensitivity and openness to diversity as additional desirable correlates of white guilt (Black 2018; Chao et al. 2015; Pinterits, Poteat, and Spanierman 2009; Poteat and Spanierman 2008; Spanierman and Heppner 2004).

Empirical research has documented the impact of educational interventions on white guilt. Findings suggest that multicultural instruction, such as diversity courses and workshops, increase white guilt (Case 2007, 233; Kernahan and Davis 2007, 50; Paone, Malott, and Barr 2015, 212). Relatedly, using randomized experimental designs, two independent research teams found significant increases in white guilt after college students viewed the 19-minute ABC *Primetime Live* special "True Colors" on racial discrimination and privilege (Garriott, Reiter, and Brownfield 2016, 165; Soble, Spanierman, and Liao 2011, 154). In short, experimental designs generally provide support for the notion that educational interventions increase students' levels of white guilt.

With regard to white guilt's consequences, evidence strongly suggests that higher levels of white guilt are associated with a variety of prosocial outcomes. For example, white guilt has been linked to greater cultural competence among white mental health trainees, such that those with higher levels of white guilt are more likely to incorporate racial and cultural factors into their case conceptualization of a client's concerns rather than attributing symptoms solely to intrapsychic factors (Spanierman et al. 2008, 85). Another example is that those white undergraduate and graduate student participants who scored higher on white guilt showed greater willingness to confront white privilege (Pinterits, Poteat, and Spanierman 2009, 415). White guilt also has been related to greater support for affirmative action (Iyer, Leach, and Crosby 2003, 125; Spanierman, Beard, and Todd 2012, 183). Taken together, these studies suggest that white guilt has the potential to be a socially productive force in challenging white privilege and disrupting the racial status quo in the United States.

My colleagues and I conducted several studies that suggest how concomitant race-related emotions influence prosocial justice outcomes (Spanierman, Beard, and Todd 2012; Spanierman et al. 2006; Spanierman, Todd, and Anderson 2009). We used cluster analysis[10] to examine the simultaneous effects of white guilt, white empathy for BIPOC, and irrational white fear of BIPOC. We identified five categories of white people that ranged from an overt racist type to an antiracist type. High levels of white guilt characterized two of the five types—*fearful guilt* and *antiracist*—which I distinguish next.

White people who score in the *fearful guilt* type are characterized by high white guilt and high white fear. Similar to left-wing critiques of white guilt, these individuals display paralyzing anxiety in response to learning about their complicity in a racist system.[11] Consequently, it is unlikely that white individuals in the *fearful guilt* type engage in productive actions to address racial injustice. In *Sister Outsider*, Audre Lorde (1984) described the intricacies of white guilt in a way that helps differentiate the *fearful guilt* from the *antiracist* type:

> Guilt is not a response to anger; it is a response to one's own actions or lack of action. If it leads to change then it can be useful, since it is then no longer guilt but the beginning of knowledge. Yet all too often, guilt is just another name for impotence, for defensiveness destructive of communication; it becomes a device to protect ignorance and the continuation of things the way they are, the ultimate protection for changelessness. (Lorde 1984, 130)

Here, Lorde emphasizes how white guilt is often employed as a defense that protects white ignorance and maintains the racial status quo. That this sort of guilt is indeed evident among individuals in the *fearful guilt* type lends some credence to the left-wing critique of white guilt as paralyzing or performative.

In contrast, Lorde also describes a sort of guilt that leads to change, which exemplifies the *antiracist* type. Research participants of the *antiracist* type reflect the highest levels of white empathy and white guilt and the lowest levels of white fear among all five types. Individuals whose score places them in the *antiracist* type report the greatest racial diversity among their friendship groups and often report meaningful relationships with BIPOC. Undergraduate focus group participants who scored in the *antiracist* type expressed a critical understanding of racial issues and white privilege (Kordesh, Spanierman, and Neville 2013, 46). They not only acknowledged white privilege; they also discussed how they personally benefit from white privilege and expressed remorse about and willingness to confront that racial privilege. White *antiracist* participants were embarrassed by white friends and family who expressed racist views. They expressed anger and frustration about racism on campus and about students' lack of involvement in social justice. *Antiracist* participants worked in solidarity with groups to fight oppression on campus. Thus, reparative possibilities for white guilt may also rely on high levels of white (racial) empathy.

Conclusion

White guilt indeed is a complex and nuanced phenomenon that may have contributed to the strong support among white people for the Black Lives Matter movement during the summer of 2020. While left-wing critiques of white guilt as paralyzing, defensive, and performative are accurate in many cases, burgeoning evidence in the field of psychology suggests that white guilt may be a socially productive force with reparative potential. Mirroring Baumeister and colleagues' (1994) interpersonal approach to understanding guilt feelings in the most general sense, white guilt, too, may serve an interpersonal (i.e., intergroup relations) function. Research findings suggest, however, that certain conditions are necessary to realize white guilt's socially productive potential. An incomplete list of such conditions includes concomitant white empathy or other race-related emotions (e.g., moral outrage), whether a sense of in-group responsibility prevails or whether the white individual easily justifies racial disparities, whether a close relationship exists that is worth repairing, whether the message about racial injustice is framed as one of unearned white privilege versus BIPOC disadvantage, who delivers the message (i.e., in-group or out-group member), and so forth.

I suspect that years of former president Donald Trump's inflammatory, racist rhetoric along with anxiety and frustration related to his mishandling of a catastrophic global pandemic also played an important role in the strong support for

Black Lives Matter among white people during the summer of 2020. Watching the brutal killing of George Floyd over and over on the news and social media, seeing police officers collude with their colleague who had his knee on Mr. Floyd's neck for nearly ten minutes, and hearing Mr. Floyd cry out for his mother just before he took his last breath might have stimulated the necessary combination of white guilt and empathy among white people to join their BIPOC brothers and sisters in the streets. Drawing from responses in an unpublished study to open-ended survey questions about their involvement in protests, white participants high in white guilt and white empathy wrote:

> As a white person, I have been extremely complacent and I'm realizing more fully about how detrimental the moderate liberal white *person* (or woke) is regarding race and racism. After Michael Brown and Freddie Gray, I thought things got better. But, all that happened was police reform, which just repackaged things so it seemed like something happened. Instead, all the liberal white people who voted for a Black president felt appeased and proud of themselves. And, that was a huge slap in the face for me. I am one of those problematic white liberal (moderates when it comes to race, apparently) queer persons who hasn't done enough. And, I don't know enough. And, I don't know how I've perpetuated racism. And, that makes me sad that I've hurt my friends of color. [31-year-old white lesbian student]

> My high school (largely white prep school) posted a list of anonymous anecdotes from Black and Latin@ students on their experiences at my school after the protests. I read them and while I don't remember exact occurrences, I realized I probably committed or perpetrated some of the racist comments these students experienced. I certainly heard and saw a lot and did nothing. I realized how much of a part I play in racism as a white woman. I need to actively change that. I will now be supporting *Black-owned* businesses whenever I can. I intend to vote for politicians who are in support of reparations and antiracist policies. I intend to speak up more about racist comments from my in-laws even though I may damage the relationship (I have offered only weak or pandering responses so far; I realize how important it is to do better now). [29-year-old heterosexual woman school counselor]

Although I have no way of knowing if white guilt was the impetus in their involvement in the protests for racial justice, these participants scored high on white guilt, with a corresponding high score in racial empathy. Their words seem heartfelt and beyond the performative antiracism described earlier. If white guilt (a particular form of collective guilt) does play a role in collective action, then we need not be so quick to dismiss expressions of guilt feelings among white people.

Instead, we need to learn how to work with guilt feelings and cultivate empathy to move from paralysis or performative allyship (i.e., playacting wokeness) to a lifelong, meaningful commitment to racial justice.

Notes

1. Referring to Black, Indigenous, and People of Color, the term BIPOC became popularized in early 2020 on social media and has been adopted in academic literature. BIPOC highlights anti-Black racism and police brutality, draws attention to the genocide of Indigenous Peoples in the United States, and emphasizes differences in experiences with oppression across racial and ethnic groups (see Raypole 2020).
2. My use of the term "performative" here and throughout this chapter refers to "performative allyship"—a superficial sort of activism that tends to serve the ally (e.g., seeking recognition as a progressive white person) more so than the group the action is supposed to support (Amponsah and Stephen 2020).
3. Tangney (1991) described guilt as an uncomfortable emotional state involving negative evaluation of one's specific behaviors.
4. See O'Keefe (2002) for a review of guilt-based social influence techniques.
5. Drawing from quantitative data from the 2016 American National Election Studies, Hawley also showed that few white Americans actually *felt* white guilt. My interpretation of the data differs slightly. I observed that respondents experienced at least "a little" guilt: 38.5% about their "association with the white race," 29% about "the privileges and benefits they receive as a white American," and 40.3% about "social inequality between white and Black Americans." Notably, "liberal" and "very liberal" respondents reported being aware of white privilege and experiencing white guilt, whereas moderates and conservatives did not.
6. Derived from African American Vernacular English and now widely used in social media, the term "woke" refers to being awake or aware of racial justice issues. "Wokesters" refers to people who exemplify awareness of racial justice.
7. There continues to be debate in the literature regarding the distinction between guilt and shame (both individual and collective; see Grzanka, Frantell, and Fassinger 2020; Miceli and Castelfranchi 2018; Tangney and Dearing 2002). It is possible that white shame, often considered a negative evaluation of oneself (rather than a specific behavior) may more closely align with white individuals' efforts not to appear racist.
8. One need not have played a role in personally harming an out-group to experience collective guilt; a group's history of having caused harm is a sufficient trigger (Doosje et al. 1998).
9. Swim and Miller's (1999) white guilt scale garnered initial psychometric support across four studies among more than 500 white university students and 51 adults at an airport. Spanierman and Heppner's (2004) white guilt scale garnered initial psychometric support from a large, college-student sample. Subsequent investigations demonstrated adequate reliability and validity primarily among undergraduate and

graduate students (Case 2007; Chao et al. 2015; Garriott, Reiter, and Brownfield 2016; Mekawi, Bresin, and Hunter 2016; Paone, Malott, and Barr 2015; Sifford, Ng, and Wang 2009; Spanierman et al. 2006, 2008; Todd, Spanierman, and Aber 2010; Turner 2011). Additional validation studies provided psychometric support among sexual minority men (Kleiman, Spanierman, and Smith 2015), and among a geographically dispersed sample of employed adults (Poteat and Spanierman 2008) and Ashkenazi Jews (Berk 2015).

10. Cluster analysis is a statistical approach that classifies data into groups to maximize similarity within a group and dissimilarity across groups.
11. The white fear scale is limited to irrational fear, anxiety, and avoidance of BIPOC, but it also is likely that white fear encompasses other domains (e.g., fear of saying the wrong thing or fear of discovering that one's achievements were not based on merit alone; see Jensen 2005; Spanierman and Heppner 2004).

References

Amponsah, Peter, and Juanita Stephen. 2020. "Developing a Practice of African-Centred Solidarity in Child and Youth Care." *International Journal of Child, Youth and Family Studies* 11, no. 2 (June): 6–24.

ANES (American National Election Studies). 2016. "Codebook and User's Guide to the ANES 2016 Pilot Study." ANES, February 23, 2016. https://electionstudies.org/wp-content/uploads/2016/02/anes_pilot_2016_CodebookUserGuide.pdf.

Baumeister, Roy F., Arlene M. Stillwell, and Todd F. Heatherton. 1994. "Guilt: An Interpersonal Approach." *Psychological Bulletin* 115, no. 2: 243–67.

Baumeister, Roy F., Arlene M. Stillwell, and Todd F. Heatherton. 1995. "Personal Narratives about Guilt: Role in Action Control and Interpersonal Relationships." *Basic and Applied Social Psychology* 17, no. 1–2: 173–98.

Baumeister, Roy F., Kathleen D. Vohs, C. Nathan DeWall, and Liqing Zhang. 2007. "How Emotion Shapes Behavior: Feedback, Anticipation, and Reflection, Rather than Direct Causation." *Personality and Social Psychology Review* 11, no. 2 (June): 167–203.

Baumeister, Roy F., Sara R. Wotman, and Arlene M. Stillwell. 1993. "Unrequited Love: On Heartbreak, Anger, Guilt, Scriptlessness, and Humiliation." *Journal of Personality and Social Psychology* 64, no. 3: 377–94.

Berk, Emile Tobias. 2015. "Construct Validation of the Psychosocial Costs of Racism to Whites Scale for Ashkenazic Jews in the United States." PhD diss., Seton Hall University.

Black, Whitney W. 2018. "An Examination of Relations among Fear, Guilt, Self-Compassion, and Multicultural Attitudes in White Adults." PhD diss., University of Kentucky.

Bonilla-Silva, Eduardo. 2003. *Racism without Racists: Color-Blind Racism and the Persistence of Racial Inequality in the United States*. Lanham, MD: Rowman & Littlefield.

Branscombe, Nyla R., and Bertjan Doosje. 2004. *Collective Guilt: International Perspectives*. Cambridge: Cambridge University Press.

Branscombe, Nyla R., Bertjan Doosje, and Craig McGarty, C. 2002. "Antecedents and Consequences of Collective Guilt." In *From Prejudice to Intergroup Emotions: Differentiated Reactions to Social Groups*, edited by Diane M. Mackie and Eliot R. Smith, 49–66. New York: Psychology Press.

Branscombe, Nyla R., Ben Slugoski, and Diane M. Kappen. 2004. "The Measurement of Collective Guilt: What It Is and What It Is Not." In *Collective Guilt: International Perspectives*, edited by Nyla R. Branscombe and Bertjan Doosje, 16–34. Cambridge: Cambridge University Press.

Brown, Rupert, Roberto González, Hanna Zagefka, Jorge Manzi, and Sabina Čehajić. 2008. "Nuestra Culpa: Collective Guilt and Shame as Predictors of Reparation for Historical Wrongdoing." *Journal of Personality and Social Psychology* 94, no. 1: 75–90.

Case, Kim A. 2007. "Raising White Privilege Awareness and Reducing Racial Prejudice: Assessing Diversity Course Effectiveness." *Teaching of Psychology* 34, no. 4 (October): 231–35.

Chao, Ruth Chu-Lien, Meifen Wei, Lisa B. Spanierman, Joseph Longo, and Dayna Northart. 2015. "White Racial Identity Attitudes and White Empathy: The Moderation Effect of Openness to Diversity." *The Counseling Psychologist* 43, no. 1 (September): 94–120.

De Hooge, Ilona E., Marcel Zeelenberg, and Seger M. Breugelmans. 2007. "Moral Sentiments and Cooperation: Differential Influences of Shame and Guilt." *Cognition and Emotion* 21, no. 5 (July): 1025–42.

Doosje, Bertjan, Nyla R. Branscombe, Russell Spears, and Antony S. R. Manstead. 1998. "Guilty by Association: When One's Group has a Negative History." *Journal of Personality and Social Psychology* 75, no. 4: 872–86.

Doosje, Bertjan, Nyla R. Branscombe, Russell Spears, and Antony S. R. Manstead. 2006. "Antecedents and Consequences of Group-Based Guilt: The Effects of Ingroup Identification." *Group Processes & Intergroup Relations* 9, no. 3: 325–38.

Ent, Michael R., and Roy F. Baumeister. 2016. "The Functions of Guilt." *Emotion Researcher*, May 2016. https://emotionresearcher.com/the-functions-of-guilt/.

Garriott, Patton O., Stephanie Reiter, and Jenna Brownfield. 2016. "Testing the Efficacy of Brief Multicultural Education Interventions in White College Students." *Journal of Diversity in Higher Education* 9, no. 2: 158–69.

Greenaway, Katharine H., Kylie Fisk, and Nyla R. Branscombe. 2017. "Context Matters: Explicit and Implicit Reminders of Ingroup Privilege Increase Collective Guilt among Foreigners in a Developing Country." *Journal of Applied Social Psychology* 47, no. 12 (August): 677–81.

Grzanka, Patrick R., Keri A. Frantell, and Ruth E. Fassinger. 2020. "The White Racial Affect Scale (WRAS): A Measure of White Guilt, Shame, and Negation." *The Counseling Psychologist* 48, no. 1: 47–77.

Harth, Nicole S., Thomas Kessler, and Collin W. Leach. 2008. "Advantaged Group's Emotional Reactions to Intergroup Inequality: The Dynamics of Pride, Guilt, and Sympathy." *Personality and Social Psychology Bulletin* 34, no. 1 (January): 11–29.

Hawley, George. 2017. "How Many People Actually Feel 'White Guilt'?" *American Conservative*, April 11, 2017. https://www.theamericanconservative.com/articles/how-many-people-actually-feel-white-guilt/.

Hughes, Coleman. 2020 "Stories and Data: Reflections on Race, Riots, and Police." *City Journal*, June 14, 2020. https://www.city-journal.org/reflections-on-race-riots-and-police.

Imhoff, Roland, Michał Bilewicz, and Hans-Peter Erb. 2012. "Collective Regret versus Collective Guilt: Different Emotional Reactions to Historical Atrocities." *European Journal of Social Psychology* 42, no. 6 (June): 729–42.

Imhoff, Roland, Michael J. A. Wohl, and Hans-Peter Erb. 2013. "When the Past Is Far from Dead: How Ongoing Consequences of Genocides Committed by the Ingroup Impact Collective Guilt." *Journal of Social Issues* 69, no. 1 (March): 74–91.

Iyer, Aarti, Collin W. Leach, and Faye J. Crosby. 2003. "White Guilt and Racial Compensation: The Benefits and Limits of Self-Focus." *Personality and Social Psychology Bulletin* 29, no. 1. (January): 117–29.

Iyer, Aarti, Colin W. Leach, and Anne Pedersen. 2004. "Racial Wrongs and Restitutions: The Role of Guilt and Other Group-Based Emotions." In *Off White: Readings on Power, Privilege, and Resistance*, edited by Michelle Fine, Lois Weiss, Linda Powell Pruitt, and April Burns, 345–61. New York: Routledge.

Iyer, Aarti, Toni Schmader, and Brian Lickel. 2007. "Why Individuals Protest the Perceived Transgressions of Their Country: The Role of Anger, Shame, and Guilt." *Personality and Social Psychology Bulletin* 33, no. 4 (April): 572–87.

Jensen, Robert. 2005. *The Heart of Whiteness: Confronting Race, Racism and White Privilege*. San Francisco: City Lights Books.

Kernahan, Cyndi, and Tricia Davis. 2007. "Changing Perspective: How Learning about Racism Influences Student Awareness and Emotion." *Teaching of Psychology* 34, no. 1: 49–52.

Klandermans, Bert, Merel Werner, and Marjoka van Doorn. 2008. "Redeeming Apartheid's Legacy: Collective Guilt, Political Ideology, and Compensation." *Political Psychology* 29, no. 3: 331–49.

Kleiman, Sela, Lisa B. Spanierman, and Nathan G. Smith. 2015. "Translating Oppression: Understanding How Sexual Minority Status Is Associated with White Men's Racial Attitudes." *Psychology of Men and Masculinity* 16, no. 4: 404–15.

Kordesh, Kathleen, Lisa B. Spanierman, and Helen A. Neville. 2013. "White Students' Racial Affect: Gaining a Deeper Understanding of the Antiracist Type." *Journal of Diversity in Higher Education* 6: 33–50.

Krauth-Gruber, Silvia, and Virginie Bonnot. 2020. "Collective Guilt, Moral Outrage, and Support for Helping the Poor: A Matter of System versus In-Group Responsibility Framing." *Journal of Community & Applied Social Psychology* 30, no. 1: 59–72.

Kuttner, Robert. 2018. "Stop Wallowing in Your White Guilt and Start Doing Something for Racial Justice: Spend Less Time Contemplating Privilege and More Time Acting, To Be Part of the Change." *American Prospect*, August 21. https://prospect.org/civil-rights/stop-wallowing-white-guilt-start-something-racial-justice/

Leach, Collin W., Aarti Iyer, and Anne Pedersen. 2006. "Anger and Guilt about Ingroup Advantage Explain the Willingness for Political Action." *Personality and Social Psychology Bulletin* 32, no. 9 (September): 1232–45.

Leonardo, Zeus. 2004. "The Color of Supremacy: Beyond the Discourse of 'White Privilege.'" *Educational Philosophy and Theory* 36, no. 2: 137–52.

Lorde, Audre. 1984. *Sister Outsider: Essays and Speeches*. Berkeley, CA: Crossing Press.

Mallett, Robyn K., Jeffrey R. Huntsinger, Stacey Sinclair, and Janet K. Swim. 2008. "Seeing through Their Eyes: When Majority Group Members Take Collective Action on Behalf of an Outgroup." *Group Processes & Intergroup Relations* 11, no. 4: 451–70.

Mallett, Robyn K., and Janet K. Swim. 2007. "The Influence of Inequality, Responsibility and Justifiability on Reports of Group-Based Guilt for Ingroup Privilege." *Group Processes & Intergroup Relations* 10, no. 1: 57–69.

Marquette University Law School Poll. 2020. "Black Lives Matter Protests in Wisconsin: Summary." *RPubs*, August 8, 2020. https://rpubs.com/PollsAndVotes/652966.

McGarty, Craig, Anne Pedersen, Collin W. Leach, Tamarra Mansell, Julie Waller, and Ana-Maria Bliuc. 2005. "Group-Based Guilt as a Predictor of Commitment to Apology." *British Journal of Social Psychology* 44, no. 4: 659–80.

Mekawi, Yara, Konrad Bresin, and Carla Hunter. 2016. "White Fear, Dehumanization, and Low Empathy: Lethal Combinations for Shooting Biases." *Cultural Diversity and Ethnic Minority Psychology* 22, no. 3: 322–32.

Miceli, Maria, and Cristiano Castelfranchi. 2018. "Reconsidering the Differences between Shame and Guilt." *Europe's Journal of Psychology* 14, no. 3: 710–33.

Neville, Helen A., Germine H. Awad, James E. Brooks, Michelle P. Flores, and Jamie Bluemel. 2013. "Color-Blind Racial Ideology: Theory, Training, and Measurement Implications in Psychology." *American Psychologist* 68, no. 6: 455–66.

Nile, Lauren N., and Jack C. Straton. 2003. "Beyond Guilt: How to Deal with Societal Racism." *Multicultural Education* 10, no. 4: 2–6.

O'Keefe, Daniel J. 2000. "Guilt and Social Influence." *Annals of the International Communication Association* 23, no. 1: 67–101.

Paone, Tina R., Krista M. Malott, and Jason J. Barr. 2015. "Assessing the Impact of a Race-Based Course on Counseling Students: A Quantitative Study." *Journal of Multicultural Counseling and Development* 43, no. 3: 206–20.

Peetz, Johanna, Gregory R. Gunn, and Anne E. Wilson. 2010. "Crimes of the Past: Defensive Temporal Distancing in the Face of Past In-Group Wrongdoing." *Personality and Social Psychology Bulletin* 36, no. 5 (April): 598–611.

Pew Research Center. 2016. "On Views of Race and Inequality, Blacks and Whites are Worlds Apart." Washington, DC: Pew Research Center. https://www.pewsocialtrends.org/2016/06/27/on-views-of-race-and-inequality-blacks-and-whites-are-worlds-apart/.

Pew Research Center. 2020. "Amid Protests, Majorities across Racial and Ethnic Groups Express Support for the Black Lives Matter Movement." Washington, DC: Pew Research Center. https://www.pewsocialtrends.org/wp-content/uploads/sites/3/2020/06/PSDT_06.12.20_protest_fullreport.pdf.

Pinterits, E. Janie, V. Paul Poteat, and Lisa B. Spanierman. 2009. "The White Privilege Attitudes Scale (WPAS): Development and Initial Validation." *Journal of Counseling Psychology* 56, no. 3: 417–29.

Poteat, V. Paul, and Lisa B. Spanierman. 2008. "Further Validation of the Psychosocial Costs of Racism to Whites Scale among a Sample of Employed Adults." *The Counseling Psychologist* 36: 871–94.

Powell, Adam A., Nyla R. Branscombe, and Michael T. Schmitt. 2005. "Inequality as Ingroup Privilege or Outgroup Disadvantage: The Impact of Group Focus on Collective Guilt and Interracial Attitudes." *Personality and Social Psychology Bulletin* 31, no. 4: 508–21.

Raypole, Crystal. 2020. "Yes, There's a Difference between 'BIPOC' and 'POC'—Here's Why It Matters." *Healthline*, September 17. https://www.healthline.com/health/bipoc-meaning.

Sifford, Amy, Kok-Mun Ng, and Chuang Wang. 2009. "Further Validation of the Psychosocial Costs of Racism to Whites Scale on a Sample of University Students in the Southeastern United States." *Journal of Counseling Psychology* 56: 585–89.

Smith, Kyle. 2020. "The White-Guilt Cult." *National Review*, June 18. https://www.nationalreview.com/magazine/2020/07/06/the-white-guilt-cult/#slide-1.

Smith, Robert C. 1995. *Racism in the Post-Civil Rights Era: Now You See It, Now You Don't.* Albany: State University of New York Press.

Soble, Jason R., Lisa B. Spanierman, and Hsin-Ya Liao. 2011. "Effects of a Brief Video Intervention on White University Students' Racial Attitudes." *Journal of Counseling Psychology* 58, no. 1: 151–57.

Spanierman, Lisa B., Jacquelyn C. Beard, and Nathan R. Todd. 2012. "White Men's Fears, White Women's Tears: Examining Gender Differences in Racial Affect Types." *Sex Roles* 67, no. 3–4: 174–86.

Spanierman, Lisa B., and Mary J. Heppner. 2004. "Psychosocial Costs of Racism to Whites Scale (PCRW): Construction and Initial Validation." *Journal of Counseling Psychology* 51, no. 2: 249–62.

Spanierman, Lisa B., V. Paul Poteat, Amanda M. Beer, and Patrick I. Armstrong. 2006. "Psychosocial Costs of Racism to Whites: Exploring Patterns through Cluster Analysis." *Journal of Counseling Psychology* 53: no. 4: 434–41.

Spanierman, Lisa B., V. Paul Poteat, Ying-Fen Wang, and Euna Oh. 2008 "Psychosocial Costs of Racism to White Counselors: Predicting Various Dimensions of Multicultural Counseling Competence." *Journal of Counseling Psychology* 55, no. 1: 75–88.

Spanierman, Lisa B., Nathan R. Todd, and Carolyn J. Anderson. 2009. "Psychosocial Costs of Racism to Whites: Understanding Patterns among University Students." *Journal of Counseling Psychology* 56, no. 2: 239–52.

Steele, Shelby. 2006. *White Guilt: How Blacks and Whites Together Destroyed the Promise of the Civil Rights Era*. New York: Harper Collins.

Swim, Janet K., and Deborah L. Miller. 1999. "White Guilt: Its Antecedents and Consequences for Attitudes toward Affirmative Action." *Personality and Social Psychology Bulletin* 25, no. 4: 500–14.

Tangney, June P. 1991. "Moral Affect: The Good, the Bad, and the Ugly." *Journal of Personality and Social Psychology* 61, no. 4: 598–607.

Tangney, June P., and Ronda L. Dearing. 2002. *Shame and Guilt*. New York: Guilford Press.

Thompson, Audrey. 2003. "Tiffany, Friend of People of Color: White Investments in Antiracism." *International Journal of Qualitative Studies in Education* 16, no. 1: 7–29.

Tignor, Stefanie M., and C. Randall Colvin. 2017. "The Interpersonal Adaptiveness of Dispositional Guilt and Shame: A Meta-Analytic Investigation." *Journal of Personality* 85, no. 3: 341–63.

Todd, Nathan R., Lisa B. Spanierman, and Mark S. Aber. 2010. "White Students Reflecting on Whiteness: Understanding Emotional Responses." *Journal of Diversity in Higher Education* 3, no. 2: 97–110.

Turner, Marcée M. 2011. "Personality Factors, Multicultural Exposure, and Cultural Self-Efficacy in a Model of White Fear." PhD diss., University of Notre Dame.

Vaish, Amrisha, Malinda Carpenter, and Michael Tomasello. 2016. "The Early Emergence of Guilt-Motivated Prosocial Behavior." *Child Development* 87, no. 6: 1772–82.

Wolf, Christine. 2020. "How to Be an Anti-racist When You're Paralyzed by White Guilt." *Medium*, June 29, 2020. https://medium.com/women-this-way/how-to-be-an-anti-racist-when-youre-paralyzed-by-white-guilt-a6509afa4350.

Zebel, Sven, Anja Zimmermann, G. Tendayi Viki, and Bertjan Doosje. 2008. "Dehumanization and Guilt as Distinct but Related Predictors of Support for Reparation Policies." *Political Psychology* 29, no. 2: 193–219.
Zimmermann, Anja, Dominic Abrams, Bertjan Doosje, and Antony S. R. Manstead. 2011. Causal and Moral Responsibility: Antecedents and Consequences of Group-Based Guilt. *European Journal of Social Psychology* 41, no. 7: 825–39.

3

From Shame to Guilt

Indonesian Strategies against Child Marriage

Nelly van Doorn-Harder

Introduction

In Indonesia, feminist Muslim activists use guilt and shame strategies to challenge and reform the widespread practice of child marriage. While we can distinguish between two types of child marriage—a union between two underage spouses, and a union between an underage girl and an older man—the latter is considered more problematic since, in most cases, the girl bride cannot consent. The battle to criminalize child marriage via legal means and by changing cultural practices and the interpretation of the Qur'an and the Hadith (also known as the Tradition) involves negotiations with powerful men, especially male guardians, who are responsible for the welfare of young girls and can therefore be held accountable for child brides' high divorce rates and, subsequently, the lifelong impoverishment of women and their children. Guardians can be shamed for their failure to protect women and children. Such guilt strategies aim to change the discourse about women's rights and misogynist practices by appealing to the need for education, which is in the long-term national interest.

This chapter discusses the strategies used against child marriage developed by two organizations that focus on women's rights: the governmental women's rights organization Komnas Perempuan and the Muslim organization Rumah KitaB. The goal of these and related organizations is to change the Indonesian mindset about child brides. They seek to empower women and to change gender-biased discourses still prevalent across Indonesia by creating new social connection models that involve male and female religious, civil, and political leaders. These activists evoke feelings of guilt and shame by referring to Islamic sacred texts and their legal interpretations. Furthermore, they appeal to Indonesia's goals toward national prosperity. Particularly in recruiting religious and civil leaders in the Indonesian legal establishment, they draw on sentiments of national pride to induce a sense of shame and guilt.

Child marriage is prevalent across Indonesia. The country ranks as having the eighth-highest number of child brides in the world, with a large number of girls

Nelly van Doorn-Harder, *From Shame to Guilt* In: *Guilt*. Edited by: Katharina von Kellenbach and Matthias Buschmeier, Oxford University Press. © Oxford University Press 2022. DOI: 10.1093/oso/9780197557433.003.0004 marrying before the age of 18.[1] Research shows that early marriage can lead to a series of health and other problems, including increased risk of maternal and infant mortality, reproductive and mental health problems, domestic violence, and life-long poverty, especially when the couple divorces. A high divorce rate is often due to an age gap between bride and groom, as an age gap can, among other detrimental outcomes, yield adverse impacts, such as incompatibility of interests, mismatches in maturity level, and hindrances to the economic independence of the younger partner.

Muslim activists have been fighting against the phenomenon of child brides for decades. At the governmental level, some of Indonesia's highest authorities, such as the Minister for Women's Empowerment and Child Protection, along with President Jokowi Widodo, agree with their arguments. With nearly 42% of Indonesia's population under the age of 24, the government has designed a national strategy, starting in 2016, to support its young population and eliminate violence against children. It is fully committed to ending the practice of child marriage.[2] Via a 2018 Presidential Decree, the government tried to enshrine the protection of child and teenage girls in the Marriage Law by raising the legal marriage age from 16 to 19 (cf. Harsono 2018). After intense lobbying from women's rights activists, the Constitutional Court ratified this decree in 2019. Muslim teenagers who want to marry at an earlier age must now appeal to a judge in the Islamic Family Court.

The new rule seems straightforward and easy to apply. However, activist groups and organizations seeking to strengthen the rights of women consider it only the beginning of an extended battle against a practice that has long been the social, cultural, and religious norm in large parts of the country. Especially since the fall of President Suharto in 1998, they have intensified their struggle for gender justice; during the post-Suharto era, extremist-minded Muslim groups emerged that propagate and encourage child marriage. Riding on the waves of a so-called moral panic (Smith-Hefner 2009; 2019, 121–3), one of the extremist arguments is that child marriage prevents sex before marriage and thereby preserves the good name of the family.

Guilt and God-Given Rights

The issue of child marriage overlaps with that of polygamy.[3] Addressing it with psychological, economic, or health arguments often fails, since those advocating it base their own arguments on spiritual traditions. Following Qur'anic references, women in a polygamous union believe that they must accept their condition and that they disobey God's will if they reject it (Nurmila 2009, 39);

meanwhile, men who practice polygamy and child marriage refer to God-given rights to do so.

The Qur'an speaks about polygamy (Q 4:3), while the Hadith supports child marriage. While the Qur'an is considered the direct Word of God, the core texts of the Hadith transmit the words and deeds of the Prophet Muhammad. According to the Hadith, the Prophet married his favorite wife, A'isyah, when she was six or seven and consummated the marriage when she was around 10. Based on this precedent, Islamic Law does not mention a minimum marriage age for girls. Nowadays, in most majority-Muslim countries, the state has set a minimum marriage age for girls at 15.

Although the Hadith is not considered divine revelation, it has gained a level of scriptural authority over time comparable to that of the Qur'an. Women's rights activists reason that there is a difference between the scriptures and their interpretations. However, according to Nina Nurmila, many Muslims "tend to equate human interpretation with the Qur'an itself and to regard this interpretation as divine and immutable" (Nurmila 2009, 39). Activists argue that since circumstances and opinions change over time, especially where it concerns social issues such as polygamy and child marriage, the holy texts need to be interpreted contextually, not literally, and that, to protect women and children, Indonesian Muslims should follow the Marriage Act rather than the Islamic injunctions (Ali et al. 2015, 25–8; Nurmila 2009).

Komnas Perempuan and Rumah KitaB activists never expressly use the words "guilt" or "shame." However, several of their strategies result in forms of transformative guilt, both personal and collective. On the personal level, this guilt relates to the objective observation that someone has committed a wrong, offense, or crime, or has failed concerning an obligation to others; alternatively, it can be the *feeling* that someone committed a wrong or failed in an obligation. When collective, the guilt that emerges becomes a kind of guilt by association based on group membership and the feeling that one's group has caused hardship or committed a wrong.

Research about guilt and shame in Indonesia (Collins and Bahar 2000; Lindquist 2004; Pakpahan 2016) indicates that in spite of local cultural differences, productive forms of guilt and shame connect to the welfare of the broader Indonesian nation, where all citizens are inclined to support macro-scale projects to reach economic and developmental goals. Hence, when playing the "guilt-shame card," organizations such as Komnas Perempuan and Rumah KitaB suggest indirectly that those involved in child marriage and polygamy obstruct the future and prosperity of the nation.

In the Indonesian landscape of emotions, the fact that thousands of different ethnic groups use their own local languages complicates speaking about shame and guilt. Local dialects have a range of words denoting and/or connecting these emotions (Heider 1991, 84, 117). In a sense, the two emotions intersect—although references to shame (*malu*) prevail over guilt (*salah*) in Bahasa Indonesia, the lingua franca. In this chapter, I limit my discussion to the use of these two words. This choice is based on the observation that, recently, researchers have started to notice a trend; namely, Indonesians are using the word *malu* more often than the one used in their local language (e.g., Stodulka 2008).

Malu and *Salah*

At the same time, research about situations in the West indicates that divorcing the two emotions can be difficult. People often refer to them interchangeably (Lickel, Steele, and Schmader 2011, 154). Helen B. Lewis has observed that "the simultaneous stirring of shame along with guilt for moral transgression may result in a fusion of shame and guilt feeling. The experience is identified, however, only as guilt, when, in fact, there are shame components operating outside awareness" (Lewis 1971, 431). She describes guilt as a force that seeks to restore a social relation that has been damaged, while shame "leads to casting outside the community, exile; a guilt morality, to suffering in order to be accepted back within" (62). And April and Mooketsi (2010, 69–70) write:

> Shame is a social emotion, related to the entire self. It is linked to ideals while guilt concerns prohibitions. Guilt and shame are internal affective states that often arise from similar situations but have different effects on the individual. Shame is oriented towards the self, while guilt is oriented towards others.

In Indonesia, shame, *malu*, is such a powerful social emotion that Goddard (1996, 432) has suggested using the term as the main lens through which to understand Malay culture. Several ethnographies have analyzed the use of the term and the register of emotions it covers (Geertz 1973; Keeler 1983; Errington 1989; Heider 1991; Collins and Bahar 2000). While many argue that emotional concepts are linked to culture, language, and place, at the basic level, *malu* means feeling bad or looking bad because something is not right or differs from what is customary. Lindquist (2004) translates *malu* as "shame, embarrassment, shyness, or restraint and propriety." According to Pakpahan (2016, 8), "[t]he Indonesian word concerns that which is customary, that which involves other people and not only the conscience of the person." In his view, shame is more relational than individual.

Meanwhile, guilt, in Indonesian society is primarily a legal term; one is guilty of a certain offense or trespass that is punishable by law. It can also involve a social judgment when something is wrong or should not be done. Researching the widespread corruption that is practiced with impunity in Indonesia, Pakpahan concludes, "the separation of guilt and shame in the Indonesian language context also plays a role in the absence of positive shame. The society will act as a social conscience if anyone breaks the rule. When shame is separated from guilt, the society is not in charge anymore in deciding whether someone is guilty" (Pakpahan 2016, 20).

Social psychology has found that the emotion of guilt relates to behavior that helps restore damaged relationships and balances situations of unequal power between individuals and groups (Baumeister, Stillwell, and Heatherton 1994; Branscombe and Doosje 2004). Guilt can thus play an important role in regulating social relations. At the same time, local cultures influence interpretations of guilt and shame. In individualistic societies such as the European and North American ones, guilt depends on external influences and is effective as a moral regulator. In shame-based societies, similar mechanisms obtain, as people need the approval of others (Pakpahan 2016, 6). Lickel, Steele, and Schmader's (2011, 156) research found that, in general, shame "may promote proactive attempts to repair the tarnished image of one's group."

Especially in the post-Suharto era, researchers have found that shame, or *malu*, in certain circumstances can become a productive force. Collins and Bahar conclude that "*[m]alu* has special power to shape human relations because it is based in deeply rooted emotional orientations that are not always accessible to rational critique." While it defines what kind of behavior is disrespectful and shameful, "*malu* can be reshaped under the pressure of social and economic changes" (Collins and Bahar 2000, 63). In the context of Indonesian migrant workers, Lindquist (2004, 489) observes that "*malu* should be understood in relation to the emergent identities connected to the nation." Researching how street youth in Java deal with feeling *malu*, Stodulka (2008, 336) noticed that "everyday life is constituted within the framework of the nation, rather than exclusively within the context of a local community in which culture, place, and language appear to easily correspond."

Collins and Bahar presented an interesting example of how, in the minds of many Indonesians, matters of everyday life and the well-being of the nation are intertwined. After analyzing what was considered *malu* during the Suharto era, they concluded that the Suharto regime mobilized *malu* in order to constrain individualism and support relations of hierarchy and deference. After the regime fell, Indonesians moved away from focusing on hierarchic relations and started to stress individual support of ethical behavior, lack of self-interest, and concern for others in order to help the nation thrive (Collins and Bahar 2000, 37, 68). Furthermore, the researchers found that social and economic circumstances seem to influence the emotions of shame and guilt, and in fact can lead to behavioral change, for example, where it concerns attitudes toward different forms of corruption that are widespread across the nation.

Komnas Perempuan and Rumah KitaB

When in the spring of 1998, during the transition period from the Suharto regime (1966–1998) to democracy, large-scale communal riots erupted, sexual assaults on women increased significantly. This was not the first time such patterns of violence against women occurred. It was an open secret that, during military operations, the regime's security forces violated human rights on a staggering scale. Igreja's chapter in this volume mentions some of the atrocities that Indonesian troops committed during the occupation of Timor-Leste (1975–1999), for example. Since troops targeted women especially, Indonesian civil society groups insisted that the state accept responsibility for this particular form of gendered violence and launched Komnas Perempuan (Komisi Nasional Anti Kekerasan terhadap Perempuan), the National Commission on Violence against Women.[4] The organization's goal was to monitor and advise on Indonesia's adherence to the International Covenant on Economic, Social and Cultural Rights, the Convention on the Elimination of All Forms of Discrimination against Women (CEDAW), and the Convention on the Rights of the Child.[5] Its mandate to report gender-based human rights abuses and create awareness among the Indonesian public is difficult to implement and requires continuous efforts to highlight the issue of violence against women. From the beginning, one of its main goals has been to change the pervasive mindset that ignores women's plight and blames the victim. (Although not specifically focused on the plight of women, Katharina von Kellenbach's chapter in this volume provides painful examples of the power of victim-blaming in Indonesian society.) At the national level, the organization monitors the application of existing laws protecting women's basic rights and advocates for the formulation of new laws (for example, raising the marriage age for women). In order to bring its message to a wider audience, it also connects with hundreds of other groups advocating for the rights of women, which translate the reports, videos, and other materials that Komnas Perempuan produces and disseminates at different levels of society, from the provincial and county levels to the local level.

While Indonesia is a Muslim-majority country, groups connected with Komnas Perempuan, depending on their location, reflect the multi-religious character of the country. One such group is Rumah KitaB (Rumah Kita Bersama), an Islamic nongovernmental organization (NGO) that operates as a think tank focusing on the interpretation of key religious texts that are used at institutes for Islamic higher education and Qur'an schools. These schools educate the future leaders of

Islam, who can become influential voices in helping either to promote or diminish early marriage practices.[6] When defending underage marriage, Indonesian Muslims refer to Islamic legal injunctions and find support for their arguments from local preachers, scholars, or judges at the Islamic Court. Since Rumah KitaB is a Muslim organization, its goal is to transform Indonesia's Muslim society by addressing the various paradigms, moral and ethical norms, values, and teachings that have shaped current ways of thinking within society.[7] Since its inception, its researchers have studied various issues, such as why women join radical-minded Muslim groups, how to alleviate poverty among divorcees, and concerns pertaining to child marriage (Marcoes-Natsir and Octavia 2014; Ali et al. 2015).

The demographic that Rumah KitaB particularly hopes to reach are the students (the *santri*) at Qur'an schools called *pesantren*. Upon graduation, many *pesantren* graduates continue their studies at an Islamic university. Then, after earning a BA or MA degree, they become part of the vast network of teachers, local imams, judges in the religious courts, and civil servants working in institutions such as the governmental offices that regulate matters of personal status law, such as marriage and divorce. Many of these professionals hold state-sponsored jobs that require them to attend post-graduate courses as well. Research among participants in such courses has shown that at the grass-roots level, village imams and other religious officials are often inspired by radical teachings about the status of women, polygamy, underage, and secret or unregistered marriage (Maufur, Hasan, and Zuhri 2014). These findings are significant since, in rural areas, the poor and the uneducated take their clues from local imams, who advise parents to give a young daughter in marriage and that an unregistered marriage, despite its negative social and economic repercussions, is theologically acceptable since it follows Islamic guidelines.

Assuming that the audience for this book might be unfamiliar with the Islamic injunctions pertaining to marriage in the Indonesian context, I shall next provide some background information before returning to the topic of guilt.

Indonesian: Not Arab!

In Muslim circles, certain districts, and rural areas, child marriage has long been an accepted practice. Socioeconomic factors such as poverty and local customs, as well as patriarchal structures and rigid gender norms that normalize male-perpetrated violence against women, support the practice.

Investigating the crowded field of radical Islamic ideals and the competition for religious authority, Rumah KitaB has found that two books that are widely available on the Internet, by Yemeni authors Abu Ammar Ali al-Hudzaifi and Arif Ibn Ahmad al-Shabri, had the most influence within extremist circles (Ali et al. 2015,

82). Child marriage is widespread in Yemen and at times causes international uproar; for example, in an internationally publicized case from 2010, a 13-year-old girl died of internal injuries after the wedding night (Mail Foreign Service 2010). However, the two Yemeni authors argue that forbidding the practice goes against Islamic law. Their writings show that resistance against abolishing the practice also emerges from rumors about Western conspiracies that ostensibly lure Muslims into unlawful practices such as fornication and homosexuality, or that are presumed to cause breast cancer and pregnancy complications (Ali et al. 2015, 82–94). Although these ideas are widespread among some conservative and radical Muslims, several ultra-conservative Indonesian platforms dedicated to defending the family have adopted these arguments too, broadcasting them in sophisticated ways via popular media. Rumah KitaB, however, stressing that these scholars read the holy texts in a literal fashion, scrutinized the various arguments they presented with the goal of convincing Indonesians (especially local Muslim leaders) that the practice of child marriage was a cultural (more than a religious) practice and is not an Islamic injunction.

We are thus observing a de facto battle about which interpretation of Islam Indonesians could or should follow. According to literalist Muslim groups, Indonesians should follow voices from the Arab world. On the other hand, the majority of Indonesian scholars and religious activists disagree with this idea; they argue that Indonesian Islam is not Arab Islam and never will be. Yet when, in 1998, the Suharto regime fell and the country's political system became more democratic, Islamic movements whose main goal was to align Indonesian Islam more closely with Middle Eastern Arab interpretations started to influence the country's public life. The country's newly found democratic freedoms not only allowed for a pluralization of Islamic ideals, but also led to a fragmentation of religious authority. Communal boundaries were redrawn, and relatively small numbers of extremist Muslim thinkers disproportionately influenced the creation of Islamic laws and regulations. In this crowded landscape, women and their bodies, roles, and rights became the battleground of cultural transformation. In particular, the face veil that few Muslim women had previously used became of primary importance to signal adherence to Islam (cf. Brenner 2011; Smith-Hefner 2007).

However, the majority of Indonesia's Muslim leaders support the idea that Indonesian Islam is firmly embedded in local cultures and that the state remains democratically governed under the ideological platform of the pluralistic model called "Pancasila," which, in principle, sanctions the full legal presence of Christian, Hindu, Buddhist, and Confucian communities (Azra 2013) alongside the Muslim one. Furthermore, a distinctive feature of Indonesian Islam is that for nearly half a century, many Indonesian Muslim leaders have allowed women to hold religious and secular leadership roles. This development is also discernible in mainstream Muslim organizations, of which Nahdlatul Ulama (NU) and Muhammadiyah are the largest and "can be seen as a perfect representation of Islamic-based civil society" (Azra 2013, 71).

Mobilizing against the Harm of Child Marriage

Child marriage exemplifies Indonesia's multilayered and complex legal and religious reality, since it finds itself at the ideological confluence of the country's cultural practices, marriage law, interpretations of the Islamic holy texts, and competing views within Indonesian society on the role and rights of women. The practice of child marriage touches on several priorities of the Indonesian and Muslim feminist agendas. Medical, social, and other research has shown that marrying young does considerable harm to a girl. It threatens her reproductive health. It also leads to economic disadvantage; with little education, young women who have children and whose marriages end in divorce face a lifetime of poverty. Based on these negative consequences, child marriage is now a prime form of violence against women.

From a legal point of view, marriages involving underage brides are contracted in secret, since such marriages cannot be registered at the Office for Religious Affairs (the Kantor Urusan Agama, or "KUA"), and Indonesian law forbids marriage arrangements that do not go through this office. In addition, a young bride often becomes part of a polygamous union.

Ever since the 1974 Marriage Law, child marriage has been expressly forbidden in the country. This law was the result of prolonged debates over the question of whether a single set of marriage-related rules should apply to all Indonesians, or if separate statutes should rule different ethnic and religious communities. Disagreements among the Muslims involved in drafting the law concerned the status of Islamic injunctions, such as the issue of polygamy, for which the Marriage Law provides strict rules. When a couple ignores these rules, the union is necessarily treated as a secret, which means that it remains unregistered. Furthermore, Indonesia's legal establishment considers children born out of unofficial unions as illegitimate, a status that affects the child's entire life.

The reality that a large part of Indonesia's Muslim population (in principle) supports child marriage requires a multilevel approach when fighting against it. Anti-child-marriage activism targets the political establishment and the judiciary (civil as well as religious law), reinterprets the Islamic holy texts, seeks fatwas (nonbinding legal advice) against the practice, appeals to public opinion, questions cultural practices, and teaches gender justice. They target local Muslim leaders and judges. Local leaders officiate illegal marriages, and couples often appeal to the Islamic court to get their union registered anyway.

On the political and legal level, Komnas Perempuan seeks to strengthen laws that protect the rights of girls. Furthermore, its members design educational campaigns to inform the public about changes in the law and the rights of women and children, the risks of marrying early, and the right to refuse. Such campaigns use national media, digital platforms, and local meetings in the marketplace, neighborhood associations, and private Qur'an study groups (Gunawan and Hayati 2019; Ali et al. 2015; Afifah 2013).

Rumah KitaB, furthermore, has produced writings that provide religious leaders with Islamic arguments against the practice. In 2015 the organization published a book reinterpreting the religious texts in the Qur'an and the Hadith with respect to the subject of child marriage (Ali et al. 2015). The book is based on extensive research and analyzes the historical and religious context of the practice. Rumah KitaB, when reading and interpreting the holy texts, takes the position that such interpretations should be placed within contemporary contexts and needs, including when considering the rules on child marriage found in Islamic jurisprudence *(fiqh)* (Ali et al. 2015, 23). Moreover, on the sociological level, Rumah KitaB emphasizes issues that are important to society, such as a girl's schooling, health, and psychological maturity.

All the activists find the key to women's liberation in reinterpreting the Qur'an and the Hadith from the perspective of gender equality. To them, while men have misread and abused these texts to subordinate women[8] throughout history, the holy texts support women's basic rights. Namely, the activists' most important religious argument against child marriage is that God created men and women from one soul to be each other's companions (Qur'an 7:189 and 4:1);[9] this argument underscores the expectation that a marriage is a union between equals, which is often not the case with child brides, who are markedly younger than their husbands (Ali et al. 2015, 28). Such an incompatibility between spouses leads to high divorce rates, unstable families, and an increase in psychological problems.

Komnas Perempuan and Rumah KitaB, furthermore, support high-level Islamic initiatives that condemn the practice. A powerful example is the international congress known as "KUPI"—Kongres Ulama Perempuan Indonesia, or, in translation, the Indonesian Congress of Female Ulama [religious authority]—which was held in April 2017. Over 500 feminists, male and female, attended this historic meeting. According to one of the organizers, its main goal was to

> build long-term perspectives on the rights of women that are currently being ignored. Many of these come down to biological and social issues that harm men as well as women. However, what is bad for men is far worse for women. The goal was to formulate what benefits religion and strengthens *the welfare of the Indonesian society.*[10]

Most of the women present were authorities of Islam, or ulama, who equaled their male counterparts in advanced understanding of the Islamic source texts. In her comment at the event, the organizer Dr. Nur Rofiah declared the congress "a universal declaration of equality, a manifesto that women have the same spiritual and mental potentials as men," thereby referencing the Qur'an verses 7:189 and 4:1.[11] Over the course of three days, the participants discussed the Islamic teachings that influence Islamic extremism, violence against women, and environmental problems. Rumah KitaB and Komnas Perempuan provided the materials in support of the foundational arguments against child marriage. The congress concluded with the passage of several statements, comparable to religious rulings of a fatwa. One of these was the decision to work toward a total ban on child marriage.

As mentioned earlier, however, apart from local leaders, it is the child's male guardian who holds primary decision-making power in her life. Initially, Rumah KitaB recruited mothers and women's organizations to advocate against child marriage, assuming that the women in a girl's family would try to prevent her from marrying early. However, their research found that the main factors influencing child marriage are structural and cultural, forces more strongly influenced by men than by women (Rumah KitaB 2018). Therefore, activists realized that they had to focus on the opinions of male guardians in order to affect change.[12]

Appealing to Male Power and Guardians' Responsibility

Is shame a productive strategy to influence guardians who support child marriage and polygamous unions? Before she marries, a girl or a woman is under the protection of her guardian, or *wali*, which means that this guardian is in charge of her life and makes decisions about her eventual marriage. The *wali* is always a man: before marriage, this is her father, brother, or uncle, and it becomes her husband when she is married. In this sense, marriage is a contract whereby the woman's *wali* transfers her to the protection of her husband. If a premarital guardian feels remorse or guilt for having harmed a girl's life and future, he would certainly not acknowledge this publicly but rather handle and address the situation quietly. And these marital failures have so far not changed the reality that a large segment of Indonesia's Muslims continues to support child marriage in principle. To change this pervasive mindset is hard and slow work, which has been ongoing for over a century (Nurmila 2009, 46–52).

Nevertheless, at the familial level, guilt and shame over harm inflicted on women and their children play an important role. When a guardian is confronted with early marriage's negative repercussions—and the even-greater harmful consequences for the children of divorce—he understands that the blame falls on him when things go wrong. In other words, he recognizes that he should feel guilty or, at a minimum, that the girl's extended family and others involved can blame him for forcing her into a lifetime of poverty.

Of course, the bridegroom is also to blame, particularly as unregistered marriages actively break the law. He is guilty in a legal sense, and if the law were to be enforced, he might be legally convicted.

But for many Indonesian Muslim men there is no feeling of guilt (or sentiment of shame). A guardian might not feel guilty, since he can always argue that he has followed the Islamic injunctions over the state's laws, and that the appeal to a higher power trumps secular law. And few guardians have ever faced jail for forcing a girl to marry early anyway. Activists can therefore not assume that guardians feel shame or guilt. Indeed, the activists are aiming for collective and intergenerational guilt.

Indeed, inducing guilt among individual male actors who feel blameless is difficult. Therefore, Komnas Perempuan's and Rumah KitaB's strategies challenge social norms, cultural practices, and religious beliefs. Their campaigns aim not primarily at individual guardians but at the cultural, religious, and social structures that uphold the practice of child marriage. The entire community, directly or indirectly, is implicated in the harm of destroying a child bride's future. Within this framework of understanding, *all* individuals who do not actively try to prevent an early marriage carry the blame.

Although, in the Indonesian context, shame trumps guilt, we can identify four levels of guilt that are at play and can lead to positive change. The first is legal guilt, as in breaking the law. Then there are the levels of collective, moral, and intergenerational guilt. Theoretically speaking, if a girl's guardian forces or allows her to marry early, his decision is punishable by law. Currently, such legal repercussions are uncommon in practice, but this situation might change if the Indonesian state manages to enforce the existing laws concerning child marriage. However, while a guardian has the last word, he seldom acts alone—especially in Indonesia's collectivist culture, where loyalty to the family is highly valued and individual actions have consequences for the good name and honor of relatives. As a result, negative behavior within the family often incurs feelings of collective or group-based guilt (see Heider 1991).

Branscombe and Doosje, in their volume on international perspectives, define collective guilt as the "distress that group members experience when they accept that their ingroup is responsible for immoral actions that harmed another group" (Branscombe and Doosje 2004, 3). Within the Indonesian context, researchers who examined the role of empathy and collective guilt in repairing Muslim-Christian relationships after the country's sectarian violence found that focusing on what connects the two religions (e.g., shared identity) motivated participants to restore fractured relationships (Mashuri, Zaduqisti, and Alroy-Thiberge 2017, 34). That said, this phenomenon is not uncommon, nor even unique to Indonesia; in her chapter on white guilt in this volume, Lisa Spanierman argues that when experiencing feelings of collective guilt, individuals may engage in actions to alleviate such feelings, and she points to the link between empathic perspective-taking and collective action. Hence, we can surmise that collective guilt is a powerful motivator, with the potential to generate very positive outcomes.

Individual group members can also experience feelings of moral guilt. Tracy Isaac (2011, 78) relates personal moral guilt to "blameworthy moral responsibility at the collective level" (71). In the introduction to this volume, Matthias Buschmeier and Katharina von Kellenbach state that while sentiments of moral guilt, in themselves, do not lead to legal responsibility, individual group members can nonetheless view themselves as accountable. Therefore, the anti-child-marriage campaigns of Komnas Perempuan are especially effective when they make individuals and groups aware that they carry part of the blame for the fate of individual group members. For instance, the organization engaged in a campaign to rehabilitate female victims of the 1965–1966 anti-communist sexualized violence, who had been outcasts of society for half a century (Doorn-Harder 2019). The rehabilitation of victims aims to reverse the habitual shaming and blaming of victims. Katharina von Kellenbach, in her chapter in this volume, examines one artistic attempt to undo the effects of the anti-communist purges in Indonesia, which for more than fifty years blamed the victims and celebrated the perpetrators as heroes. And the Indonesian film co-director and film crew of *The Act of Killing*, fearing retribution, must still remain anonymous. Komnas Perempuan's campaign to remember and help female victims of this purge targeted its message toward the younger generation, who questions how Indonesian society could condone these atrocities (Doorn-Harder 2019). Due both to increased Internet access and to government efforts to raise the population's level of education, many members of younger generations have started to question the treatment of those women who were accused of having communist sympathies. Generational guilt felt by the younger generation has become one of the strongest forces in restoring justice for the female victims.

Conclusion: Playing the Guilt-Shame Card

Changing deeply engrained patterns of behavior requires new forms of awareness. Komnas Perempuan and Rumah KitaB aim to insert new information about child marriage into the spaces where the official Marriage Law and Islamic law intersect. At this juncture, guardians still act with impunity when they disregard child brides' well-being. While they assume that their power over these girls grants them legal immunity, the women's rights activists aim at showing them that, while legal according to Islamic Law, their actions are nonetheless in violation of the Marriage Law and, most of all, are shameful. They are guilty of destroying not just a child bride's life, but also that of her children—thus, those who support, facilitate, or condone child marriage obstruct the future of the nation.

Since Indonesian identities are bound to family, the actions of close relatives (and not just of one's own) can also evoke emotions of guilt and shame (Collins and Bahar 2000, 41). The social relations connected to the concept of shame (*malu*) emphasize harmony and group solidarity and enforce cultural expectations of acceptable or proper behavior (Collins and Bahar 2000, 42). However, according to Talal Asad (1993, 134), emotions are not just feelings but also "processes of power." Men hold power over women. Showing them the negative consequences of exercising this power might evoke emotions of *malu* that lead to behavioral changes.

Men who allow or force a girl to marry at a young age also must face the immense force of collective guilt when communities must absorb the poverty and obstacles these girls encounter after their marriages end in divorce. Furthermore, when a younger generation questions an older generation's actions, the latter must deal with forces of intergenerational guilt as well (Baumeister, Stillwell, and Heatherton 1994, 251). Younger generations of women will question why their grandmother, mother, aunt, or older sister is living in precarious circumstances that could have been avoided had she received a better education and not been taken out of school upon being forced to marry. Indeed, social and economic changes reshape feelings and expressions of *malu* also—not just of guilt (Collins and Bahar 2000, 63). These changes force individuals to question the consequences of their actions and open questions of responsibility and accountability.

This is the juncture where Indonesia's activists for women's rights, by playing on the interconnected emotions of shame and guilt, hope to bring about further change: by providing explanations, education, and information in the intellectual, cultural, and legal spaces that are opened by the social changes that have already begun. Activists for women's rights strive to find a middle way between following the holy scriptures, ensuring the well-being of girls, respecting Indonesia's many cultures, and securing the needs of the country. This is not a new challenge. During the 1990s, the prominent Muslim leader Muhammad Ali Yafie (born 1926) saw education for men and women as the main secret to the country's progress. His advice was that

> young people should be encouraged to study longer so that they achieve better employment and the quality of human resources in Indonesia will increase. The

tradition of marrying young belongs to a certain period and can no longer be seen as the general marriage pattern (Ali et al. 2015, 110).

The activists working in Rumah KitaB and Komnas Perempuan all raise questions about the type of society that Indonesians envision for future generations. As is the case in many other countries, in Indonesia, child marriage has become one of the symbolic touchstones in Islamic debates. Fusing religious interpretations of the Islamic texts with contemporary ideas of human rights, activists argue that Indonesia cannot afford to destroy the futures of generations of women and children. In the mind of KUPI leader Nur Rofiah, religion, social context, biological realities, and women's rights are all intertwined:

> [W]hen we improve the rights of women and women's issues are no longer being ignored, we not only look at our Holy texts with fresh eyes, but we strengthen the policies about these issues and strengthen our nation.[13]

With the government on their side, the activists are already witnessing modest results with respect to their efforts to raise the marriage age. Now that the minimum age is 19, however, Islamic judges have seen an increase in appeals to register unregistered marriages.[14] While immediate outcomes in changing patriarchal mindsets might be hard to measure, the activists will know that they have succeeded in changing the nation's mindset when, one day, the family and community members of a child bride will say "Shame on you!" to her guardian and cast blame on him for allowing the marriage—and when, with the government on their side, his legal guilt can finally result in an arrest or a fine.

Notes

1. According to the 2015 report by Coram International, 7.8% of Indonesian brides were 12–14 years old, and 30.6% were 15–17 at the time of marriage. (According to Indonesian law, the minimum age for girls is 16 and for boys is 19.) (Statistics from: PLAN International and Coram International, 2015, *Getting the Evidence: Asia Child Marriage Initiative Summary Report*. In *girlsnotbrides.org*, 7, accessed August 10, 2021, https://www.girlsnotbrides.org/learning-resources/resource-centre/getting-the-evidence-asia-child-marriage-initiative-summary-report/) These numbers are higher than the numbers given by UNICEF in 2014 that estimated that 21% of Indonesian women between the ages of 20–24 (inclusive) married before the age of 18, of whom 3% were under the age of 15. See: https://www.girlsnotbrides.org/articles/tackling-child-marriage-in-indonesia/
2. See the website of the international network End Violence against Children, at https://www.end-violence.org/impact/countries/indonesia.

3. In this chapter, I will use the word "polygamy" to describe the union structure of a man married to multiple spouses. Although, technically speaking, the word "polygyny" is more appropriate for this type of marriage, I follow the Indonesian custom of using the word "polygamy."
4. https://www.komnasperempuan.go.id/.
5. Indonesia is a party to the International Covenant on Economic, Social and Cultural Rights, the Convention on the Elimination of All Forms of Discrimination against Women (CEDAW), and the Convention on the Rights of the Child. Under these core human rights instruments, Indonesia is legally obligated to protect the rights of girls and women, including their right to freedom from discrimination; to the highest attainable standard of health; to education; to free and full consent to marriage; to choosing one's spouse; and to be free from physical, mental, and sexual violence.
6. https://rumahkitab.com/en/.
7. For the Rumah KitaB goals, see: http://rumahkitab.com/en/.
8. Some of the foundational Qur'anic teachings that feminists refer to are justice for all and equality between men and women (Q. 33:35); women and men were both created from the same spirit (*nafs*) (Q. 4:1, 6:98, 7:189); equality, especially concerning religious observance (Q. 33:35); and husband and wife as "each other's garments" (Q. 2:187).
9. Qur'an 7:89: "It is He who created you from one soul and created from it its mate that he might dwell in security with her," and Qur'an 4:1: "O mankind, fear your Lord, who created you from one soul and created from it its mate and dispersed from both of them many men and women."
10. Nur Rofiah interview, Jakarta, July 6, 2018.
11. Nur Rofiah interview, Jakarta, July 6, 2018.
12. While the other religions present in Indonesia (Christianity, Hinduism, Buddhism, and Confucianism) do not have the institution of the *wali*, culturally they follow a similar pattern that allows a male relative to make decisions on behalf of a girl or woman.
13. Nur Rofiah interview, Jakarta, July 6, 2019.
14. Lies Marcoes-Natsir, communication via email, August 26, 2020.

References

Afifah, Wiwik. 2018. "Pencegahan Perkawinan Dini Menggunakan Publik Space di Taman Bungkul Kota Surabaya." *Jurnal Masyarakat Mandiri JMM* 2: 189–96.

Ali, Mukti, Roland Gunawan, Jamaluddin Mohammad, and Ahmad Hilmi. 2015. *I, You, End: Re-Reading Religious Texts about Child Marriage*. Jakarta: Rumah KitaB.

April, Kurt, and B. Boipelo Mooketsi. 2010. "Dealing with Guilt and Shame after Breaking the Glass Ceiling: The Case of South African Executive Women." *Effective Executive Journal* 13, no. 8: 69–70.

Asad, Talal. 1993. *Genealogies of Religion: Discipline and Reasons of Power in Christianity and Islam*. Baltimore: John Hopkins University Press.

Azra, Azyumardi. 2013. "Distinguishing Indonesian Islam: Some Lessons to Learn." In *Islam in Indonesia: Contrasting Images and Interpretations*, edited by Jajat Burhanudin and Kees van Dijk, 63–74. Amsterdam: ICAS/Amsterdam University Press.
Baumeister, Roy F., Arlene M. Stillwell, and Todd F. Heatherton. 1994. "Guilt: An Interpersonal Approach." *Psychological Bulletin* 115, no. 2: 243–67.
Branscombe, Nyla R., and Bertjan Doosje, eds. 2004. *Collective Guilt: International Perspectives*. Cambridge: Cambridge University Press.
Brenner, Suzanne. 2011. "Private Moralities in the Public Sphere: Democratization, Islam, and Gender in Indonesia." *American Anthropologist* 13, no. 3: 478–90.
Collins, Elizabeth Fulle, and Ernaldi Bahar. 2000. "To Know Shame: Malu and Its Uses in Malay Societies." *Crossroads: An Interdisciplinary Journal of Southeast Asian Studies* 14, no. 1: 35–69.
Doorn-Harder, Nelly van. 2019. "Purifying Indonesia, Purifying Women: The National Commission for Women's Rights and the 1965–1966 Anti-Communist Violence." *CrossCurrents Journal* 69, no. 3: 301–18. https://onlinelibrary.wiley.com/doi/full/10.1111/cros.12380.
Errington, Shelly. 1989. *Meaning and Power in a Southeast Asian Realm*. Princeton, NJ: Princeton University Press.
Geertz, Clifford. 1973. *The Interpretation of Cultures*. New York: Basic Books.
Goddard, Cliff. 1996. "The 'Social Emotions' of Malay (Bahasa Melayu)." *ETHOS* 24, no. 3: 426–64.
Gunawan, Roland, and Nur Hayati Aida, eds. 2019. *Fikih Perwalian: Membaca Ulang Hak Perwalian untuk Perlindungan Perempuan dari Kawin Paksa dan Kawin Anak*. Jakarta: Rumah KitaB.
Harsono, Andreas. 2018. "Indonesian President Jokowi to Ban Child Marriage: Pending Presidential Decree Will Raise Minimum Marriage Age." Human Rights Watch, April 23. https://www.hrw.org/news/2018/04/23/indonesian-president-jokowi-ban-child-marriage#.
Heider, Karl. 1991. *Landscapes of Emotion: Mapping Three Cultures of Emotion in Indonesia*. New York: Cambridge University Press.
Isaac, Tracy. 2011. *Moral Responsibility in Collective Contexts*. Oxford: Oxford University Press.
Keeler, Ward. 1983. "Shame and Stage Fright in Java." *ETHOS* 11, no. 3: 152–65.
Lickel, Brian, Rachel Steele, and Toni Schmader. 2011. "Group-Based Shame and Guilt: Emerging Directions in Research." *Social and Personality Psychology Compass* 5, no. 3: 153–63.
Lindquist, Johan. 2004. "Veils and Ecstasy: Negotiating Shame in the Indonesian Borderlands." *Ethnos* 69, no. 4: 487–508.
Mail Foreign Service. 2010. "Child Bride, 13, Dies of Internal Injuries Four Days after Arranged Marriage in Yemen." *Daily Mail.com*, April 9. http://www.dailymail.co.uk/news/article-1264729/Child-bride-13-dies-internal-injuries-days-arranged-marriage-Yemen.html.
Marcoes-Natsir, Lies, and Lanny Octavia, eds. 2014. *Kesaksian Para Pengabdi: Kajian tentang Perempuan dan Fundamentalisme di Indonesia* [Testimony of the servants: Research concerning women and fundamentalism in Indonesia]. Jakarta: Rumah KitaB Bersama.
Mashuri, Ali, Esti Zaduqisti, and Daphne Alroy-Thiberge. 2017. "The Role of Dual Categorization and Relative Ingroup Prototypicality in Reparations to a Minority Group: An Examination of Empathy and Collective Guilt as Mediators." *Asian Journal of Social Psychology* 20: 33–44.

Maufur, Noorhaidi Hasan, and Syaifuddin Zuhri. 2014. *Modul Pelatihan Fiqh dan HAM*. Yogyakarta: LKiS.
Nurmila, Nina. 2009. *Women, Islam and Everyday Life: Renegotiating Polygamy in Indonesia*. London: Routledge.
Pakpahan, Binsar. 2016. "Shameless and Guiltless: the Role of Two Emotions in the Context of the Absence of God in Public Practice in the Indonesian Context." *Exchange* 45, no. 6: 1–20.
Rumah KitaB. 2018. "Discussion: *Wilayah* (Guardianship) and *Qiwamah* (Protection) of Females." Unpublished Activity Report. Jakarta: Rumah KitaB.
Smith-Hefner, Nancy J. 2007. "Javanese Women and the Veil in Post-Soeharto Indonesia." *Journal of Asian Studies* 66, no. 2: 389–420.
Smith-Hefner, Nancy J. 2009. "'Hypersexed' Youth and the New Muslim Sexology in Contemporary Java." *Review of Indonesian and Malaysian Affairs* 43, no. 1: 209–44.
Smith-Hefner, Nancy J. 2019. *Islamizing Intimacies: Youth, Sexuality, and Gender in Contemporary Indonesia*. Honolulu: University of Hawai'i Press.
Stodulka, Thomas. 2008. "'Beggars' and 'Kings': Emotional Shame among Street Youths in a Javanese City in Indonesia." In *Emotions as Bio-Cultural Processes*, edited by Birgitt Röttger-Rössker and Hans J. Markowitschm, 329–49. New York: Springer.

4

Historical and Survivor Guilt in the Incorporation of Refugees in Germany

John Borneman

Guilt and Refugee Incorporation

Upon their arrival in Germany, Syrian refugees interact with Germans amid a landscape that is marked by two kinds of guilt: historical guilt concerning culpability, and survivor guilt concerning victimization and agency. In postwar Germany, these two kinds of guilt were assumed to be mutually exclusive, with the Germans themselves ascribed only historical guilt. New arrivals have difficulty grasping their relation to this landscape, yet for those who stay, incorporation into the full body of the nation demands an awareness of what they are to become a part of.[1] Even for native Germans themselves, what this relation should be is highly contested—entailing questions such as who actually inherited guilt from the past and whether this guilt should be collectively shared.

In the summer of 2015, I began an ethnographic project on what in Europe tends to be called "integration," but I soon turned to the thicker concept of what some call the "incorporation" of foreigners/outsiders/strangers into German society (McKowen and Borneman 2020). During that summer and through the subsequent fall and winter, 1.1 million refugees and migrants entered Germany; 441,899 applied for asylum, of whom 298,000 were from Syria, 79% were male, and 67% were under the age of 30 (UNHRC 2018). This chapter explicates how the demands of incorporation are experienced by Syrian refugees through the lens of historical and survivor guilt. One of its aims is to broaden the definition of guilt by drawing on its actual experience. Based on current research conducted with Parvis Ghassem-Fachandi on refugee experiences of incorporation, I examine here how guilt arises in Syrian encounters with Germans, how it is communicated, and how these encounters provide an opening into the process of incorporation.[2]

Parvis and I are not only anthropologists but also have an intimate familiarity with Germany, which is home for Parvis and a long-term fieldsite for myself. As friends of our German and Syrian research subjects, we are engaged in a multifaceted, counter-transferential relationship, helping them understand occupational

John Borneman, *Historical and Survivor Guilt in the Incorporation of Refugees in Germany* In: *Guilt*. Edited by: Katharina von Kellenbach and Matthias Buschmeier, Oxford University Press.
DOI: 10.1093/oso/9780197557433.003.0005

structures, rental markets, and contacts they may need, and facilitating the translation of notions of generosity, obligation, eroticism, and debt as these arise in conflicts with Germans. For the first three years of research, we took it upon ourselves to introduce them to local ideas, places, and activities of which they were unaware: for one, to issues about which Germans have a particular sensitivity (such as history and the Holocaust), and for another, to those things readily available to residents that Syrians may find pleasurable or interesting but do not know. This situates us within the process of their incorporation, and enables us to expedite their knowledge of place and attachments to potential locations within it. Such an approach contributes to more ethical research and also facilitates our learning from them.

On the Productivity of Guilt in Interactions between Germans and Syrians

Many features of guilt make the subject productive for analysis beyond any particular temporal-cultural context:

- Guilt has a temporal durability that exceeds the time of the individual, as it is passed on through various acts of transference from generation to generation.
- It moves between conscious and unconscious states, and between actual and imagined wrongdoing, thus providing a connective between what is explicitly known and motives of which one is unaware.
- It moves between individuals and collectives, and thus imbricates both.
- Its source could be either an internal feeling or an external ascription, though internal and external are always related.
- It grows out of action involving a relationship with an "Other;" in other words, it emerges from the interpersonal.

This final feature distinguishes guilt from shame, with which it is often paired or confused. Shame, by contrast, involves what one is rather than what one does, and thus is experienced as a narcissistic injury to the ego. Unlike guilt, shame is not about a relationship or about an empathic response to a wrong, but grows out of an inner inadequacy, often traceable to a failure of attachment that leads to blaming the self. It is even more widespread than guilt. Needless to say, shame and guilt are also frequently present at the same time.

Beyond its generative power as an analytic, guilt is socially productive only when it carries with it supplements such as responsibility, accountability, atonement, rectification, redress, or reparation—an argument that Klaus Günther makes in this volume. If guilt does not include a reckoning with these supplements, then one can become guilt-ridden about some wrong. To be guilt-ridden is not only nonproductive but also counterproductive. In that case, guilt does not motivate action but instead paralyzes and overwhelms; it appears to define being, similar to shame, thus inhibiting action that might transform the guilt into reparative action. Thus, to avoid this paralysis and become productive, shame, when present, must be converted to guilt, and the feeling and ascription of guilt must be transformed into thought and action.

Many scholars have characterized German culture as marked by guilt (Buruma 2015; Benedict 1946) and Arab societies as cultures of shame (Crapanzano 1980; Eickelman 1976). In the anthropology of the Mediterranean of the 1960s, the entire region was framed by a focus on gender that was structured by the shame of women and by the honor of men, with little consideration of guilt (Peristiany 1966). Such culturalist accounts may in fact accurately describe the predominant mode in which wrongful behavior is felt, expressed, or censored. But they elide the question of guilt's presence and its productivity in one place, and that of shame in another. Instead, I argue we should acknowledge the frequent co-presence of shame and guilt, the specific nature of learning from transference in a cross-cultural relation of guilt, and how the feeling of shame can become a major source of responsibility when converted to guilt (cf. Buschmeier's nuanced argument in this volume on the particular balance of shame and guilt in postwar Germany).

Syrians are a diasporic group in Germany. There, Germans and Syrians think with and about each other in primarily three languages: English, German, and Arabic. Some Syrians also speak Turkish with Turkish-speaking Germans. Guilt as a legal judgment and guilty as a state of being are readily translatable into each of these languages, though the semantic overlaps and frequency of use vary. The German concept of *Schuld* has a richness that is lost in English translation, or, rather, that is parsed into different words to designate a wide range of feelings and states of being. These include feelings and states of remorse, fault, wrong, blame, debt, and responsibility. In Arabic, there is also no single word for all the meanings of *Schuld*, but at the same time there are many ways to express the different dimensions contained within the German concept. This is to say that the linguistic concepts do not present an undue limitation to understanding; the diverse meanings of *Schuld* can be readily translated into Arabic expressions. These expressions allow for both the ascription of guilt (accusing, suspecting, tormenting, admitting, confessing) and its internalization (intending and feeling) (Hermans 1999, 303–4).

Despite this translatability, linguistic concepts alone are insufficient to analyze real or imagined feelings or behavior that are judged as wrongful or shameful. Direct translation can flatten the diverse forms of expression and their relation to experiences of shame and guilt. Shame, for example, is actualized through external moral codes whose utility rests largely in social control functions. It is most often expressed not in language but in blushing, a reddening of the face. It would be wrong to assume that there is no process of internalization of these codes. When moral codes of shame are internalized, they may become feelings of guilt and work in the superego to deter certain behavior, in which case it is unnecessary to initiate shaming in order to change behavior that has been deemed wrong. Moreover, social groups cannot function over time with external controls alone; they also necessarily rely on forms of compliance that are dependent on emotions (such as the feeling of guilt) that motivate individuals from sources not external to but rather within the self.

Among Syrians, the feeling of shame and its ascription are ubiquitous, closely related to both family-oriented morality and to Islamic moral codes. Yet what frequently begins with an external ascription of shame is subsequently internalized as guilt. Philip Hermans makes this very point in refuting the argument that Moroccan adolescents in Belgium do not feel guilt: "The gaze of the other turns the offenders' attention to his inner self. He is forced to look at himself, to take his responsibility and make amends" (Hermans 1999, 314). Guilt, then, is often the outcome of a sequence of events that begin with shame.

In interactions between Germans and refugees that I have observed, many feelings of guilt are communicated without being expressed in language. The lack of linguistic articulation in interactions has some relation to the uneven competence of interlocutors in the use of multiple languages (at least during the time of my fieldwork, 2015–2019). Speakers often resort to simplified language for the purposes of translation, losing some of the understandings of guilt and the subtleties of its inner expression in their mother tongues. Nonetheless, these subtleties can be found in dreams and in interactions in the form of gestures, composure, hesitations, and pregnant verbal omissions. For the more attentive interlocutor, traces are indeed observable.

The Syrian Experience of German Historical Guilt

Before their arrival, Syrian refugees knew little about the German landscape of guilt, yet their personal sense of guilt is now being shaped through the collectively formed landscape in which they live (on a landscape of memory, see Kirmayer 1996). Initially, Syrians saw and heard only an emphasis on historical guilt.[3] In visiting German cities, new arrivals no longer see the destruction and suffering of the war, but they cannot avoid seeing the extensive memorialization culture that today revolves around Germany's memory of and responsibility for political crimes: two world wars and the Holocaust. Germans themselves usually explain to refugees their sense of guilt as arising only from a history of descent from perpetrators who partook in these events.[4] Refugees in turn sense this abiding historical guilt that haunts Germany's present in their everyday encounters with places and people. It appears frequently as an unspoken awkwardness in initial meetings between Syrians and Germans.

Also unspoken are the dominant cultural perspectives that frame encounters. For Germans and Europeans generally, there is the legacy of cultural concepts that view difference in essentialist or ontological categories (Pew Research Center 2017). From the mid-9th-century binaries of Muslims and Christians to the contemporary Middle Easterners and Europeans, there has been remarkable continuity (Mastnak 2010). Even when the participants in an exchange today may not think of themselves in such exclusive, oppositional terms, they tend to think of the other in those terms. For Syrians, of more concern in their encounters with Germans is what they know of the Holocaust, and what Germans expect them to know. Most Syrians knew of past German crimes even before they arrived in Germany, but many were convinced that those crimes were exaggerated, or that conspiracies connecting Arabs to Jewish victimization were the real problem. They also did not know much about the German reckoning with its own criminal past.

Current refugees enter this landscape of historical guilt inhabited by people who are three-to-four generations removed from the original crimes.[5] Germans who engage Syrians about it are not generally the ones who refuse to admit to this history. They often try to share details with Syrians of the complicity and culpability of their grandparents. It is fair to say, then, that current refugees have a different experience of Germany, of acknowledgment of guilt and debt, than they would have had living among the generations with direct experience of Nazi or war crimes; today's discourse on culpability for Nazi crimes only superficially resembles the discourse of the 1950s or 1960s elaborated by Jaspers (1948). The denial of guilt in the past by those who lived through Nazi crimes was linked to a denial of Nazi crimes themselves. Today this link is severed. Even leaders and followers of Alternative für Deutschland (AfD), the anti-EU, anti-immigrant, anti-Islam political party, acknowledge the reality of Nazi crimes, though they minimize these crimes' historical significance and disavow their own guilt. Also, given the history of reckoning with Nazi crimes, and the awareness of apartheid policies targeting Palestinians in Israel (where Jews become perpetrators), the grandchildren and great-grandchildren of the World War II generation in no way feel the same sense of guilt or responsibility for Jewish victimization as did the generation of 1968.

In my experience of nearly 40 years of fieldwork in Germany, most Germans share an understanding, even if they dispute specific facts, of what has happened in the nearly 80 years since their *Stunde Null* (Hour Zero) in 1945. Facilitated by Allied occupation and division, history has been critiqued and revised; older narratives of the homogeneous Volk have been undermined; and German personhood has been reconstructed not in opposition but in relation to neighbors, and it welcomes some degree of diversity within Germany. While reactionary thought continues to find public expression, that represented in its most radical form by the neo-Nazi National Democratic Party has a national political constituency of less than 1%.

The terms *Täter* and *Opfer,* those who persecute and those who suffer, have framed the understanding of guilt's relation to *Vergangenheitsbewältigung.* Defined as the struggle to overcome the negatives of the past, this entailed the wish to identify with those who suffered and to acknowledge the role that Germans played in this. It generated ongoing attempts to redress individual survivors and their descendants for war crimes and the Holocaust. This psychic and material work has generated increasingly nuanced insights into the nature of guilt, complicity, and generational responsibility. In the process of generational reckoning, guilt has been continually redefined, as has the meaning of responsibility for the Other. In this sense, guilt has been a motivator for transformative action.

One of the most recent transformations was the sudden emergence of a newly dubbed *Willkommenskultur* in 2015. Although this term initially described the welcome in Munich, where the first large waves of refugees arrived, the refugees were immediately dispersed, with welcome signs and receptivity to those in flight quickly appearing in all parts of the country. In contrast, Italy and Greece, which were ports of first arrivals, waived the refugees on to the north, while most other countries in the north refused any cooperative effort at resettlement.[6] Many people throughout Germany experienced their own collective action with a strange relief, as the many small acts of kindness were seen as a vigorous rejection of past xenophobia and affirmation of a more open country today. People in many settings have told me of familial experiences of flight and of their own guilt as motivation for participation in initiatives to welcome refugees (Borneman and Ghassem-Fachandi 2016).[7] This transformative action—from a country known for death to a country offering life—can be traced both to the overcoming of an original destructive impulse (Klein [1937] 1975) and to a changed sense of responsibility.

This does not mean that refugees thought of their welcome as the result of deep transformations in German society. Despite the many opportunities to do so, Syrians, like many recent immigrants to Germany, have been unmotivated to learn about German history (cf. Assmann and Schwartz 2013; Huyssen 2003; Özyürek 2018). Of more concern to them has been their own immediate past—of the Syrian uprising; its evolution into civil war; and the originary violence of the regime that forced them into resistance, flight, and what is likely to be a permanent exile. While such experiences surface continually in dreams, life in

Germany offers few settings in which refugees are encouraged to work through their immediate past—and when encouraged to do so, they quickly became attuned to media demands to tell standardized narratives of suffering instead of their own unique stories.

Consciously, it is the future, not the past—neither their own nor that of their German hosts—that remains a focus. To think or fantasize about the future has been essential to survival during the last several years and exercises different mental functions than thinking or dreaming about *history*. Much like the dream described in the chapter by Ghassem-Fachandi in this volume, it is Syrians' unconscious activity to which I am drawing attention, more like having a vision or hallucinating than activating memory to reconstruct the past.

This focus on the future is reinforced by an inability to comprehend what was experienced in Syria before arrival in Germany. The implosion of Syria is indeed very complicated, even for historians. And refugees know too little of the larger stories of contemporary Syrian and Levantine history to make sense of their own place in the current implosion. History was not a major part of the curriculum in Syrian schools or universities. In 2004, my year teaching in Syria, I asked students what they wanted to study. Most wanted, above all, engineering or medicine, and linguistics, which is an employable skill as is, though of lesser status, translation—practical endeavors of the present. In Syrian universities linguistics was a popular and rigorous major for those talented enough and uninterested in the hard sciences. It was one humanities discipline with expansive employment possibilities. This was in part the result of Syria seeing itself as the "beating heart" of the Arab world, and thus cultivating a spoken vernacular Arabic more readily understandable than in, for example, Northern Africa. Translation studies was one track of linguistic study. Nobody expressed a desire to study history and politics. In television and films, Syrian authorities selectively emphasized a grandiose past and present unity, avoiding conflicts such as the recent 15 years of the Lebanese Civil War and Syrian occupation. The Lebanese, by comparison, bring that history into contemporary critical films. Partly because the Syrian regime subsumed the stories of individual citizens and groups into its own progressive narrative of overcoming all obstacles, discrete personal experiences of loss were incapable of being coherently linked (Borneman 2007). In the first two years of exile in Germany, the most common frame Syrians used to make their histories cohere was a conspiracy narrative. I heard similar narratives while visiting Syrians in exile in Turkey, in 2012–2014 (Borneman 2019).

After more than three years in Germany, however, many Syrians have turned away from conspiracies. The legal security that most have obtained has contributed to the license to pursue other explanations for their changing presents. Factors such as individual effort and achievement, institutional luck, fortuitous friendships, or the help of strangers at key moments began to differentiate the causal reasoning of those who obtained formal legal status from those who did not. This does not mean they will all construct individualist accounts, which were largely foreign to Syrians I had known in Syria. One of the more difficult tasks for me in supporting applications to universities and apprenticeship programs in Europe and North America between 2012 and 2017 was to help craft their lives into a convincing, coherent personal narrative. Those who made quick resort to fate (in a *longue durée* religious perspective) to explain events could not at the same time situate their own experiences as uniquely personal within their own narratives.

When confronted with German narratives of reckoning with political crimes, Syrian refugees could not initially imagine such accountability in their country of birth. It is so improbable—the dictator Bashar al-Assad's brutality and power have only increased since they left—that to enter into such fantasies is more likely to lead to depression than hope. This may change, as Berlin has become the center of refuge for people fleeing the countries of the Arab Spring, from Palestinians to members of the Syrian resistance and Egyptian opposition. Syrians-in-exile have begun to ally with German legal authorities to prosecute Syrian perpetrators of war crimes in German courts, which may create a belief in the possibility of political accountability (Borneman 2011).

The Syrian Experience of Survivor Guilt

Most Syrians who have fled to Germany have direct experience with forms of helplessness and victimization that are related to the *other* kind of guilt that marks Germany's landscape: survivor guilt. To obtain asylum they must initially objectify, even exaggerate, their helplessness, in standardized narratives (on such narratives of Hutu refugees from Burundi in exile in Tanzania, see Malkki 1995). Obtaining refuge and legal asylum is, of course, a key to incorporation, but it is an ambivalent accomplishment. Survivors are also haunted by those they left behind, those who were either killed or unable to escape the unending war. Hence shame at having capitalized on their helplessness accompanies the guilt of survival, and some I know have indeed returned.

The concept of survivor guilt emerged in the early writing of Freud and his followers and was prominent in Sandor Ferenczi's ([1933] 1980) explication of the guilt of survivors of child abuse. As Ruth Leys (2007) makes clear in her excellent reconstruction of the origins of the concept and its dismantling, following World War II, psychoanalysts and psychiatrists needed the concept of "survivor syndrome" to assist those who escaped the Nazi camps, specifically, in obtaining recognition from the German government for suffering due to the camp experience. Although those who had survived the camps were not alone in this experience of guilt, analysts had to make their suffering singular, deserving of redress because of this experience alone. Why was the concentration camp the source of psychological problems after the war? Why not prior childhood experiences, or the war more generally? Psychoanalysts employed the concept to specify the effects of camp life as delayed and long-term, expressed in the symptoms of trauma. They also used the concept to account for how the dilemma of helplessness leads to an uncanny identification of survivors with those who persecuted them. The concept proved essential in providing survivors with a diagnosis of injuries (traumata) singularly caused by camp experience that could justify their demands for reparations from the West German government.[8]

At the same time, the Allied occupiers and Europeans generally defined Germans as perpetrators, leading to a privatization, if not repression, of forms of German victimization and suffering, in particular of the firebombing of German cities, the mass rape of German women, and the experience of flight and expulsion. Many German leaders encouraged this repression because they feared that acknowledgment would lead to a revanchist drive to reclaim lost territories and re-establish the former Reich. The occupiers—the Soviet Union, the United States, Britain, and France—cooperated in this and enforced the repression despite the Cold War divide among them. This repression of public acknowledgment, however, did not preclude the private sharing of experiences of suffering and loss, the private mourning of the dead, or the growing generational break embracing the notion that Germans deserved punishment (e.g., Mitscherlich and Mitscherlich [1967] 1984). While the landscape is changing now, these changes still occur in the context of a "perpetually recurrent [pattern]," as Aleida Assmann (2006, 194; cf. Schmitz 2007) argues. "The perpetrator renounces his guilt by asserting the memory of himself as a victim."

Central to this story of suffering contra guilt is the fate of the *Vertriebene* (expellees) from the former eastern territories, which were ceded after the war to those Central European countries that had suffered under German occupation. Between 1944 and 1948, some 11,935,000 expellees fled or were coerced into leaving their homes, a population transfer legitimated by the Potsdam Accords in 1945. People resettled in what was to become a territorially shrunken Germany of occupied zones (Bauerkämper 2008, 477–485). Fear that the refugees themselves would unite behind a neo-Nazi party led, in 1946 in the Western Allied zones (with similar policies enacted by the Soviets), to forbidding the political organizing of refugees; expellees were allowed only cultural recognition. Repression of this stew of victimization and unacknowledged political crime (complicity in Nazi crimes) distorted the discourse about war crimes and responsibility, with effects to this day.[9]

How the entrance of Syrian refugees fits into this picture of victimization and political crime is by no means obvious. My observation is that survivor guilt arises among Syrian refugees in Germany when it provides a transferential opportunity for Syrians and Germans to connect unconsciously. Syrian experience with Germans began with their arrival following flight from the civil war. The initial xenophilic reception quickly turned more ambivalent, soon followed by intermittent xenophobic reactions (Borneman and Ghassem-Fachandi 2017). Assignment to cities and regions all over Germany and lengthy processing of documents and status kept refugees dispersed, without a concentration of listeners. This dispersal discouraged them from sharing the experiences they had in Syria and Turkey that predated their arrival in Germany. What they most disliked in this early period after arrival, as explained to me, was the pity that such experiences evoked when Germans saw them as refugees.

Experiences in war and flight, and reasons for flight, are so diverse that they do not unite Syrians with each other. The Syrian regime fomented distrust among citizens, and Syrians often fought on or identified with different sides in the conflict before fleeing. If they had remained in Syria, this recent past would be felt daily in their environment, amid the fear and the destruction of the war. Conflicts would have to be confronted and managed, and new solidarities, as well as homicidal impulses, might arise out of wishes for revenge, reciprocity, or mutual protection. The failed uprising would today be literally part of the air they breathe. But in Germany they breathe a different air; the uprising is seldom part of everyday discussion, even when refugees are among other Arabs. Also, the regime of Bashar al-Assad has sent assassins to Europe, making exiles who are still politically active feel unsafe. In this climate, trust is difficult to build. Past experience is left to dreams and to internal conversations, intrapersonal dialogues that Syrians have with themselves. Germans rarely hear these dialogues voiced.

Some of this opaque experience is revealed in recurring dreams or, more specifically, in nightmares. Two have been repeated to me often.

- In one, the young men are suddenly back in Syria and pursued by the Mukhabarat (state security) and other government agents; they flee and are barely escaping until, suddenly, they are not—or they wake up.
- In the second nightmare, the young men are reunited in Syria with their entire families—parents, siblings, sometimes cousins and grandparents, sometimes neighbors and friends, including those who have since died—when, suddenly, these intimate others vanish and they are again alone, or in trouble. It tends to focus on reunion with the mother, but may also include many other details of estrangement, danger, or death.

The first dream is about fear of persecution and survival in the past. It says: *I can survive. Barely. But, in fact, I did it. I appear to have escaped.* The second is centrally about a relationship with those left behind and is set in the future (Ward and Styles 2012). It tries to repair what has been damaged by the war, by the flight and escape. It says: *I have not abandoned you. But I want to create unity where there has been separation.* Although working through the guilt of surviving may be the exterior motive for the dream, it is transposed into a reparative mission of atonement in the dream. What cannot be undone consciously is contemplated in the unconscious.

Dream Versions

The first nightmare of pursuit and escape may repeat many other details of death or near-death experience. For example:

- They have returned to Syria, and the authorities show up at their homes to forcibly abduct them to serve in the military, or to arrest them. The fear of being coerced into indiscriminately shooting the Assad regime's enemies—namely "the people"—is one major reason many young men fled Syria. Many had indeed escaped a close call, actually fleeing from Syria days before authorities arrived. (Although all dreams are dreamt in the present, this one places a past experience in the future.)
- They are drowning on a boat, crossing the Mediterranean from Libya to Lampedusa; or on a ferry from Tripoli, Lebanon, to Mersin, Turkey; or on a raft from Turkey to a Greek island. In waking life someone else on the boat had already drowned, but the person is still alive in the dream. Or, in waking life, someone with whom they were close (e.g., cousin, uncle, niece, sibling) had already drowned, and they witness or are told of this in the dream. Or scenes of waking life are repeated accurately in the dream: they are lost at sea, and the waves rock the boat wildly, causing vomiting; or a raft for 15 contains 50 people, and it is filling with water, which mixes with gasoline and vomit.
- They are packed into a prison cell where there is standing room only. They are taken out individually and tortured. In waking life, some were in fact imprisoned, most for a matter of days, weeks, or months. There they learned how to sleep standing up. Some were also, in fact, tortured—for example, burned with cigarettes, hung by their wrists, beaten, given electric shocks to the genitals. In the dream, those who were not tortured also place themselves at the scene of the torture.
- Their papers are not in order, and they are not allowed to leave Syria (or, in some dreams, to leave Turkey) because of missing documents. In waking life, most escaped without all the personal documents they needed but were nonetheless able to recover sufficient proof of their pasts to request and obtain asylum.

The second nightmare stages different outcomes of reunion:

- The reunion with family is filled with joy, much weeping, kissing, and hugging. In one telling, the dreamer cannot cry in waking life since he became a refugee; he can only cry in this dream (though he admitted that he sheds a few tears when he watches Turkish soap operas). His mother, with whom he particularly connects in the dream, died from a sudden cancer within a year of his escape.
- The reunion is disappointing. In one telling, the returnee keeps trying to unite his family, but they have nothing in common; they cannot find a line of discussion that brings them closer. His father appears in this dream, even though he died after being tortured in prison.
- After they return to Syria, authorities will not allow them to return to Germany, offering them no reasons or justifications. In some tellings, the reunion is planned but at the last minute not approved by German authorities. In other versions, Syrian authorities refuse to let them leave and return to Germany. Sometimes Syrian authorities claim that documents are missing. In waking life, none of these young men I know has yet been approved to travel to Turkey, Lebanon, or Jordan (which is where the family might reunite). I do know adolescents, however, who have been approved to make the trip.

Dream Interpretation and Context

Many refugees have talked to me about psychological problems, including clear signs of depression, yet very few have been willing to accept professional help when given the opportunity. Their dreams stage problem scenarios that they cannot access or address in their everyday lives. The first dream concerns reawakened fears of persecution and suffering in a return to the landscape from which they fled; the second stages a reconciliation and a unity that ultimately fails.

It is paradoxical that the young men in the examples above have difficulty sleeping, because they are in fact dreaming their own experience (Ogden 2017, 2004). Usually, dreaming would process disturbing experiences, thereby making more restful sleep possible. But they are putting into their dreams affective experiences that make little sense to them—what Bion (1962, 1959) called "beta elements," or *undigested* bits of experience. Because this experience is frightening, the unconscious re-presents the beta elements that they cannot deal with consciously. The question is whether these elements are being successfully metabolized through what Bion termed an "alpha-function": the mental and emotional resources to link discrete sensory data into narratives that enable thinking. This inability to link and make sense in the dream is likely what prevents them from getting restful sleep.

Guilt enters into the second dream when one asks what the men are atoning for. What do they imagine they did wrong? Why are they responsible for family reunification? The German government has largely frozen such reunification, and it rarely grants to refugees of the age group with which I am interacting visas to visit Lebanon, Turkey, or Jordan, where many have relatives in exile. Most of the young men played some kind of role in the uprising against Assad, often against the advice of their parents. Rather than bring about any kind of political reform or opening, the uprising has miserably failed; instead, the country has been destroyed to preserve Alawite rule and is now totally dependent on Russia and Iran (Borneman 2019, 2012). Refugees are haunted by the idea that they are responsible for this destruction and, because of this, many today disavow any interest in politics. But even more disturbing, as the dream suggests, is how their parents have experienced the attempted revolution: loss of homes, businesses, and children (Borneman 2020b, 2019). The suspected actions of children and youths has directly led, in some cases, to the interrogation, torture, arbitrary confiscation of property, dismissal from jobs, imprisonment, and even murder of their parents.

The context most relevant to the present in which dreaming takes place is the abandonment of family (Olwig 2011). Thus, family reunification is the wish that refugees most frequently articulate, but also an idea that generates a great deal of shame. They have abandoned family to escape the war *and* to obtain a better life; the two are often inseparable and create a great deal of ambivalence. Most fled Syria only through the financial support of relatives—for such things as food and lodging, bribes, transport, and fake documents. Support ranged from several hundred to ten thousand euros. From the time they began to receive payments from the German state—the amount varies by state, but in Berlin it is approximately €365/month plus housing subsidies for a single person—most refugees have sent remittances to family members or friends elsewhere. Some relatives who lent them money demand repayment with interest. One man told me he recently repaid his uncle €2,500, having saved €100 a month over two years. The uncle, who fled to Germany in 2011 with his wife and their one child, has learned no German in the interim and does not work. Since arrival he has had three more children and now lives well from the state's child benefits. When the uncle said the €2,500 was not enough, his nephew broke off all contact.

Living in diaspora with no hope of return makes it impossible to address a strong Oedipal wish, which is to assume the responsibilities of an adult as one grows up. Many men consider the care of their parents or the fulfillment of parental wishes an ethical imperative for themselves (Loewald 1980). One day I received a call from a friend who asked me to inform his younger brother that their father had died. He said he could not do this himself. The first thing he said to me was: "No. No. No," before quickly lamenting the inability to help his father fulfill his most important wish: "He always wanted to go to Hajj."

Within Arab kin groups, generally, there is a strong sense of caring for each other in emergencies. This care is often a literal transformation of shame at survival into *Schuld*—guilt, debt, responsibility. Caring can extend beyond the domain of kin relations or primordial ties to redressing indebtedness to and responsibility for the larger collectives of refugees, neighborhood, city, nation, or the *umma* (the community of Muslims) through alms to the poor and charity (Borneman 2012). However, young men in exile are often unable to fulfill this responsibility to others, though many try; because survivors have been widely scattered in many different countries in Europe, North America, and the Near or Middle East, they are unable to assess accurately what others in exile elsewhere actually need. Moreover, especially those kin who are not successful in completing education or finding work not only ask for money frequently but also often prey on the guilt feelings of their stronger or more established kin. Given this dynamic, many of my interlocutors have even stopped communicating with some family, while still extending help to non-kin or within friendship circles with other Arab migrants in Germany.

On Productive Guilt and Incorporation

The motivations of Germans and Syrians to act, to take responsibility for the past, are informed by guilt of different kinds. For Germans, the origins of historical guilt go back several generations, to the 1940s and the Nazi regime; for Syrians, to a failed revolution that began as an uprising in 2011. Complicating the German position is that the guilt of survivors was defined in opposition to their role as perpetrators, excluding important elements of the German experience of victimization. Complicating the Syrian position is the recency of their diasporic condition and the difficulty of locating themselves in time and place with respect to both their own and German history.

Even though Germans and Syrian refugees are most often unable to access each other's histories, it is not as if the dynamics of historical and survivor guilt fail to intersect. They do, continually, but largely through unconscious communication that, at least at this point in time, never enters speech. And they do so largely through a shared experience of survivor guilt; the German welcoming culture of 2015, and the continued vigorous defense of asylum rights by courts, grew directly out of what has been learned from German experiences as both perpetrators and victims during and after the war, including the importance of atoning for past wrongs and debts. Syrians benefit from this culture without understanding where it comes from.

Contemporary Syrian refugees share an experience with other strangers who have visited and stayed in Germany over the last seven decades. That experience is of attempted incorporation into a German society that has, minimally since the Nazi period, struggled with a conscious recognition of its ethnic and cultural heterogeneity. The sentiment of homogeneity has been widely criticized but is remarkably resilient, finding a permanent place in televised discussions of immigration. It actually prevents people from developing an accurate view of their own society, which has become a country of immigration. Incorporation now occurs in this more recognizably diverse society with many people with migration backgrounds, each required to live within a landscape of guilt that is also changing. Economic integration was and is often thought of as the foundation of all other forms of "integration" of strangers into European societies. It may, in fact, be fundamental, especially in welfare states that support strangers with the expectation that they later contribute monetarily to the general well-being (Bertoli, Brückner, and Moraga 2013; Borneman 2020a, 2020b; Brückner et al. 2016). But employment is not the sole goal of life for a refugee, nor is it sufficient to create a sense of belonging to a group and a place. Nor is citizenship, the attainment of rights in collective membership, sufficient for group incorporation. Belonging exceeds the demands and expectations of labor integration or the acquisition of citizenship. It becomes meaningful only through encounters that force imaginative work with disparate pasts within and of the collective body. It entails physical and metaphorical ingestion: digestion as in the desire to incorporate, indigestion as in the tensions of integration, and excretion as in the desire to expel and deport. In this process, the boundaries of the physical, social, and political body are reconfigured. The feeling and ascription of guilt affects the digestive capacities of those involved, influencing the imagination of groups and aspirations regarding self-worth and belonging. Guilt can act as a productive force in creating a social whole.

Notes

1. Incorporation of strangers/foreigners into Germany means into peoplehood (Volk and nation), territory, and polity (democratic state), all written into law (for a discussion of concepts, see McKowen and Borneman 2020). Germany is a corporate body with borders determining entrances and exits imagined in relation to the human body. Strangers become part of the imagination and reality of the collective body even if their stay is brief (on links between individual and social bodies, see Douglas 1996; Lefort 1986; and Merleau-Ponty 1962, 58–84, 185–217). Successful incorporation would be the capacity to care for oneself and others (Borneman 2000a, 2000b).
2. We concentrate on young men from Syria between ages 18 and 35 who arrived in Germany around 2015. They have contact with refugees from other countries, migrants from Syria who are not refugees, and refugees who are either younger or

older, or are married with families. Each group of new entrants is positioned differently with respect to Germans, and thus with the emotional transference of guilt.

3. The public reaction that greeted the (mostly) Syrian refugees in 2015 created an unusually strong xenophilic sentiment, which was then inverted through the vehemently xenophobic anti-foreigner debate mobilized against it (Borneman and Ghassem-Fachandi 2017). Germany has always had to incorporate strangers (Göktürk, Gramling, and Kaes, 2007), and xenophobic and racist reactions have been part of its history. But the patterns of German immigration in the 20th century are unlike those among its European neighbors, in that extraordinary migration (or huge spikes in numbers) frequently followed ordinary migration (Héran 2016, 246). Political mobilization against Muslims and against immigration have been more effective in this most recent spike, aided by social media's widespread coverage of two new radical parties—Pegida (Patriotic Europeans Against the Islamization of the Occident) and the AfD (Alternative for Germany)—contributing to their electoral success (Ceylan 2010). Nonetheless, the fact that many Germans have personal experience with migration and refugees helps to explain why there is also a xenophilic reaction. Most important in this regard is the experience of the expellees and displaced persons resettled after World War II; the experience of the 2.5 million foreign workers (known as *Gastarbeiter*) who were recruited for industry between 1955 and 1973, most of whom have not returned to their countries of origin; and the growth of asylum applications and refugees from the Balkan Wars in the early 1990s.
4. My focus in this paper on Syrian reactions to German historical guilt leaves the German point of view less developed. Although the Allies began in 1945 with legal retribution and punishment (which the Germans followed with trials in the 1950s), the initial form of redress by Germans was instead monetary compensation, followed by performative redress (e.g., events of cultural accountability, such as apologies, commissions of inquiry, historical commissions) and rites of commemoration (e.g., days of mourning, monuments, museums). These modes of accountability all worked with an idea of becoming conscious of guilt and its ascription, punishing the culprits, acknowledging the victims, and redressing wrongs (Borneman 2011).
5. Arendt's ([1945] 2000) distinction between guilt and responsibility holds at most only for the immediate generation of actors with whom she was concerned. With the generational inheritance of guilt and responsibility from a prior time, these degrees do not absolve the living from all historical responsibility for the collective today.
6. Germany remains the preferred destination in Europe for people in flight from the Global South (see the excellent review of these complex issues by Münkler and Münkler 2016, 81–114).
7. The fact that refugees sought aliveness in Germany countered the post-war narrative ascribed to Germany as a land of death. Their arrival presented Germans with an existential experience in which both internal and external environments could be transformed. As "transformational objects," they were no longer the cause of a crisis to control; instead, through welcoming, they offered "an experience that promised to radically alter German collective self-definition" (Bollas 1987, Borneman and Ghassem-Fachandi 2017, 118f.).

8. The early public discourse on victims that shaped the German landscape of guilt arose out of early writings by Holocaust survivors of the concentration camps (cf. Sebald 2002, 66–77). Among the most insightful were the accounts of Primo Levi, who made very nuanced arguments that supported the psychoanalytic theory of the nature of guilt for survivors of extreme violence. Central to survivor guilt is the paradoxical notion that because of helplessness, the victim or survivor of violence mimetically and unconsciously identifies with the aggressor, hence becoming contaminated, if not complicit, in the violence to which he or she had initially been subjected. Many survivors of the camps, including Levi, reported feeling guilty for having survived when so many others had died, and, in order to survive, they had to cooperate and "identify" with the Nazis on some level. Leys argues that, more recently, several other theoretical arguments from very different fields with dissimilar interests, critical of this way of viewing camp survivors, have converged. Above all, the psychoanalytic frame is accused of blaming the victims; these scholars argue that guilt arising from a mimetic identification with the Other—that is, in this case, the perpetrator—is more accurately characterized as shame. In shame the victim simply reacts to being seen or harmed. Violence is an assault from *without*, of which one can be fully conscious; the victim need not experience any internal conflicts of significance. Leys concludes, "This has the advantage of portraying the victim as in no way mimetically collusive with the violence directed against him" (Leys 2007, 9). Such a framing, while appearing to be less judgmental and more sympathetic to survivors, is plausible only by ignoring the widely documented ambivalence in the actual experience of victims of traumatic violence. Statements and testimonies of those who have experienced trauma—and not only camp survivors—frequently elaborate the intense pressure to identify with the aggressor in order to survive psychically as an integral self. In the extreme, such victims can suffer from dissociative disorders that result in schizophrenia (Borneman 2015; Howell 2005; Sullivan 1927).
9. One of these effects is the emotional arousal that accompanies recurrent attempts to bring the repressed into discussion and debate. When these attempts are successful, what appears is not only a demand to recognize the facts surrounding German helplessness alongside the acknowledgment of having been perpetrators, but also a sudden emergence of repressed and contagious affect (Assmann 2006; Maier 1988; Peter 1995). This affect erupts in ways that far exceed expectations of its significance. There are many such moments in the last thirty years, such as the 1979 broadcast in Germany of the American miniseries *Holocaust*, which focused on the human dimension of Nazi mass murder; the "historians' dispute" from 1986–1989, which debated Germany's special burden and the singularity of the Holocaust (original documents in Knowlton 1993); Daniel Goldhagen's (1996) public tour of Germany to talk about his study of the willingness of ordinary Germans to act as Hitler's executioners; the traveling Wehrmacht exhibition, an original 1995 exhibition (revised and re-exhibited from 2001–2004) seen by an estimated 1.2 million visitors, which showed that the regular German army (not just the SS) had been involved in the planning and implementation of the Holocaust and other war crimes; and the 2002 publication of Jörg

Friedrich's ([2002] 2006) emotion-laden historical account of German suffering in the Allied bombing campaign of German cities. This affect follows the transmission of memory, as Assmann (2006, 197) states, from a generation "that actually experienced the events ('*Erfahrungsgeneration*') to a generation that identifies with them ('*Bekennt nisgeneration*')."

References

Arendt, Hannah. (1945) 2000. "Organized Guilt and Responsibility." In *The Portable Hannah Arendt*, edited by Peter Baehr, 146–56. New York: Penguin Books.

Assmann, Aleida. 2006. "On the (In)compatability of Guilt and Suffering in German Memory." *German Life and Letters* 59, no. 2: 180–200.

Assmann, Aleida, and Anja Schwartz. 2013. "Memory, Migration, Guilt." *Crossings* 4, no. 1: 51–65.

Bauerkämper, Arnd. 2008. "Deutsche Flüchtlinge und Vertriebene aus Ost-, Ostmittel- und Südosteuropa in Deutschland und Österreich seit dem Ende des Zweiten Weltkrieges." In *Enzyklopädie Migration in Europa*. Vol. 17, *Jahrhundert bis zur Gegenwart*, edited by Klaus J. Bade, Pieter C. Emmer, Leo Lucassen, and Jochen Oltmer, 477–85. Paderborn: Leibniz Institut für Sozialwissenschaften.

Benedict, Ruth. 1946. *The Chrysanthemum and the Sword: Patterns of Japanese Culture*. New York: Houghton Mifflin Harcourt Publishing Company.

Bertoli, Simone, Herbert Brückner, and Jesus Moraga. 2013. *The European Crisis and Migration to Germany: Expectations and the Diversion of Migration Flows*. Berlin: Bundesamt für Migration und Flüchtlinge (BAMF).

Bion, Wilfred R. 1959. "Attacks on linking." *International Journal of Psycho-Analysis* 40: 308–15.

Bion, Wilfred R. 1962. "A Theory of Thinking." *International Journal of Psycho-Analysis* 43: 306–10.

Bollas, Christopher. 1987. "The Transformational Object." In *The Shadow of the Object: Psychoanalysis of the Unthought*, 13–30. New York: Columbia University Press.

Borneman, John. 2007. *Syrian Episodes: Sons, Fathers, and an Anthropologist in Aleppo*. Princeton: Princeton University Press.

Borneman, John. 2011. *Political Crime and the Memory of Loss*. Bloomington: Indiana University Press.

Borneman, John. 2012. "Und nach den Tyrannen? Macht, Verwandtschaft und Gemeinschaft in der Arabellion." Translated by Martin Zillinger and Daniele Saracino. *La Lettre International* 98 (Fall): 33–48.

Borneman, John. 2015. *Cruel Attachments: The Ritual Rehab of Child Molesters in Germany*. Chicago: University of Chicago Press.

Borneman, John. 2019. "The Syrian Revolution: The Political Subject." In *Crowds: Ethnographic Encounters*, edited by Megan Steffen, 23–39. New York: Bloomsbury Academic.

Borneman, John. 2020a. "The German Welfare State as a Holding Environment for Refugees: A Case Study of Incorporation." In *Digesting Difference: Ethnographies of Sociocultural Incorporation in Migrant Europe*, edited by Kelly McKowen and John Borneman, 22–35. New York: Palgrave.

Borneman, John. 2020b. "Witnessing, Containing, Holding? The German Social Welfare State (*Sozialstaat*) and People in Flight." In *Spaces of Care*, edited by Lorraine Gelsthorpe, Perveez Mody, and Brian Sloan, 219–40. Oxford: Hart.
Borneman, John, and Parvis Ghassem-Fachandi. 2017. "The Concept of *Stimmung*: From Indifference to Xenophobia in
Germany's Refugee Crisis." *Hau: Journal of Ethnographic Theory* 7, no. 3: 105–35.
Brückner, Herbet, Tanjo Fendel, Astrid Kunert, Ulrike Mangold, Manuel Siegert, and Jürgen Schupp 2016. "Geflüchtete Menschen in Deutschland: Warum sie kommen, was sie mitbringen und welche Erfahrung sie machen," *IAB-Kurzbericht* 15: 1–12. Nürnberg: Institut für Arbeitsmarkt- und Berufsforschung (IAB).
Buruma, Ian. 1994. *The Wages of Guilt: Memories of War in Germany and in Japan*. New York: New York Review of Books.
Ceylan, Rauf. 2010. "Muslims in Germany: Religious and Political Challenges and Perspectives in the Diaspora." In *The Many Sides of Muslim Integration: An American-German Comparison*, edited by American Institute for Contemporary German Studies, 43–50. Washington, DC: AICGS.
Crapanzano, Vincent. 1980. *Tuhami: Portrait of a Moroccan*. Chicago: University of Chicago Press.
Douglas, Mary. 1966. "Secular Defilement and External Boundaries." In *Purity and Danger: An Analysis of Concept of Pollution and Taboo*. New York: Praeger, 141–59.
Eickelman, Dale. 1976. *Moroccan Islam: Tradition and Society in a Pilgrimage Center*. Austin: University of Texas Press.
Ferenczi, Sandor. (1933) 1980. "Confusion of Tongues between Adults and the Child: The Language of Tenderness and Passion." In *Final Contributions to the Problems and Methods of Psycho-Analysis*, 156–67. New York: Brunner/Mazel.
Friedrich, Jörg. (2002) 2006. *The Fire: The Bombing of Germany, 1940–1945*. Translated by Allison Brown. New York: Columbia University Press.
Göktürk, Deniz, David Gramling, and Anton Kaes, eds. 2007. *Germany in Transit: Nation and Migration 1995-2005*. Berkeley: University of California Press.
Goldhagen, Daniel. 1996. *Hitler's Willing Executioners: Ordinary Germans and the Holocaust*. New York: Alfred Knopf.
Héran, Francois. 2016. "De la 'crise des migrantes' `a la crise de l'Europe: Un éclairage démographique." In *Migrations, réfugiés, exil*, edited by Patrick Boucheron, 239–60. Paris: Odile Jacob.
Hermans, Philip. 1999. "The Expression of Guilt by Moroccan Adolescents: Ethnographic Interpretations by Western Teachers and Social Workers." *International Journal of Education Research* 31: 303–16.
Howell, Elizabeth. 2005. *The Dissociative Mind*. Hillsdale, NJ: Analytic Press.
Jaspers, Karl. 1948. *The Question of German Guilt*. New York: Dial Press.
Kirmayer, Laurence. 1996. "Landscapes of Memory: Trauma, Narrative, Disassociation." In *Tense Past: Cultural Essays in Trauma and Memory*, edited by Paul Antze and Michael Lambek, 173–98. New York: Routledge.
Klein, Melanie (1937) 1975. "Love, Guilt, and Reparation." In *Love, Guilt, and Reparation, and Other Works, 1921–1945*, 306–43. London: Hogarth Press.
Knowlton, James. 1993. *Forever in the Shadow of Hitler? Original Documents of the Historikerstreit, the Controversy concerning the Singularity of the Holocaust*. New York: Prometheus Books.

Leys, Ruth. 2007. *From Guilt to Shame: Auschwitz and After*. Princeton: Princeton University Press.
Loewald, Hans W. 1980. "The Waning of the Oedipal Complex." In *Papers on Psychoanalysis*, edited by Hans W. Loewald, 384–04. New Haven, CT: Yale University Press.
Maier, Charles. 1988. *The Unmasterable Past: History, Holocaust and German National Identity*. Cambridge, MA: Harvard University Press.
Malkki, Liisa. 1995. *Purity and Exile: Violence, Memory, and National Cosmology among Hutu Refugees in Tanzania*. Chicago: University of Chicago Press.
Mastnak, Tomaz. 2010. "Western Hostility toward Muslims: A History of the Present." In *Islamophobia/Islamophilia: Beyond the Politics of Enemy and Friend*, edited by Andrew Shryock, 29–52. Bloomington: Indiana University Press.
McKowen, Kelly, and John Borneman. 2020. "Digesting Difference: Migrants, Refugees, and Incorporation in Europe." In *Digesting Difference: Migrant Incorporation and Mutual Belonging in Europe*, edited by Kelly McKowen and John Borneman, 1–21. New York: Palgrave Macmillan.
Mitscherlich, Alexander, and Margarete Mitscherlich. (1967) 1984. *The Inability to Mourn*. New York: Grove Press.
Münkler, Herfried, and Marina Münkler. 2016. *Die Neuen Deutschen: Ein Land vor seiner Zukunft*. Reinbek: Rowohlt.
Ogden, Thomas. 2004. "On Containing and Holding, Being and Dreaming." *International Journal of Psychoanalysis* 85, no. 6: 1349–64.
Ogden, Thomas. 2017. "Dreaming the Analytic Session: A Clinical Essay." *Psychoanalytic Quarterly* 86, no. 1: 1–20.
Olwig, Karen Fog. 2011. "'Integration': Migrants and Refugees between Scandinavian Welfare Societies and Family Relations." *Journal of Ethnic and Migration Studies* 37, no. 2: 179–96.
Özyürek, Esra. 2018. "Rethinking Empathy: Emotions Triggered by the Holocaust among Muslim Minority in Germany." *Anthropological Theory* 18, no. 4: 456–77.
Peristiany, J. G., ed. 1966. *Honour and Shame: The Values of Mediterranean Society*. Chicago: University of Chicago Press.
Peter, Jürgen. 1995. *Historikerstreit und die Suche nach einer nationalen Identität der achtziger Jahre*. European University Studies, Political Science, 288. Frankfurt am Main: Peter Lang.
Pew Research Center. 2017. "Europe's Growing Muslim Population." Last modified January 10, 2018. http://www.pewforum.org/2017/11/29/europes-growing-muslim-population/.
Schmitz, Helmut, ed. 2007. *A Nation of Victims? Representations of German Wartime Suffering from 1945 to the Present*. New York: Rodopi.
Sebald, W. G. 2002. "A Natural History of Destruction." *New Yorker*, November 4.
Sullivan, Harry. 1927. "The Onset of Schizophrenia." *American Journal of Insanity* 84, no. 1: 105–34.
UNHRC (United Nations Human Rights Council). 2018. "Syrian Emergency." Last modified March 5, 2018. http://www.unhcr.org/syria-emergency.html.
Ward, Catherine, and Irene Styles. 2012. "Guilt as a Consequence of Migration." *International Journal of Applied Psychoanalytic. Studies* 9, no. 4: 330–43.

PART II

TRANSFORMING GUILT INTO (RESTORATIVE) JUSTICE

5

The Productivity of Guilt in Criminal Law Discourse

Klaus Günther

Foucault's Observation

In his seminal study, *Discipline and Punish*, published in 1975, Michel Foucault makes the observation that recent penal law does more than just continue along the well-known path of panoptical control in order to establish a regime of (self-) discipline on a prisoner's body and soul. What is, according to Foucault, much more important than this function is that, with the *help* of panoptical control (and as a result of it), the modern welfare state has generated a universe of discourse that has transformed a perpetrator from a person who has violated the law into a "delinquent" (Foucault 1977, 251). It is a discourse about and around the delinquent and his unlawful behavior; about his character, inclinations, and dispositions; and about possible treatment and therapies, the combined effect of which, in fact, serve to lessen the importance of incarceration itself. Participating in this discourse are not only judges and lawyers, but also social workers, criminologists, psychologists, psychiatrists, and the corresponding disciplines in the humanities, such as medicine and pedagogy, which create a new episteme about the subject (303).

According to Foucault, the delinquent himself plays an important role in the creation of this discourse. The delinquent's most important activity is his confession in the courtroom, as well as in the process of the therapeutic interventions he undergoes (Foucault 2014). As the welfare state transforms into one of neoliberal governance, the defendant's courtroom confessions gain a new function: now, in addition to the *questio facti* of whether the defendant really was the perpetrator, the question of his or her internal mental state, subjectivity, motivation, disposition, and biography, which made this person into a dangerous individual—a perpetrator—becomes relevant. Defendants are expected to reveal their soul, to perform a "truth-telling of criminal subjectivity," or what Foucaults characterizes as an "avowal" (Foucault 2014, 228). What is striking about this observation, perhaps beyond Foucault's own intention, is the fact that all parties who inhabit this universe of discourse act to produce and perpetuate it: they are

Klaus Günther, *The Productivity of Guilt in Criminal Law Discourse* In: *Guilt*. Edited by: Katharina von Kellenbach and Matthias Buschmeier, Oxford University Press. © Oxford University Press 2022.
DOI: 10.1093/oso/9780197557433.003.0006

continuously performing by communicating and acting in accordance with their position in that universe, thereby constituting the new empisteme.

Taking this observation as a starting point, I shall, in the following, generalize and expand upon it. Foucault helps to show that the modern criminal justice system does not just incarcerate delinquents in order to restrain them, as passive objects who should be left alone or exploited for their labor. Instead, this system is, above all, highly *productive* when it activates the delinquent—just as it is productive in how it activates members of the criminal justice system, welfare institutions, universities, and civil society. A crime is not only an outrageous event that attracts public attention and resentment; a crime becomes an engine that drives a variety of activities across society at large. These activities are articulated in a language governed by the central signifiers of guilt or responsibility.[1] The terms are present in the search for a perpetrator and in the resulting questions of accountability, intentions and motives, justifications and excuses, mitigating circumstances, and ultimate culpability. They are also present in public discourse, in welfare institutions, in decisions about parole and rehabilitation, in victim support and therapeutic interventions, as well as in political debates over criminal law and legislation. In the following, I shall elaborate on the hypothesis that guilt in modern criminal law is productive in a self-generative way: there is no guilt in criminal law without productive discourses *of* guilt.

This is already apparent in the public setting of criminal trials, where defendants deny or admit guilt (legal and/or emotional), and in impact statements from victims (or their relatives) who are coping with trauma (or survivor guilt). In the case of public interest trials, society at large engages in debate over their own complicity and contribution to a crime. Take the example of child sexual abuse cases that implicate entire institutions (e.g., the church, sports), sometimes even across national boundaries, or the case of crimes against humanity, which raise global concerns of complicity and guilt.

The Productivity of Power and/or of Guilt

There is one grave *prima facie* objection against my hypothesis of the self-generative character of guilt in modern criminal law: from a Foucaultian point of view, it is not guilt but rather power that drives the discursive crime-guilt machine. It is primarily the establishment of a regime of observation, control, and discipline that is supposed to "normalize" the delinquent at a micro level, as well of a regime of subjectivation and justification to control the people at a macro level: "We are in the society of the teacher-judge, the doctor-judge, the educator-judge, the 'socialworker'-judge; it is on them that the universal reign of the normative is based" (Foucault 1977, 304). Part of that story is a shift from the perpetrator's individual responsibility or guilt for a criminal act to the risks inherent in the person's subjectivity, which poses a danger to the security of society. Briefly said, guilt becomes replaced by risk.

While the discourse of guilt revolves around questions of what might have made delinquents dangerous and why they began violating legal norms—as these questions help with designing particular rehabilitative interventions and life-management plans aimed at a law-abiding social reintegration—the criminal law discourse remains primarily concerned with *individual* criminal responsibility. Popular opinion demands that human beings can and should be held responsible for their actions (in particular, for breaking the law). Therefore, society cannot be entirely guided by an interest in the delinquents' subjectivity (i.e., the causes and effects that guide their life choices); otherwise, the famous phrase attributed to Madame de Staël, *tout comprendre c'est tout pardonner* (to understand everything is to forgive everything) would be the outcome of criminal law discourse. This is not the case, as shown by a short look at the operations of the criminal justice system and the public debates about crime.

The notion of personal responsibility is a key concept, and its significance has only increased in the last decades. Indeed, the productive nature of responsibility has really began to flourish in that time, as already observed by Foucault in his studies on government and biopolitics, as well as by Gilles Deleuze (1992), Nikolas Rose (2000), and David Garland (2001), who have been concerned about the emergence of the new paradigm of control. The neoliberal transformation of the welfare state into an incentive-based social policy of activation, self-management, and self-improvement operates not by forming but by *modulating* the subject through control (Deleuze 1992, 4; Rose 2000, 325). As Rose has maintained, the paradigm of discipline was about "seeking to mould conduct by inscribing enduring corporeal and behavioural competences and persisting practices of self-scrutiny, and self-constraint into the soul." The paradigm of control is characterized by a "never-ending *modulation*" of subjectivities, practices, and institutions: "One is always in continuous training, life-long learning, perpetual assessment, continual incitement to buy, to improve oneself, constant monitoring of health and never-ending risk management" (Rose 2000, 325). Such control requires agencies and procedures of evaluation, justification, benchmarking, and implementation. Different from the traditional welfare state, the control paradigm does not operate primarily by central state administrations but instead is enacted by individuals and communities intent on promoting autonomy and independence (as opposed to passivity and dependence). Therefore, an essential tool of the paradigm of control is the "responsibilisation" (328) of all subjects who participate in such practices. It becomes the *subjects'* task to improve themselves and to self-manage effectively. Thus, responsibility, as a notion, comes back even stronger than before.

The attribution of responsibility is no longer guided primarily by an abstract belief that all human beings have a free will deployed to constrain their passions on the basis of normative codes (e.g., the law), except for those who are incompetent because of mental illness. Rather, in the control paradigm, responsibility is more concerned with external and internal factors that contribute to the formation—and the actual behavior—of a dangerous individual. Now, responsibility consists of learning how to manage and control the factors that shape, determine, and change subjectivity, and, more importantly, how to determine *who* exactly is responsible for taking the steps that might adequately prevent individuals from becoming dangerous or committing a crime. The task of managing, diminishing, or eliminating these factors becomes a matter of responsibility, not only of the dangerous individual himself, but also of the entire community, from family members to experts. The prevention of and reaction to danger, risk, and crime becomes a matter of the "professional responsibility of a host of professionals [. . .] under the threat of being held accountable for any harm to the 'general public'—'normal people'—which might result" (333).[2]

Two Dimensions of Responsibility

To sum up, one can observe a shift in emphasis and relevance in the broad and complex meaning of "responsibility." It is not a wholesale change in meaning, but more a change in the relevance of the notion's different properties. These properties fall along two dimensions, which, though mutually dependent, can and should be distinguished.

On the one hand, agents have *ex post facto* responsibility for their actions, in particular for conduct that violated the law and/or caused harm. Though "responsibility" here refers to the accountability and culpability of the offender in the moment of misconduct, criminal responsibility is attributed *ex post facto* by way of a formalized process. Part of that process is the clarification and, subsequently, the defendant's affirmation or negation of the charges. Therefore, *ex post facto* responsibility is received passively as a judgment and accepted as a burden. Of course, if offenders do attribute responsibility to themselves, they may experience a bad conscience or feelings of guilt.

On the other hand, responsibility entails a forward-looking dimension, directed at future acts and conditions, a *terminus ad quem*. (A well-known example is parents' responsibility for their children.) Here, responsibility is understood as a commitment undertaken on behalf of others, although this responsibility also requires a commitment to oneself, in the sense that one must manage oneself in order to be sufficiently ready and able to fulfill that commitment. Such *terminus ad quem* responsibility requires "competence or the capacity of responsible ethical self-management" (Rose 2000, 333), an activation and self-activation, which demands that the bearer *do something* in order to honor the commitment. At the very least, one must be mindful of possible risks (for oneself and others) in order to take measures to avoid harm and damage.

The shift to *terminus ad quem* responsibility does not mean that *ex post facto* responsibility ceases to play a role; the latter remains present (in the background), because taking the task-oriented, *terminus ad quem* responsibility for future actions always implies that an *ex post facto* responsibility would arise were agents to fail at fulfilling their task. Viewed the other way around, *ex post facto* responsibility always presupposes a *terminus ad quem* responsibility: the duty to undertake one's task and manage oneself in a way that avoids norm violations. The practices, techniques, and institutions of *responsibilization* can be characterized as an attempt to implement more *terminus ad quem* responsibility in order to avoid or minimize danger, damage, and harm, which would carry *ex post facto* responsibility. As an essential part of the general political project of the transformation of the welfare state into a globalized liberal market economy, *terminus ad quem* responsibility serves as a tool to transfer the responsibility for the prevention of risks and dangers from society and its welfare institutions to individuals and local communities.

With regard to criminal law and the criminal justice system, the significance of this shift between the two dimensions of responsibility is not so obvious at first glance. Criminal law is mainly concerned with *ex post facto* responsibility; the criminal justice system does not start an investigation unless a suspicion arises that someone has committed a crime. From the first police investigation up to the trial, and then to conviction and punishment, criminal law consists of a system of legal norms, procedural rules, doctrinal rules for the interpretation of legal norms and court precedents, tools for legal imputation, and other measures centered around the conditions and justifications for the attribution of responsibility for a crime to a perpetrator. This has changed over the course of the last decades for several reasons.

A foundational rationale for the modern state, since Thomas Hobbes's political theory in the 17th century, has been the aim of ensuring security by preventing crime. This is also one of the central tasks of the police. Furthermore, crime prevention has always been an essential justification for punishment. The threat of punishment deters potential perpetrators, and the execution of punishment signals to others that it is in their best interests not to commit a crime either. The welfare state emphasizes criminal rehabilitation through punishment or therapeutic intervention as measures of prevention on the micro level, along with welfare policies and the amelioration of severe economic inequality (as a possible contributor to crime) on the macro level (e.g., Garland and Sparks 2000, 8; Garland 2001, 48).

With the emergence of the control paradigm, crime prevention becomes even more important, but at the same time diversified and decentralized. It is not only—and not even primarily—a task of the state, but also of civil society and of each individual. This responsibility is distributed across several public and private institutions and people, including dangerous individuals themselves—all should actively contribute in their own way(s) to ameliorating the internal and external factors that might otherwise increase a potential perpetrator's chances of causing damage or harm (Garland 2001, chap. 7). This concerted prevention effort is orchestrated by ethical narratives about community values and virtues, which place ethical demands on the individual for self-management and risk management.

In criminal law, there is a turn toward legal moralism, according to which norms of criminal law are justified by fundamental moral principles. The moral turn is often supplemented by an expressivist or communicative turn, which emphasizes the moral messages conveyed in trials and punishment. Consequently, victims of crime receive more public attention because they suffer moral harm, and perpetrators are forced to bear moral blame. In committing a crime, perpetrators reveal deficits in their ethical self and capacity for self-management, but also implicate their families and upbringing, their schools and communities, the welfare institutions and society at large, all of which failed to prevent and control the risk efficiently. The task-and-commitment dimension of responsibility switches to *ex post facto* responsibility, which must ascertain who should take responsibility for the failures to prevent risk and harm to others.

With regard to the delinquent, the expansion of *terminus ad quem* responsibility is noticeable in several phenomena. Defendants on trial—and then as convicts—are expected to present themselves as *ethical* selves who actively comply with the ethical ideals of the community. Before and during trial, especially trials that attract great public attention, the defendant faces increasing moral pressure to tell the truth, confess, show regret, and apologize to victims and survivors of the harm he or she caused. The trial itself becomes a public stage for the victim, not only as a witness (in keeping with conventional roles), but as a sufferer of trauma who must live with the consequences of crime. Procedural norms have changed in order to give victims a formal role and the right to give a "victim impact statement" (Booth 2015; Ashworth 1993). In Germany, victims have also been given the formal position of joint plaintiff alongside the public prosecutor (*Nebenklage*) since the 1980s, and the numbers of these cases have been growing steadily (Barton 2011). The defendant's right to remain silent (*nemo tenetur se ispe accusare*), meanwhile, is put under increasing pressure, because victims and their relatives (e.g., in cases of genocide and crimes against humanity) demand to hear the truth.

In order to offer incentives for defendants to confess in support of the search for truth, alternatives to a formal trial have been proposed and, in some cases, already been established and put in place: truth commissions and similar procedural settings aim at full disclosure of the facts of the crime, including additional participants (e.g., accomplices, instigators, assistants, superiors, and subordinates), as well as of the psychological factors and ideological motivations. One of the reasons given for the need for facts is that this knowledge might help victims and their relatives to cope better. Furthermore, these facts are supposed to help the defendant to understand and empathize with the victim's situation, take *ad quem* responsibility, undertake reparation and restoration, and seek forgiveness as a condition for reconciliation. In cases of genocide and crimes against humanity, this is considered a promising path toward the restoration of peace and security. The incentive that truth commissions and similar approaches offer to defendants include the total waiver of punishment, mitigation, or postponement. A recent example is the Special Jurisdiction for Peace in Colombia (Jurisdicción Especial para la Paz).[3] The program's name itself outlines the necessary steps: "Comprehensive System of Truth, Justice, Reparation and Non-Repetition" (in the original Spanish: Sistema Integral de Verdad, Justicia, Reparación y No Repetición).[4]

The defendant (and in the case of a guilty verdict, the convict) is expected to show remorse and repent, and make efforts at repairing and restoring the pre-crime state of affairs. The concept of *restorative justice* was invented as an alternative to imprisonment and other measures of hard treatment, or at least as a complement to them (Braithwaite 1989; for a more detailed account see Zisman's chapter in this volume). This aim is clearly articulated in the Council of Europe's *Recommendation on Restorative Justice*, as adopted by the Committee of Ministers in 2018:

> "Restorative justice" refers to any process which enables those harmed by crime, and those responsible for that harm, if they freely consent, to participate actively in the resolution of matters arising from the offence through the help of a trained and impartial third party. (Council of Europe 2018, Appendix II.3)

Restorative justice is not limited to compensation for victims but comprises an entire agenda of activities for the defendant, beginning with confession and a willingness to take *ex post facto* responsibility for the legal violation and for the victim's consequent suffering, before also accepting *terminus ad quem* responsibility for initiatives such as communication with victims, expressions of empathy, and symbolic acts of repair and restoration, culminating in a request for forgiveness. The transition from *ex post facto* to *terminus ad quem* responsibility requires and assists a perpetrator's self-transformation, which contributes to "ethical reconstruction in the attempt to instill the capacity of self-management" (Rose 2000, 336), which, in turn, allows for a law-abiding life that does no harm to others. The Council of Europe's *Recommendation* is based on the consideration of "the importance of encouraging the offenders' sense of responsibility and offering them opportunities to make amends, which may further their reintegration, enable redress and mutual understanding, and encourage desistance from crime" (Council of Europe 2018). In summary, if guilt has been replaced by risk, then the search for those responsible for controlling risk before its realization in the form of a crime becomes paramount. Therefore, the meaning of responsibility shifts from *ex post facto* to a stronger emphasis on *terminus ad quem* responsibility.

The Consequence of *Terminus ad Quem* Responsibility: Punishment and/or Restoration?

The above elaboration on *terminus ad quem* responsibility explains the increasing importance of the role of restoration, repair, and reconciliation, as well as the decreasing emphasis on hard treatment and punishment in criminal law. Hard treatment treats the prisoner as a passive object who is more or less left alone for a certain time, sometimes for many years. Even if the prisoner is obligated to work—involved in economic production inside or outside prison walls—this is usually merely an exploitation of labor; at best, it *could* serve rehabilitation, particularly if it leads to some professional qualification for future jobs (although the prison system has always been characterized by a huge gap between ideal and reality).

But such hard treatment does not serve *terminus ad quem* responsibility, which aims to reconstruct the ethical self and imposes activities to repair community values. Instead of "passivating" inmates, so to speak, it endeavors to *activate* convicts. According to Deleuze, control-focused societies try hard "to find penalties of 'substitution'" for the prison system (Deleuze 1992, 7). This leaves only those who can never be expected to turn into ethical selves (by way of taking *terminus ad quem* responsibility) to languish incarcerated. Exclusionary confinement is reserved for "'monstrous individuals' who either cannot or do not wish to exercise the self-control upon conduct necessary in a culture of freedom" (Rose 2000, 333). For the majority, possible corrective measures range from less intrusive interventions that curtail one's right to liberty (e.g., electronic ankle cuffs) to the aforementioned activities of restorative justice. These alternatives to hard treatment and incarceration communicate the symbolic message that responsibility has been taken. Imposing service requirements and community work are supposed to strengthen *terminus ad quem* responsibility (cf., e.g., the discussion of communal work in Duff 2001, 99–106). Thus, the criminal justice system and criminal law discourse become more and more a communicative performance of public affirmation, as well as a corroboration of the concept of subjectivity and the distribution of responsibility.

The growing importance of *terminus ad quem* responsibility in criminal law discourse adds support to conventional objections against punishment as counterproductive. The first, which was already raised by Beccaria and Mill, points out that responding to harm with another round of harm, in the form of hard treatment, only multiplies harm and suffering and does little to remedy the original suffering. The second objection refers to the popular assumption that holds that victims supposedly have a right and a natural desire to see their tormentors suffer, and that only punishment provides compensation and satisfaction; this justification for punishment, however, fails to recognize that for most of those involved—the victims and their relatives, more than anyone—such a desire is frequently about something *other* than seeing their tormentors suffer. Besides the fact that this may not be a legitimate aim of *public* punishment, and that subjective needs can hardly be objectified in the form of punishment that follows the principle of equality, the need for compensation is often misdirected when it is articulated by the desire to exercise hard treatment (see also Zisman's chapter in this volume). As the Council of Europe's *Recommendation on Restorative Justice* states, it is more important to care for "the legitimate interest of victims to have a stronger voice regarding the response to their victimisation, to communicate with the offender and to obtain reparation and satisfaction within the justice process" (Council of Europe 2018). As von Kellenbach points out in this volume, when it serves to cultivate mindfulness, compassion, and respect for self and others, this might indeed also include an element of pain and suffering. But then pain and suffering are already understood as a part of restorative justice and *terminus ad quem* responsibility. Finally, there is the objection that the prospect of being met with hard punishment prompts defendants to seek self-protection—to flee, suppress evidence, and pursue strategies that question and undermine the truth as well as their *ex post facto* responsibility. While the individual motive to escape and minimize punishment grounds the defendant's constitutional right to remain silent, this same motive blocks *terminus ad quem* responsibility—and its inherent productivity.

Does this mean that nothing should happen after a defendant has been found guilty of a crime with *ex post facto* responsibility? The concept of responsibilization by *terminus ad quem* offers a negative answer. At the same time, it presents an alternative to punishment: passivation by imprisonment is substituted (or at least complemented) by activity and productivity. Only *outside of* and *apart from* punishment can opportunities for cultivating a positive outcome for crime be found; that is, not until punishment ceases to be the primary focus and culmination can the productive potential of *terminus ad quem* responsibility become visible and possible.

But what could be the rationale for the attribution of *ex post facto* responsibility for having committed a crime if it provides, at best, a necessary but not *sufficient* justification for punishment? The answer to this question should be clear: *ex post facto* responsibility becomes a justification for *terminus ad quem* responsibility. Its core implication is that there is a proper course of action, which the (convicted) defendant will be required to "pay" for having chosen an *improper* course of action. The operations of the criminal justice system do the preparatory work to turn *ex post facto* responsibility into *terminus ad quem* responsibility; they are productive insofar as they elaborate on and prepare a judgment about the *terminus ad quem* responsibility for criminal harm.

Responsibility between Domination and Empowerment

Responsibility is a Janus-faced concept. As Foucault and, in the same vein, Deleuze, Rose, and others have pointed out, "responsibility" is an essential signifier in a society of control, particularly in the function *of* responsibilization, which creates a "new system of domination" (Deleuze 1992, 7). The validation and reproduction of society's dominant norms with its social hierarchies, structural inequalities, and normative "morality" involves the continual assertion that any deviance is immoral. Such a system results in practices of inclusion and exclusion. In civil society, individuals are held accountable for activities necessary for the reproduction of normative order. A society of control establishes and institutionalizes mechanisms for supervision, evaluation, self-assessment, and other control-focused procedures, to ensure that each individual fulfills his or her *terminus ad quem* responsibility and engages in reconstructing and optimizing the ethical self. In so doing, dominant patterns of responsibility are reproduced, as the "ethical" self—as an ideal of self-*control*, self-management, and self-improvement—is activated to successfully pursue projects and activities that validate the community's ethical values (for a similar example in the case of war veterans, see Derwin in this volume). The criminal justice system's reproduction of dominant patterns of responsibility has been accurately described by David Garland with regard to penal practices in the welfare state:

> Penal practices, discourses, and institutions hold out specific conceptions of subjectivity and they authorize specific forms of individual identity. In its routine practices—as well as in its more philosophical pronouncements—penalty projects define notions of what it is to be a person, what kinds of persons there are, and how such persons and their subjectivities are understood. Through its

> procedures for holding individuals accountable, penalty defines the nature of normal subjectivity and the relationship which is generally assumed to hold between individual agents and their personal conduct. (Garland 1990: 268)

Although Garland refers to punishment and its preconditional *ex post facto* responsibility, the same is true for *terminus ad quem* responsibility, by which defendants are expected to confess, reach out to victims, and engage in repentance. These activities define the ethical self, which reconstructs itself in order to improve its self-management abilities.

Such a description treats normative orders as mechanisms that give rise to normalization by producing behavioral regularities, including the institutionalization of the responsible subject. External observers are more keenly aware of the existence and maintenance of specific normative systems than are internal participants. But any practice of domination that relies on rule conformity exhibits gaps between so-called normativity and reality itself. Discrepancies between rules and factual practice give rise to criticism. According to H. L. A. Hart, norms presuppose critical reflective attitudes. Norms can be criticized on the basis of other norms (Hart [1961] 1997, 57), and they provide the standards that allow accusation and judgment, reproach and censure, criticism and sanction. Whether any such criticism is *sufficiently* justified by normative reasons is a matter of discourse and continuous contestation.

Participants in normative orders react with resentment, as Peter Strawson ([1974] 2008) has demonstrated, when they have reasons to believe that agents who violated a norm have acted voluntarily and can be held responsible for their conduct. Once there are reasonable doubts as to whether a perpetrator acted willingly and voluntarily and could have avoided a norm violation, one switches to the observer's point of view, treating agents as objects whose behavior can be explained by causes beyond their control (e.g., internal incapacities, such as mental illness, or various external factors). In some cases, these explanations are accepted as legitimate excuses; for instance, when an agent did not deliberately step on another's foot but rather stumbled involuntarily.

The practices of responsibilization are Janus-faced, too; they have an internal and an external dimension, and they serve both as a mechanism of domination and as a justification for criticism. On the one hand, the responsible subject is a necessary presupposition of the whole project of "embark(ing) on the enterprise of submitting human conduct to the governance of rules" (Fuller [1964] 1969, 162). Control societies institute practices of criticism, evaluation, assessment, or justification as techniques of responsibilization. Control societies even invite, encourage, and reward criticism. Thus, the difference between norm and reality, which constitutes the participant's critical reflective attitude, can also contribute to the confirmation of domination and normalcy. This is true as well for critical evaluation of the ethical self who fails at the ideals of self-management and self-improvement. Criticisms of conduct with regard to norms, values, and ideals become customary. According to Boltanski and Thévenot, societies of control establish different, coexisting "common worlds," each with their own specific values and ideals (e.g., the domestic world, the civic world, the market world, etc.). These worlds constitute a "critical matrix" (Boltanski and Thévenot [1991] 2006, chap. 8), which determines the type of criticism, justification, and examination that is practiced with regard to the specific values.

The current normative order of responsibilization can be criticized on two counts: first, practices of responsibilization overburden the subject due to the fact that the social conditions for the exercise of *terminus ad quem* responsibility are deficient (e.g., unequal access to higher education can reduce one's capacity for satisfying *terminus ad quem* responsibility if this lack of access prevents the actor from achieving sufficient, relevant knowledge). Responsibilization then becomes both a matter of empowerment and of discipline and oppression (Günther 2002). Second, it is possible to criticize the normative order, its values and ethical ideals (which are, of course, enforced and reproduced by responsibilization), as unjust and unfair in themselves (e.g., when they violate minorities' fundamental human rights or attempt to legitimize an arbitrarily unequal distribution of power, wealth, or recognition).

Despite these criticisms, the current normative order of responsibilization that accompanies the rise of control societies necessitates new perspectives on guilt as a transformative force. The finding of *ex post facto* responsibility in a criminal trial should no longer be considered a necessary condition for punishment in the form of hard treatment, but rather can justify the cultivation of *terminus ad quem* responsibility. Punishment is no longer the only and natural consequence of *ex post facto* responsibility. The passive suffering of punishment as hard treatment is, of course, still a possible consequence of guilt, but its very passivity blocks the performative, productive power of guilt; in this sense, hard treatment restricts guilt's transformative potential. By contrast, the withholding of punishment could unleash guilt's performative power, as in the case of the tragic heroes of antiquity who embraced their own hard treatment (dramatically and with pathos)—and *accepted their terminus ad quem responsibility*. In decoupling guilt from hard treatment, the force of guilt could unfold its performative and productive power in multiple directions, including repair and restoration.

The Community's Responsibilities

Communities do not experience or treat a crime as a private conflict between an offender and a victim. Crimes (at least those that consist of a serious offense and cause severe harm to a victim) affect the community's fundamental norms and values on a deep level. These norms and values are shared by each member and consist of intertwined affective and cognitive elements that are more or less the same for everyone and shape the basis of their social identity. For this reason, the violation of such a norm not only offends an individual victim but also the community as a whole. This is even more true if the community transforms its fundamental norms and values into positive criminal law by political legislation and an independent judiciary. Crime represents a grave threat to the society—a figurative pandemic.

In her book *AIDS and Its Metaphors* (1989), Susan Sontag described the public interpretation of pandemic diseases in ancient societies before they turned from an issue that concerns the whole community into a matter of individualized behavior in modernity: "Diseases insofar as they acquired meaning were collective calamities, and judgments on a community" (Sontag [1989] 1991, 131). Writing amid the HIV/AIDS crisis, Sontag strongly opposed the view that illness has any metaphorical meaning at all, particularly arguing against its interpretation as a punishment imposed for any fault or sinful behavior in the past.

Her analysis seems appropriate with regard to crime. Different from an illness, a crime is a norm violation by a perpetrator who offends a victim, generally on the individual level, while it is simultaneously also a "judgment on a community." Interpreted as a judgment, crime opens something like an empty space that causes unrest and craves to be filled, in the sense that everybody in the community knows something must be done in response to the norm violation that has occurred. Crime itself is a judgment on the *terminus ad quem* responsibility of the community, which must respond on three levels: in the social, temporal, and material dimensions.

These three dimensions can best be articulated with three questions. In the social dimension, the question is: *Who* should do something? In the temporal, it is: *When* should something be done? And in the material dimension, the question is: *What* should be done? These questions and the answers to them constitute the primordial *terminus ad quem* responsibility of the community to respond to the crime. If it did not, society itself would become guilty; that is, society would be attributed with *ex post facto* responsibility.[5]

One way in which a community practices its *terminus ad quem* responsibility is to find the individual who must bear the *ex post facto* responsibility for the crime. This is why criminal law discourse centers on the attribution of *ex post facto* responsibility: to find out who is responsible or guilty for the offense, and to what degree. But this search for criminal responsibility has an additional meaning: it is how the community discharges its primordial *terminus as quem* responsibility to react to crime. Criminal law is anchored in the collective *terminus ad quem* responsibility. This is easily missed because it seems natural to say that only the perpetrator carries *ex post facto* responsibility for "his" offense—but the possessive adjective, "his" (or "her") is a *petitio principii*. First, the crime must be investigated, and then it must be confirmed. The community must have first already established the normative order that can support the justification(s) for assigning individual responsibility for specific conduct. As previously mentioned by David Garland, such normative order consists of reasons, procedures, and penalties "for holding individuals accountable," and they "define the nature of normal subjectivity and the relationship which is generally assumed to hold between individual agents and their personal conduct" (Garland 1990, 268). The community's *terminus ad quem* responsibility will then be implemented by institutions and procedures, beginning with the legislation of criminal law (see Ristoph 2011), and continuing with the institutionalization of criminal trials and a criminal law discourse that defines (and justifies) the conditions under which a perpetrator could be found guilty.

Included in this process is positive criminal law and what Garland describes as the definition of "the nature of normal subjectivity." The central notion of a responsible subject must be understood as a political and legal construction, which implements the community's primordial *terminus ad quem* responsibility to react to crime (for more on the rational reconstruction of the notion of the responsible subject, see Günther 1985). Individual responsibility is constructed within the subject's relationships and dependence on nature, community, and other people. The whole point of attributing *ex post facto* responsibility to an individual consists of the (organized and justified) transfer of the community's primordial *terminus ad quem* responsibility to the convicted defendant on the basis of criminal law's definitions of the notion of the responsible subject. Furthermore, this transfer means that it is now the convict who must take the *terminus ad quem* responsibility for responding to the crime. It is critically important to keep in mind that the responsibility lay with the community first and the perpetrator second. The community's collective responsibility for a crime is not an archaic remnant that has been overcome by modern, secularized societies; rather, the fact that societies have *lost* their sense of collective responsibility over time is a consequence of individualist biases. The individualization of responsibility and the corresponding transfer of its locus to the guilty perpetrator is just one, but not the only, possible solution—and so is the productive-guilt reaction to a crime.

However, society's concern with *terminus ad quem* responsibility is not only about the fact that a crime harms both a victim *and* a community's fundamental norms and values. The notion borrowed from Susan Sontag about the "judgment on a community" also helps one to realize another kind of involvement: no crime takes place in pure isolation, on a lonely planet inhabited by perpetrator and victim alone. It always takes place in a community of which perpetrator and victim are members. The norms that are violated are ultimately norms that define the position and the status of the community's members, their relationships to each other, the basic structure of cooperation, and the distribution of resources and products. A crime, therefore, is not something external that crashes into a well-ordered community out of the blue, but rather something *internal*, which cannot be separated from the norms, institutions, and structures that shape the conduct of all its members. There is no community so sufficiently well ordered that crimes do not and will never happen. Even when the causes of a crime are primarily located in the individual, that individual grew up in a family (often dysfunctional), within a community, and in permanent interaction with these, which are all shaped by particular distributions of wealth, power, and recognition that vary in their egalitarianism and fairness. All of this contributes to the formation of any individual, who may become dangerous to others under certain circumstances. Part of this complex social constellation includes the peculiar relationship between perpetrator and victim, which provides impetus and reason for an offense. Thus, taking all of these influences into consideration, a crime is indeed a judgment on a community, which comes close to Susan Sontag's insight.

Therefore, a community can never completely shed its general responsibility for the crimes that take place within it. For this reason, a crime is a judgment on a community's *terminus ad quem* responsibility, and the community must therefore pay attention to the crime and take it seriously—because the crime refers to conditions and circumstances that contribute to the commission of crimes and therefore could be changed, either by political reform on the macro level or by minimizing criminogenic factors on the micro level. Each crime is also a judgment on the community's *ex post facto* responsibility, for its share in the reasons that shaped one of its members into a dangerous individual. Again, it is the criminal law discourse and the criminal justice system that make judgments about the shares of the society, perpetrator, and victim (and their relative weight) in the case of each crime. When *ex post facto* responsibility is attributed to a perpetrator in a criminal trial according to the criteria of the criminal law discourse, it claims to be justified within this triangle. The more guilt is attributed to the perpetrator, the less to the victim and society—and vice versa.

In conclusion: crime is always a communal matter, and the community has an *ex post facto* responsibility for its share in the conditions under which offenses happen *and* a *terminus ad quem* responsibility to respond to such offenses. Modern societies implement their *terminus ad quem* responsibility by establishing criminal law and justice systems, which attribute *ex post facto* responsibility to individuals, according to assumptions about the theoretical responsible subject. Thereby, they judge who shall take the *terminus ad quem* responsibility and "do something about" the crime. Punishment by hard treatment is the conventional way to realize this responsibility, but this may well obstruct the productive unfolding of *terminus ad quem* responsibility.

The Communication of *Ex Post Facto* Responsibility by Criminal Law Discourse

Returning to Foucault's observations about the productivity of criminal law discourse and the justice system, we can give a more detailed and differentiated description. The productivity of criminal law discourse consists of performative, communicative acts, which are guided by the interrelationship between *ex post facto* and *terminus ad quem* responsibilities and are (re)distributed between community, perpetrator, and even victims. As a discourse on responsibilization, criminal law is part of a system of domination but also a communicative space of criticism and contestation. It is important to note that all performative acts and communications that take place beginning immediately after a crime has been committed—from trial to verdict and its resulting punishment, restoration, and/or repair—are driven by the productive force of the *community's terminus ad quem* responsibility. In this communicative process, the community discharges its *terminus ad quem* responsibility by transforming it into an *ex post facto* responsibility to be borne by the convict. This occurs in a series of communicative messages that are sent by performative acts by all of the participants in this criminal law discourse. The productive force of *terminus ad quem* responsibility is manifest already in the performative activities that compel the exchange of such messages among defendant, community, and victim. I will rely on recent theoretical scholarship on the communicative or expressive nature of punishment, which explains the primary function and justification of criminal trials (and punishment) as communication between these three parties (see Duff and Garland 1994; Günther 2014, 2020.). I shall focus on two meanings of the propositional content of these communicative messages: the unlawfulness (or norm-violation) component and the responsibilization and guilt-related component. Both parts are equally relevant to all three parties, but I will focus in particular on the relevance of the guilt-related part, not only directed at defendants but also at victims and society. Finally, I will conclude with an ideal version of a criminal law discourse that is not overly repressive or part of a system of domination (although it could become such under certain circumstances, as mentioned previously).

Under ideal conditions, a community assumes and implements its *terminus ad quem* responsibility by democratic procedures of legislation and the establishment of an independent judiciary. It must be presupposed that a community's fundamental norms and values are codified into positive criminal law by democratic legislation (i.e., legislation based on citizens' equal rights to and opportunities for political participation). Citizens can conceive of themselves as co-legislators of the norms of criminal law and the criminal justice system; it is "their" community, and consequently also "their" *terminus ad quem* responsibility. The presumption is that the productive force of guilt can only unfold its full potential if citizens of a community are able and willing to conceive of the *terminus ad quem* responsibility (in this case, the responsibility to do something in reaction to a crime) as their own.

The Perpetrator

It may sound trivial to affirm that a trial, verdict, and punishment are conducted in order to send messages to a perpetrator. But the defendant is not the only recipient of these messages. While an offender commits an individual act in pursuit of personal desires and on the basis of individual motivations, every unlawful act is also communicative by nature (although offenders are usually only concerned with their personal intentions). A crime is a *performative* act that declares publicly that that murder, rape, etc. is possible. Hence, it becomes a normative statement, too. It contains both the descriptive and normative propositions that human life is not safe from harm. As Hegel said, crime asserts a normative facticity of its own, "which would otherwise be positively established" if it were not negated and its corresponding, violated right not restored (Hegel [1821] 2001, sect. 99, 90). This "false norm" has a communicative effect upon society as well as upon the victim; for society, specifically, a crime denies the validity of the violated legal norm and presents an alternative, which cannot be justified in a public discourse of legislation. That is, by presenting the alternative norm, perpetrators assert themselves as authoritarian legislators, unique and sovereign sources of law, imposing their particular will (authoritarian in the sense that their crimes deny their fellow citizens' equal status as co-legislators). Offenders subjugate victims under the norms of their own making, thereby exercising domination without justification. Impunity could therefore have devastating effects on a society's trust in the validity of the legal system (see Hofmann's chapter in this volume).

Most offenders are not conscious of this communicative messaging when planning and committing a crime. But what is relevant is not whether and how perpetrators *construe* their own actions so much as the normative impact that these actions have on other people, particularly the victim(s) and society. Nevertheless, there is some empirical evidence that some perpetrators do provide legitimizing narratives for their lawbreaking and use techniques such as victim-blaming in order to neutralize the normative impact of their conduct (Sykes and Matza 1957).

Ideological justifications and excuses are typical of political and collective human rights abuses, such as crimes against humanity or genocide. They are usually embedded in ideological narratives, containing large and comprehensive stories that validate and justify atrocities conducted by members of the government or the military (for ideological justifications of this kind, see Lotter in this volume). These narratives provide political, religious, or racist reasoning that defends violence as necessary for the survival of the community and the defense of core values—that is, as part of an apocalyptic battle between Civilization and Barbarism, Good and Evil. In these cases, wrong is cast as right, and the victims' suffering appears deserved and inevitable. Unlawfulness is denied. When prominent and trusted leaders spout these narratives, many people listen and are seduced into believing them. In fact, such narratives often become a central part of the founding narrative and legitimacy of a state. If they remain unchallenged, they can prevail and structure a community's collective memory, identity, and public ethos in the present and the future. In the worst case scenario, such narratives become "legitimate" sources of law. In such instances, therefore, it becomes even more important to empower victims' voices and to deny perpetrators' presumed "right" to be treated as authentic and trustworthy narrators. Otherwise, the victors will write history and shape not only a community's collective memory but also its sovereign sources of legitimate legislation. Criminal trials and convictions send messages that negate these narratives; the verdict, in particular, serves as a performative act of public rejection.

The other essential component of the communicative procedure of criminal trials consists of the activation and responsibilization of the defendant. In judicial practice, establishing the facts of a case, including most prominently the fact of culpability beyond reasonable doubt, takes the majority of the time and attention. However, more than the facts are necessary in order to establish the degree of *ex post facto* responsibility. The convict must also be treated as a responsible person, responsive to *reasons*, and compelled to accountability—required to account for the reasons that led to a violation of the community's norms (and, thereby, to challenge the validity of the community's laws). Therefore, the public trial sets the stage to convey messages of *ex post facto* responsibilization directed at the perpetrator. Convicts must recognize their responsibility for validating norms and laws and renounce their previous usurpation of normative authority. In international criminal cases, this entails the explicit de-authorization of those who previously held legislative power (and, thereby, responsibility for the crimes) as valid sources of law.

The Victim

The message of a guilty verdict, for victims and/or their relatives who are present as witnesses or as "joint plaintiffs" (*Nebenkläger*), powerfully affirms the vindication of their suffering as well as condemns the misconduct that violated their rights to life, bodily integrity, dignity, property, and so on. Furthermore, victims are told explicitly and publicly that someone else is responsible for the harm they suffered, which was neither bad luck, nor fate, nor an inevitable natural event. Finally, because *ex post facto* responsibility is linked to the *terminus ad quem*, a guilty verdict opens the door to either punishment or repair and reparation. These communicative messages can help victims, especially of traumatic, violent crimes, to explain suffering and to undo the effects of violation and degradation. Unless the offender's narratives of neutralization (i.e., justifications and excuses) are publicly repudiated, they serve to reinscribe the power of perpetrators to define the victim as a victim.

Often, it is presumed that the victim was in the wrong, be it at the collective macro level of ideological conflict or at the micro level of individual behavior (which supposedly provoked the offense). Such victim-blaming is internalized, and victims are forced to absorb *ex post facto* responsibility and begin to blame themselves. This phenomenon has at least two related consequences. Firstly, on a psychological level, victims experience a severe loss of control over their own lives; they live in constant fear of recurrence and of a powerlessness that they could not and will not be able to prevent in the future. This may lead to a condition of "learned helplessness," which, once ingrained and habitual, causes depression, passivity, and lack of self-confidence (Seligman 1975; Abramson, Seligman, and Teasdale 1978; see also von Kellenbach in this volume). Secondly, the injury is not only psychological but also normative: victims who blame themselves confirm the normative position of the criminal, who was able to impose his or her will. In the case of international crime, victim-blaming confirms the justificatory ideology that has caused the mass atrocities while simultaneously denying the rights of citizens as coequal legislators; victims are relegated to objects of control and domination, deprived of agency. Victimization that is not exonerated deprives people of their ability to claim political rights and to resist injustice.

In declaring a perpetrator guilty of *ex post facto* responsibility, the justice system absolves the victim of wrongful suffering and redistributes the blame to the perpetrator. With the declaration of guilt, the justice system affirms victims' rights and recognizes them as integral citizens with equal rights in the normative community.

The Community

Finally, with the public determination of guilt, society repudiates complicity and communicates its recognition and rejection of injustice. Without such rejection, society remains implicated in the (unjust) normative reasons or "shared wrongs" (Duff 2001, 63) that found expression in the crime; that is, if it were to neglect its responsibility to justice, society would essentially accommodate wrongdoing into its normative order and renounce its solidarity with victims. The declaration of guilt affirms the validity of the legal norm (notwithstanding its former violation) and restores victims' and society's shared belief in justice.

This performative act is important not only with regard to low-level offenses but even more so in the context of systemic human rights abuses and state crimes. Declarations of guilt refute perpetrators' implicit claims to the validity or assertion of their authority to legislate arbitrary normative codes of conduct. For citizens, this message helps to restore and reaffirm faith in human rights and democratic legislation as legitimate authority—and it validates those who resisted the usurpation of legislative authority and declares them correct in doing so. Ultimately, a determination of wrong is only complete *when attended by* the determination of *ex post facto* responsibility or guilt. Otherwise, denial and dismissal, and revisionism and neutralization, remain widespread—such as when public officials assert that they were unable to act differently and that they were powerless to avert atrocity and injustice. Society must reject these excuses. It is necessary to oppose assertions and narratives that aim to excuse or to absolve guilt, in order to make clear how much freedom was indeed available to individual perpetrators and to what extent the wrong was the result of their individual lack of motivation to avoid it. Finally, the justified responsibilization of the perpetrator must be the result of a close elaboration of society's possible share in the crime. Only then may society be exonerated; otherwise it must acknowledge its own *ex post facto* responsibility (on the difficulties that a society faces in entering into a discourse on its own guilt, see—with regard to Germany after 1945—Buschmeier's chapter in this volume).

Conclusion

By now it seems clear that the productivity of criminal law discourse is driven by the community's *terminus ad quem* responsibility that grounds the need to react to crime. In order to meet this responsibility, a community must come up with a normative order and a criminal law discourse that allows for calling a certain deed a "guilty" deed. In other words, the society must create a system that recognizes and justly assigns for *ex post facto* responsibility. But the function of criminal law discourse is only to prepare and to justify the reallocation of *terminus ad quem* responsibility from the community itself to the perpetrator, who has been convicted in a fair trial, precisely according to the conditions established by criminal law and its rules and procedures for the attribution of *ex post facto* responsibility for a crime. Criminal law discourse is, therefore, a kind of proactivity.

In criminal law, the conventional consequence of a conviction is punishment by hard treatment, notwithstanding the exceptions that mitigate or substitute punishment by other measures. Punishment, however, as I have argued, does not incentivize convicts to accept any productivity inherent to the *terminus ad quem* responsibility. Therefore, restorative justice and similar alternatives are more suited to unleashing its inherent productivity. As the Council of Europe says:

> The core principles of restorative justice are that the parties should be enabled to participate actively in the resolution of crime (the principle of stakeholder participation), and that these responses should be primarily oriented towards addressing and repairing the harm which crime causes to individuals, relationships and wider society (the principle of repairing harm). (Council of Europe 2018, Appendix III.13)

Recent abolitionist movements argue for a complete substitution of the criminal justice system in general, and hard treatment in particular, by favoring the idea of a "transformative justice," which encourages the perpetrator, victim, and members of the community to engage in the realization of the productive potential of *terminus ad quem* responsibility. This would be accomplished through communal conflict resolution, which includes the attribution to and acceptance of *ex post facto* responsibility by the perpetrator, as well as the examination of the community's responsibility for its own share. Such an approach would help to restore intersubjective relationships between perpetrators, victims, and the community (Morris 2000; Langer 2020). These alternatives depend on an important condition, as stated explicitly in the Council of Europe's *Recommendation*: all participants must "freely consent" (Council of Europe 2018, Appendix II.3). This involves *terminus ad quem* responsibility, which requires consent and cannot be imposed and enforced by criminal law. At this point, criminal law discourse comes to its limit. But if it were to conceive of itself as an essential part of the community's *terminus ad quem* responsibility, it could do much more than it actually does to improve the conditions under which all participants could freely consent to release the productive forces of guilt (for a similar conclusion with regard to restorative justice, see Zisman's chapter in this volume).

Notes

1. Of course, it is important to disguise different meanings associated with the broad notion of guilt and the even broader notion of responsibility. The German term *Schuld* (guilt) is still used in criminal law, comprising *mens rea* and culpability, notwithstanding much criticism of its religious, psychological, and moral connotations. Therefore, some German legal scholars prefer the term "criminal responsibility" (e.g. Roxin [1997] 2005, 847–74). The English language provides different expressions, such as accountability, liability, and culpability, which altogether belong to the family of responsibility. In the following, I shall use the term "responsibility," and I will suggest some distinctions that are necessary for my argument.
2. This can be observed with the increasing number of cases where police officers or psychiatrists who are working in the law enforcement system or in forensic psychiatry are prosecuted because of negligence (often, negligent homicide) because they permitted imprisonment facilities or limited releases that were abused by the prisoner to commit a new crime (often murder). See, e.g., the Bundesgerichtshof judgment of Nov. 26, 2019: BGH 2 StR 557/18, https://www.hrr-strafrecht.de/hrr/2/18/2-557-18-1.php), or the Bundesgerichtshof judgment of Nov. 11, 2003: BGH 5 StR 327/03, https://www.hrr-strafrecht.de/hrr/5/03/5-327-03.php3.
3. https://www.jep.gov.co/Paginas/Inicio.aspx (February 22, 2021).
4. https://www.jep.gov.co/Infografas/SIVJRNR_EN.pdf (February 22, 2021).
5. This could perhaps be a possible interpretation of an enigmatic paragraph about criminal justice in Kant's *Metaphysics of Morals* from 1797/98: "Even if a civil society were to be dissolved by the consent of all its members (e.g., if a people inhabiting an island decided to separate and disperse throughout the world), the last murderer remaining in prison would first have to be executed, so that each has done to him what his deeds deserve and blood guilt does not cling to the people for not having insisted upon this punishment; for otherwise the people can be regarded as collaborators in this public violation of justice" (Kant [1797/98] 2017, 116). Often quoted as a bad example and irrational affirmation of an archaic theory of retribution, it nevertheless emphasizes the link between a crime and the community's *terminus ad quem* responsibility to do something about it after the perpetrator has been found guilty; although for Kant, the execution of the death penalty was the only activity which he considered sufficiently justified, not to mention necessary.

References

Abramson, Lyn Y., Martin E. Seligman, and John D. Teasdale. 1978. "Learned Helplessness in Humans: Critique and Reformulation." *Journal of Abnormal Psychology* 87, no. 1: 49–74.

Ashworth, Andrew. 1993. "Victim Impact Statements and Sentencing." *Criminal Law Review*, July: 498–509.

Barton, Stephan. 2011. "Nebenklagevertretung im Strafverfahren." *Strafverteidiger Forum (StraFO)* 5: 161–68.

Boltanski, Luc, and Laurent Thévenot. 2006. *On Justification: Economies of Worth*. Translated by Catherine Porter. Oxford: Oxford University Press.

Booth, Tracey. 2015. "Victim Impact Statements, Sentencing and Contemporary Standards of Fairness in the Courtroom." In *Crime, Victims and Policy: International Contexts, Local Experiences*, edited by Dean Wilson, and Stuart Ross, 161–83. London: Palgrave.

Braithwaite, John. 1989. *Crime, Shame and Reintegration*. New York: Cambridge University Press.

Council of Europe. 2018. "Recommendation CM/Rec(2018) 8 of the Committee of Ministers to Member States concerning Restorative Justice in Criminal Matters." Strasbourg: Council of Europe. https://search.coe.int/cm/Pages/result_details.aspx?ObjectId=09000016808e35f3.

Deleuze, Gilles. 1992. "Postscript on the Societies of Control." October 59: 3–7.

Duff, R. Antony. 2001. *Punishment, Communication, and Community*. Oxford: Oxford University Press.

Duff, R. Antony, and David Garland, eds. 1994. *A Reader on Punishment*. Oxford: Oxford University Press.

Foucault, Michel. 1977. *Discipline and Punish*. Translated by Alan Sheridan. New York: Vintage Books.

Foucault, Michel. 2014. *Wrong-Doing, Truth-Telling: The Function of Avowal in Justice*. Edited by Fabinne Brion and Bernard E. Harcourt. Translated by Stephen W. Sawyer. Chicago: University of Chicago Press.

Fuller, Lon L. (1964) 1969. *The Morality of Law*. 2nd ed. New Haven, CT: Yale University Press.

Garland, David. 1990. *Punishment and Modern Society*. Oxford: Clarendon.

Garland, David. 2001. *The Culture of Control*. Oxford: Oxford University Press.

Garland, David, and Richard Sparks. 2000. "Criminology, Social Theory and the Challenge of Our Times." In *Criminology and Social Theory*, edited by David Garland, and Richard Sparks, 1–22. Oxford: Oxford University Press.

Günther, Klaus. 1985. *Schuld und kommunikative Freiheit*. Frankfurt am Main: Vittorio Klostermann.

Günther, Klaus. 2002. "Zwischen Ermächtigung und Disziplinierung: Verantwortung im gegenwärtigen Kapitalismus." In *Befreiung aus der Mündigkeit*, edited by Axel Honneth, 117–40. Frankfurt am Main: Campus.

Günther, Klaus. 2014. "Criminal Law, Crime and Punishment as Communication." In *Liberal Criminal Theory: Essays for Andreas von Hirsch*, edited by A. P. Simester, Antje du Bois-Pedain, and Ulfrid Neumann, 123–40. Oxford: Hart Publishing.

Günther, Klaus. 2020. "Positive General Prevention and the Idea of Civic Courage in International Criminal Law." In *Why Punish Perpetrators of Mass Atrocities?*, edited by Florian Jeßberger and Julia Geneuss, 213–27. Cambridge: Cambridge University Press.

Hart, Herbert L. A. (1961) 1997. *The Concept of Law*. 2nd ed. Oxford: Oxford University Press.

Hegel, Georg W. F. (1821) 2001. *Philosophy of Right*. Translated by S. W. Dyde. Kitchener, ON: Batoche Books. https://socialsciences.mcmaster.ca/econ/ugcm/3ll3/hegel/right.pdf.

Kant, Immanuel. (1797/8) 2017. "Public Right Section I: The Right of a State." In *The Metaphysics of Morals*, edited by Lara Denis, translated by Mary Gregor, 97–148.

Cambridge: Cambridge University Press. Published online 2018. https://doi.org/10.1017/9781316091388.

Langer, Máximo. 2020. "Penal Abolitionism and Criminal Law Minimalism: Here and There, Now and Then." *Harvard Law Review Forum* 134: 42–77.

Morris, Ruth. 2000. *Stories of Transformative Justice*. Toronto: Canadian Scholars Press.

Ristoph, Alice. 2011. "Responsibility for the Criminal Law." In *Philosophical Foundations of Criminal Law*, edited by Richard A. Duff and Stuart P. Green, 107–24. Oxford: Oxford University Press.

Rose, Nikolas. 2000. "Government and Control." *British Journal of Criminology* 40: 321–39.

Roxin, Claus. (1997) 2005. *Strafrecht Allgemeiner Teil*. Vol. 1. 4th ed. Munich: C. H. Beck.

Seligman, Martin E. P. 1975. *Helplessness: On Depression, Development and Death*. San Francisco: Freeman.

Sontag, Susan (1989) 1991. "AIDS and Its Metaphors." In *Illness as Metaphor and AIDS and Its Metaphors*, 89–180. London: Penguin.

Strawson, Peter F. (1974) 2008. "Freedom and Resentment." In *Freedom and Resentment, and Other Essays*, 1–28. New York: Routledge.

Sykes, Gresham, and David Matza. 1957. "Techniques of Neutralization: A Theory of Delinquency." *American Sociological Review* 22: 664–70.

6

Making Guilt Productive

The Case for Restorative Justice in Criminal Law

Valerij Zisman

Introduction

In recent debates in criminal law, the idea that legal punishment can be justified as a form of secular penance[1] has gained prominence (Duff 2001, 2003; Garvey 1999, 2003; Radzik 2009). However, this justification might come as a surprise, especially to Western and liberal theorists. Penance, repentance, and atonement are all taken to be religious concepts, and it is widely held among these theorists that religious concepts should not play a role in the criminal codes of liberal societies unless there is also an independent secular justification for integrating such concepts into criminal law. Relying on concepts that would only be justifiable from a certain religious rationale would violate the commonly held liberal demand that the justification for coercion be universal. The adherents of penance accounts that I want to discuss in this chapter are aware of this worry and have tried to offer secular justifications that might do justice to the idea that offenders should indeed endure some sort of penance, without committing to any contested religious views but still borrowing from the same core ideas.

In this chapter, I have two central aims. First, I will argue that legal punishment cannot be justified by invoking the penance rationale, even with a secular account of penance that focuses on citizens as constituting a community with shared values. The two central problems of such justifications, which I will discuss, are that they grant the state too much authority over the "inner feelings" of offenders, and that they lead to disproportionate punishments. The second aim of this chapter is to argue that the penance rationale gives a plausible justification for restorative justice as a *procedure* for addressing criminal wrongdoing. This gives penance a more limited role to play in criminal law than some of the above-mentioned theorists assumed, but it is nonetheless an important takeaway attendant to taking penance seriously, especially for its productive potential with respect to various aims in criminal law.

This chapter will be structured as follows: First, I will briefly present the justification for punishment based on the penance rationale. Next, I will discuss two

Valerij Zisman, *Making Guilt Productive* In: *Guilt*. Edited by: Katharina von Kellenbach and Matthias Buschmeier, Oxford University Press. © Oxford University Press 2022. DOI: 10.1093/oso/9780197557433.003.0007 main objections against said justification. Lastly, I will defend penance—not as a justification for punishment, but as a rationale that justifies approaching criminal law *via* restorative justice. The defense of restorative justice will proceed in three steps, and each step will explore, one by one, the relevance of a specific aspect of the penance rationale and restorative justice: feelings of guilt, reparation, and reconciliation.

From Religious Penance to a Secular Community

The adherents of penance theorists that I want to discuss here accept that penance needs to be justified in secular terms, if we are to consider penance a legitimate justification for criminal law at all. Thus, instead of relying on concepts such as "sin" or "offenses against God," such authors try to understand wrongdoing as an offense against the victim, the community, and the relationship of all affected parties to one another. Stephen Garvey, for example, asks us to imagine a (near-) ideal community and how such a community would respond to wrongdoings performed by people within the group. His own answer is that

> [p]unishment in such a community would [. . .] be a form of secular penance aimed at the expiation of the wrongdoer's guilt and his reconciliation with the victim and the community. (Garvey 1999, 1802)

Antony Duff, similarly, argues that the aim of punishment with regards to the community should be understood along these lines:

> I have argued [. . .] that we should understand criminal punishment as, ideally, a kind of secular penance. The culpable commission of a crime involves wrongdoing that violates the central values of the political community, as expressed in its criminal law; the crime thus damages or threatens the offender's normative relationships not only with the direct victim, but also with her fellow citizens. (Duff 2003, 300)

Linda Radzik (2009, 4) also reasons that the perpetrator-victim and perpetrator-community relationships are paramount; she asks what "morality demands of wrongdoers themselves," and how they can make amends with their victims and the broader community in order to atone.

For a first approximation, we can thus say that, according to all of these theories, punishment is justified because criminal wrongdoing violates the central values of the community and the rights of the victim, and morality *demands* that the offender somehow account for said violation. The state's job, in this context, is to enforce the offender's compliance with that expectation. The specifics may vary regarding how penance should be understood and defined within each of these different penance approaches, but all of these approaches share some common aspects that will guide our discussion, which I want to highlight now.

Radzik argues that the appropriate response of an offender to their own wrongdoing consists of reconciliation with the victim and the community. Reconciliation requires that the offender:

1. make moral improvements,
2. express that the victim holds value (this response also entails that the offender feel guilty, remorseful, or ashamed), and
3. take reparative actions towards the victim. (Radzik 2009, 85)

For Garvey (1999, 1813), two stages of penance are important: expiation and reconciliation. Expiation, in particular, consists of repentance, including feelings of guilt, apology, reparation, and some measure of intentional harsh treatment toward the wrongdoer.

For Duff, wrongdoing deserves censure, and punishment-as-penance is a way to communicate this censure, in addition to being a way for offenders to restore their moral standing within the community by accepting the punishment and expressing remorse (Duff 2001, 107).

The commonalities across these three perspectives from Garvey, Duff, and Radzik consist, first, of the importance of recognition (of the wrongdoing's "wrongness") and the role played by feelings of guilt.[2] In each of these perspectives, it is important that offenders show a certain emotional response that not only acknowledges their wrongdoing but also signals that (1) the gravity of the action has been understood, and (2) they will not violate the victim's rights or the community's values in the future. The three authors also agree that the offender must provide reparation for the victim.

Two of the three authors also agree that penance requires that the offender be punished. For Garvey, punishment constitutes the penance stage of atonement (here is where he deviates from my suggested terminology); as penance must be burdensome, punishment is needed. Similarly, Duff argues that punishment is the only adequate means of expressing censure for the offender's act, and accepting a sentence offers a way for offenders to signal their assent to undergoing some form of penance, as well as their desire to reconcile themselves with their communities. Radzik, by contrast, does not agree that penance requires punishment.

However, all three agree that one of the *aims* of penance is reconciliation between the stakeholders in a conflict. It is thus important to stress that the penance rationale can aim to offer a justification for punitive treatment (as in Duff's and

Garvey's accounts), but that it can also aim to justify nonpunitive treatment of people in accordance with other penance demands, such as reparation, feelings of guilt, and reconciliation (as in Radzik's account).

In short, all three authors emphasize the importance of recognition and the experience of guilt, punishment (or reparation), and reconciliation.

Objections to Penance Theories

I want to discuss two objections to penance theories thus understood. The first objection challenges the strong communitarian assumption in secular penance accounts, along with the state's right to intrude upon the "inner feelings" of wrongdoers. The second raises the concern that penance theories open up the possibility of disproportionate punishment.

The Liberal Objection

This first objection aims at one of the central assumptions that has just been mentioned; namely, that the justification for punishment-as-penance can be built on the relationship between the offender and the victim, as well as between the offender and the broader community. The liberal objection will go on to argue that such a justification of punishment grants the state too much authority over the "inner feelings" of offenders.

Andrew von Hirsch and his colleagues have raised versions of the liberal objection several times in debates over Duff's penance account, but these objections can be applied to Garvey's and Radzik's accounts as well. The criticism concerns Duff's (alleged) overestimation of "how deeply the censurer may properly involve himself in the feelings and attitudes of the offender, in order to bring about a morally appropriate response" (Hirsch and Ashworth 2005, 95). I shall use the term "inner feelings" henceforth to refer to the aspect of penance that involves feelings of guilt, along with core moral commitments and values—all, arguably, deeply personal, internal experiences and perspectives. According to this objection, penance theories would justify the state's punishment of offenders, along with its attendant aim of making them regret violating certain values—even if the offenders did not share those values, and even if they had sound reasons for rejecting them. Such an imposition on the offenders seems morally wrong, and as penance theories entail the right to such intrusions, penance theories ought to be dismissed. (So the reasoning goes.)

To illustrate their argument that the liberal state lacks such a right, Hirsch and Ashworth start by highlighting examples of parties who (they believe) are indeed justified in demanding penance because of their close and intimate relationship to each other. It might very well be morally appropriate, for instance, that a religious leader demand (and enforce) penance from his followers whenever his followers commit sins, because he, as their religious leader, stands in the appropriate relationship to make this legitimacy claim. To add another example: when a family member or a close friend wrongs me, I might be entitled to expect (among other things) that she shows remorse in response to the wrongdoing. In cases like the aforementioned, an insistence on regret, feelings of guilt, and penance would be justified, because the stakeholders in the conflict stand in the appropriate relation to each other for making such demands. In the case of the state and its citizens, however, Hirsch and Ashworth insist that the relationship is of a very different kind—one that lacks the closeness and intimacy that intrusions upon inner feelings arguably require.

So far, Hirsch and Ashworth only lean on the intuition that the respective relationships are different in kind, without offering a clear-cut argument. So, what precisely are the relevant differences? One additional reason to think that the relationship between citizens and the state is different from the relationships within religious communities (besides the elements of closeness and intimacy) is that, in the case of states, citizens arguably give neither explicit nor implicit consent to the laws governing their behavior (Simmons 1979, secs. 3–4). Admittedly, this is sometimes also true for religious and familial relationships. But that observation only helps us to nail the point down. If we never agreed to the values of a certain religious community, the intuitive assertion that leaders overstep their authority by demanding penance grows strong. Even though communitarian authors such as Duff argue that communities are typically governed by shared values, the theorists do not invoke consent theories of punishment to make a case for the legitimacy of demanding penance. Such consent theories have been widely dismissed in the debate on the justification of punishment (Boonin 2008, 156–170).

It is plausible to concede that most states do not enjoy the same kind of relationship as (many) families or religious communities do. It is also reasonable to assume that, for example, if I were to harm a stranger, this person could not justifiably expect any right to change my core attitudes, feelings, or convictions—nor could the person expect to count on the state's help in doing so.

Yet it is perhaps inaccurate to view other citizens as complete strangers with whom we lack connectedness; as Duff emphasizes, citizens of liberal communities typically share, at the very least, "central liberal values [such] as freedom, autonomy, privacy, and pluralism, and [. . .] a mutual regard that reflects those values" (Duff 2001, 47). Even so, we should ask: Why is it that being connected through these basic values would justify demanding penance from offenders? The answer is not self-evident. The penance theorists need a further justification for this claim.

To defend the state's concern with the inner feelings of offenders, Duff attacks the liberal conception of freedom held by such critics of penance theories, along with their understanding of harms and wrongs:

> The harm suffered by the victims of central *mala in se* crimes (such as murder, rape, theft, violent assault) consists not just in the physically, materially, or psychologically damaging *effects* of such crimes but in the fact that they are victims of an *attack* on their legitimate interests—on their selves. The harmfulness and wrongfulness of such attacks lie in the malicious, contemptuous, or disrespectful intentions and attitudes that they manifest, as well as in their effects. The agent's intentions and practical attitudes (those directly manifest in his conduct) are thus relevant as conditions of liability—conditions for holding him liable for conduct that causes or threatens to cause harm. (Duff 2001, 128)

Duff, in other words, argues that the inner feelings and moral commitments of an offender do concern the state after all, as the state should be concerned with wrongs—and wrongs, in turn, can only be adequately understood when we take a look at the offender's attitudes. Thus, if it hopes to address (adequately) the wrongs as wrongs, the state should be concerned with an offender's inner feelings.

There are at least two problems with this response. First, as Duff mentions, his description concerns *mala in se* crimes—not all crimes in general. If that is true, his account would be plausible for a good range of criminal wrongdoings but not as a comprehensive approach to justify punishment for *all* wrongdoings, as not everything that the state criminalizes relies on *mala in se* wrongs.

Second, Duff's justification might not even be plausible in cases of *mala in se* crimes. It is true that we need to pay attention to the disrespectful intentions of offenders in order to be sure that we capture all dimensions of their wrongdoing—but even then, one could argue that the inner feelings of offenders (or anyone else) are beyond the state's jurisdiction. There is a morally relevant difference between taking into account the malicious intentions in order to determine an adequate amount of punishment from a retributive or consequentialist perspective versus doing so with the goal to change these attitudes through the imposition of punishment. We can hold as morally justifiable the taking into account of offenders' intentions without thereby also admitting that some form of penance is necessary. Duff's attempt to counter the liberal objection by pointing to the necessity of taking offenders' intentions into account thus fails to show that the state ought to be concerned with the inner feelings of offenders more broadly.

Without further argument as to why the state should be concerned with the inner feelings of offenders over and above the intentionality of their wrongdoing in order to assess the severity of the wrong, the liberal objection still stands against penance accounts.

The Proportionality Objection

Another problem is that the penance rationale alone cannot guarantee the proportionality of punishment, in either of two senses: first, it cannot guarantee that the punishment is proportional to the gravity of the wrongdoing, and second, it cannot guarantee that similar crimes would receive similar punishments. Meanwhile, both of these proportionality concerns are often seen as fundamental to justice in criminal law.

According to the penance approach, a hard-headed offender might, for example, receive a punishment that is disproportionate to the gravity of the wrong, as the amount of punishment would be determined by what was necessary to evoke penance. In other words, a particularly hard-headed person might "need" a fairly harsh punishment for a relatively minor offense, simply by virtue of his or her stubbornness. Not only this, but furthermore (and by extension), if two different offenders were to commit a similar wrong (in kind, gravity, and culpability) but were hard-headed to unequal degrees, this would mean that each of them would require different amounts and maybe even different *types* of punishment in order to understand that what they did was wrong. Of course, given that every account of criminal law should respect the principle of proportional and equal punishment (so the argument goes), we have to dismiss justifications of punishment that are based on the penance rationale.

Duff explicitly addresses the proportionality objection in his work. He argues that what justifies the imposition of punishment upon offenders is the deserved censure, which takes the form of penance. When the deserved censure is the primary focus, then the proportionality worry can be addressed. Censure must be proportional to the gravity of the wrong; therefore, similar actions must always receive similar types and amounts of censure. The problem with this perspective, however, is that the proper justification for proportional punishment in Duff's account stems from his retributive commitment, not from the penance rationale. In other words, what demands proportionality is not the remorse that offenders ought to experience, but the censure that their actions deserve. In turn, the censure should take a form of secular penance, but this comes into play independently of the proportionality concern.

Combining proportionality with penance in this way might be convincing in principle, but this move would essentially require us to discard the attempt to offer a justification of punishment solely based on the penance rationale and instead go with a predominantly retributive theory. It is perfectly legitimate to combine penance with other theories, but that departs from the question I aimed to answer in this chapter: Can the penance rationale represent plausible grounds for criminal punishment? If the penance rationale needs help from retributivism to address central objections, then the answer to that question is "no."

How Penance Informs Criminal Law

In the following sections, I will argue that penance has important implications for criminal law, regardless of its failure to justify the imposition of punishment, because it can be used to argue for restorative justice in criminal law.

The Importance of Feelings of Guilt for Various Aims of Criminal Law

When talking about guilt, I want to refer to the definition described in social psychology, in the tradition of Roy Baumeister and colleagues:

> By guilt we refer to an individual's unpleasant emotional state associated with possible objections to his or her actions, inaction, circumstances, or intentions. Guilt is an aroused form of emotional distress that is distinct from fear and anger and based on the possibility that one may be in the wrong or that others may have such a perception. [. . .] Guilt can be distinguished from shame on the basis of specificity. Guilt concerns one particular action, in contrast to shame, which pertains to the entire self. (Baumeister, Stillwell, and Heatherton 1994, 245)

Why should guilt, thus understood, be valuable for criminal law? We can argue for guilt in two ways. First, we can show that guilt facilitates the achievement of goals in criminal law that are widely considered justifiable; I want to refer to this first point as the argument that is based on the productivity of guilt. Second, we can show that pursuing guilt in the restorative framework does not fall prey to the objections presented above.

Let me begin by turning to the argument based on the productivity of guilt. I cannot offer a full account of the empirical benefits of feelings of guilt in this chapter; rather, I want to summarize what I take to be the key findings in the relevant research (for a detailed account, see Spanierman in this volume).

First, "(g)uilt often motivates reparative action (e.g., confession, apology, efforts to undo the harm)" (Tangney, Stuewig, and Hafez 2011, 710; see also

Tangney et. al. 1996; Ketelaar and Au 2003; Lindsay-Hartz 1984; Wallbott and Scherer 1995; Wicker, Payne, and Morgan 1983). Second, "[f]eelings of guilt go hand in hand with other-oriented empathy" (Tangney, Stuewig, and Hafez 2011, 710; see also Tangney 1991; Leith and Baumeister 1998; Stuewig et al. 2010). If offenders start to empathize with their victims, we can hypothesize that the former are also more likely to acknowledge the wrong that was done. Third, "guilt-prone individuals are inclined to take responsibility for their transgressions and errors" (Tangney, Stuewig, and Hafez 2011, 210; see also Tangney, Stuewig, and Mashek 2007). Furthermore, guilt proneness "negatively predicted arrests and convictions" in a long-term study of children (Tangney, Stuewig, and Hafez 2011, 712).

On the face of it, many theorists of criminal law should welcome these benefits. Various consequentialists (i.e., those who think that punishment is justified in terms of its deterrent or rehabilitative effect) should welcome the beneficial consequences that feelings of guilt might have. And even retributivists (i.e., those who think that punishment is justified because it is intrinsically morally valuable) can allow for guilt to play a role, because of its beneficial consequences; retributivism does not allow such consequences to play into the justifications for punishment, but once we have a retributive justification, we can at least allow the knowledge of guilt's correlation with beneficial outcomes to influence how we address wrongdoing.

In the following subsection, I argue that the productivity of guilt feelings, as well as the productivity of penance as a whole, can best be captured by adopting restorative procedures, and that it can address the objections mentioned in the previous section.

Addressing the Objections against Feelings of Guilt

June Tangney and her colleagues agree that the research on guilt lends validity to restorative justice as an approach to criminal law:

> Restorative justice approaches emphasize the need to acknowledge and take responsibility for one's wrongdoings and act to make amends for the negative consequences of one's behavior. But the restorative justice approaches eschew practices aimed at shaming offenders, ascribing bad behaviors to a bad defective self. Restorative justice interventions are consistent with RST [the reintegrative shaming theory] [. . .] and with psychologists' self vs. behavior distinction [. . .] but they often do not refer to the emotions of shame and guilt explicitly. Such interventions may be enhanced by the addition of components aimed explicitly at transforming problematic feelings of shame about the self

> into adaptive feelings of guilt about behaviors and their negative consequences for others. (Tangney, Stuewig, and Hafez 2011, 718)

Restorative justice as a philosophical theory is, in fact, a fairly new approach in the debate within criminal law. However, while people argue over precise understandings of restorative justice, they agree roughly on some central themes. First, a wrongdoing is not understood as an offense against the state or the rule of law, but as an offense against an individual (or a community), which creates a conflict (Christie 1977; Zehr 1990). According to restorative justice, then, the aim of the criminal justice system should be to resolve the conflict that has emerged from the wrongdoing, and it is crucial for this that the conflict remain in the hands of the stakeholders (i.e., offender, victim, respective friends and family members, broader community, etc.). The central idea of restorative justice is that the stakeholders enter a dialogue about the conflict. In this dialogue, the victim's voice is heard and, in the best case, the offender apologizes to the victim.

According to proponents of restorative justice, moreover, the stakeholder ought to agree on some sort of compensatory payment or reparation that helps victims cover the losses they have suffered, both materially and emotionally. Further to this concern, in some versions of restorative justice, the state checks whether the amounts and types of reparation are adequate, and it reserves the right to amend these decisions.

Reparative efforts might go beyond the aforementioned and also include the offender's willingness to take steps against factors that contributed to the wrong, such as addiction, aggression, etc. In some versions of restorative justice, what the offender and victim agree upon is understood as constituting punishment (Hirsch, Roberts, and Bottoms 2004). Despite the potential for verbal disputes over whether this agreement represents punishment or compensation, I think there is value in emphasizing that the version of restorative justice I am interested in here significantly differs from how punishment itself is traditionally conceptualized. Restorative justice does not aim at the infliction of hardship on the offender for its own sake, nor because the offender supposedly "deserves" to endure it. Hardship, rather, is justified in restorative justice only as a foreseeable side effect of the necessary reparative action; in other words, the reparative element itself takes primacy.

Research offers promising data for the relationship between restorative approaches and the evocation of guilt in the offender. As John Braithwaite, a leading figure in the restorative justice movement, argues, it is to be expected that guilt might arise naturally in face-to-face communication with the victim(s) and the other stakeholder(s).[3] Evidence shows that "74 per cent of victims get an apology from the offender in [restorative] conferences compared to 11 per cent with cases randomly assigned to court" (Braithwaite 2000, 123). If we take apology to be a proxy for guilt, then these figures support our expectations that restorative justice holds promise as a guilt-evoking method for addressing criminal wrongdoing (for a more detailed discussion of the benefits of restorative justice, see Saulnier and Sivasubramaniam 2018).

I have argued here that the penance rationale favors the restorative approach, as it promises to harness the productive *potential* of feelings of guilt. But should we be worried about the objections introduced above, even with the more limited role given to the penance rationale?

The first worry over penance rationales was expressed by the liberal objection; namely, given that the state is not a close-knit community with shared values that go beyond minimal liberal principles, the state has no right to intervene in the inner feelings of the offender. The restorative justice approach mentioned above is untouched by this worry, as restorative justice does not directly aim to make the offender feel guilty in the first place. Nonetheless, one of the reasons why restorative justice is appealing for criminal law is that offenders are more likely to feel guilty in restorative justice settings, as alluded to above.

Solving the liberal challenge by resorting to a procedural claim alone, rather than to a justification of punishment, might seem like a cheap trick: the idea that, instead of aiming *directly* at feelings of guilt and making these the justification for criminal law, we propose a strategy that aims only *indirectly* at such feelings and thereby believe that we have solved the problem.[4] In fact, on the contrary, this shift toward the concept of restorative justice is more than a mere rephrasing; it has important implications for the state's role in the matter, specifically with respect to whether the state is justified in imposing its will on offenders. For this reason, the liberal worry does not apply here.

Up until now, my suggestion has been merely a *procedural* one. Because of this, such a defense of penance, of course, presupposes that there is a plausible justification for punishment in the first place; otherwise, the thesis concerning how to approach criminal law would be vacuous. The suggestion is to give the offender the opportunity to express remorse within the personal setting of restorative justice but not to force him to express remorse. Furthermore, the state need not be involved in the restorative setting at all; thus, we also bypass the worry that the state will intrude into the offender's inner feelings. In sentencing, the state needs to have the last word, but the victim-offender mediation—and even a suggestion for an adequate sentence—can come from the conflict's stakeholders themselves. The worry that the state might intrude into the inner feelings of offenders within the procedural suggestion is therefore unjustified.

What about the worry concerning proportionality? Let us assume here that proportionality of punishment is something that every approach to criminal law must guarantee. As explained above, the risk that is run by making penance a direct aim of criminal law is that two people who commit equal crimes but are unequally "hard-headed" might be sentenced to different kinds and degrees of punishment, in order that they both arrive at the same "end goal" of conscience, so to speak. But this worry does not apply to the limited account of penance; the fact that such an account does not aim to offer a justification for punishment makes the proportionality worry irrelevant. Rather, if implementing restorative justice fails at any point to evoke feelings of guilt, then no justification remains for pushing the offender further. Unlike the account of penance that tries to justify punishment, the procedural thesis does not, therefore, trigger worries about proportionality. On the other hand, if we still have good reason to expect feelings of guilt to occur generally—as I have argued that we do—then the social psychology literature can be used to argue for restorative justice as an approach to criminal law.

Punishment or Reparation?

In this section, I want to defend another element of the penance rationale: reparation. The claim that derives from the penance rationale is that, to expiate their wrongs, offenders must provide some form of compensatory payment or reparative action for their victims. Taken together with the arguments above, this would mean that the penance rationale not only favors restorative approaches (e.g., victim-offender mediation) from a procedural perspective, but would also favor reparative sanctions to punitive ones.

I have already argued against penance-based justifications for punishment. That said, as for the procedural thesis concerning restorative justice, we can still ask whether the penance rationale justifies reparation, even if we must reject punishment on the basis of penance. The objection I want to consider here states that penance without punishment is necessarily inadequate penance. The procedural thesis, which emphasizes reparation instead of punishment, and which I wish to defend here, would thus not be a penance account in any interesting sense, as it lacks a commitment to punishment.

To substantiate this objection against the limited account of penance, we can point to a differentiation that is commonly invoked in the debate in criminal law and ethics: when we violate criminal or moral law, we not only (typically) harm people but also *wrong* them. The harm consists of the material loss that a person suffers, while the wrong refers to the demeaning message (about a victim's moral worth) that is conveyed by the offender's choice to treat the victim in such a way. This is where the argument against nonpunitive penance approaches begins: harm can be accounted for simply by providing restitution for a victim—but to counter the demeaning message regarding the victim's moral worth, the offender must be punished (Duff 2001; Garvey 1999; Hampton 1991). Penance is only achieved if the offender undergoes some form of punishment that cancels out the demeaning message of his or her wrongdoing.

Let us suppose that wrongs are distinct from harms, in that the former entail such demeaning messages, and let us further assume that the penance approach must be able to account for such messages. I think that the best response to the claim that nonpunitive penance approaches are not adequate penance approaches is to show how reparative approaches can also account for wrongs.

For this, we first need to distinguish between compensation by third parties or by the offender, versus what is more properly thought of as "reparation." To understand the difference, consider Linda Radzik's discontent with purely compensatory approaches; namely, she warns that they would allow for the state or other third parties to assume the duty of compensating victims for their losses (Radzik 2009, 45–50). However, although it is important to be sensitive to victims' practical needs, it is also important to take offenders' responsibility seriously; it is not just that victims deserve to be compensated, but also that offenders ought to take responsibility *for* that compensation. If a third, non-offending party simply compensates a victim, the demeaning message conveyed through the offender's wrong goes effectively uncountered. Hence, for wrongs to be adequately addressed, it is necessary that the offender take responsibility for the compensation.

But while this is a necessary condition, it is not a sufficient one. According to the compensation rationale, it would, in principle, suffice if the offender wrote a check covering the victim's material losses in order to fulfill the demands of penance—but this would seem inadequate, for the reasons mentioned above. In fact, simply writing a check and assuming that, thereby, one has atoned seems to add insult to injury. Here, we can grant the critics of nonpunitive approaches to penance their intuition that something about such an approach to handling wrongdoings is inappropriate and represents an inadequate penance.

Nonetheless, I do not think that compensatory payments are, *in principle*, problematic. Imagine the following case: after a burglary, the victim and the offender participate in a mediation wherein they talk about what happened and about the impact the wrongdoing had on the victim. In this setting, the offender ultimately acknowledges the wrongdoing and agrees to pay the victim a certain amount in compensation for the material loss and psychological distress incurred. The victim is happy with the outcome. In such a case, the assertion that compensatory payment adds insult to injury loses its thrust. And if it really loses its thrust, then the insult-to-injury objection would not ultimately be grounded in the compensatory approach itself, but rather the *kind* of compensation provided, as well as the process of getting to the verdict or agreement.

What is necessary for compensatory payments to establish adequate reparation, then? Geoffrey Sayre-McCord, who has defended such reparative approaches, shares some worries about the shortcomings of compensatory approaches, arguing that it is not enough for the offender simply to write a check that compensates for a victim's losses. Such a response would fail to capture the nature of the wrongdoing itself as an infringement on the victim's rights, more than just a matter of material loss (Sayre-McCord 2001, 504–10). To resolve the issue of the victim's rights, the offender must work through a process of making amends, not by simply writing a check, but by engaging in more substantial forms of reparation (which Sayre-McCord leaves somewhat open, although victim-offender mediation and community service arguably fall under this category). What precisely is entailed by adequate reparation is thus not clearly defined, but expressing respect for the victim seems essential. Writing a check, in most cases, seems inadequate, as this action alone does not express respect for the victim, whereas respect *is* expressed when the offender sincerely apologizes to the victim; after a sincere apology, however, writing a check ceases to feel inadequate. Adequate reparation, therefore, must involve an act that expresses respect for the victim—which, arguably, leaves the precise understanding of reparation somewhat vague.

Nonetheless, the important difference that sets reparation apart from punishment and from some versions of the penance theories mentioned above is that the suffering of the offender is *not* necessary to render a response to a wrongdoing adequate. Rather, the reparation perspective does not insist that offenders deserve to suffer—only that they owe victims reparation of a certain kind. This difference in moral perspective has important implications for the kinds of sentences that could be justified.

In emphasizing the relevance of this distinction, let us address Garvey's following point:

> On this view, the difference between restorative justice and retributive justice boils down to the particular forms of punishment that restorativism [restorative justice] and retributivism are said to prefer. Restorativism is said to prefer milder punishments like restitution and community service, while retributivism is said to prefer harsher punishments like imprisonment. (Garvey 2003, 309)

The difference here is more substantial than Garvey seems to suggest; he himself takes it to be, at best, a practical difference, but not a philosophically interesting one. However, the difference between restorative justice and retributivism is a difference in justification, and not only in the types of sanctions indicated. One classical mark of the definition of "punishment" that has been important in philosophical debates is that punishment is an *intentional infliction of harm* or of hardship on offenders for their wrongs, which also expresses condemnation of the action (Benn 1958; Feinberg 1965; Flew 1954; Hart 1960). According to the purely reparative perspective, the *intentional* infliction of harm is unjustified—which thus means that only certain kinds of actions can be justified in turn. If the infliction of suffering or harsh treatment itself is not morally valuable, then the state ought always to opt for the methods of reparation that are the least burdensome to the offender yet still sufficient for the reparative effort.

I have argued that reparation suffices for penance, even without punishment. But we should also at least briefly return to the proportionality worry: as some versions of restorative justice advocate that the stakeholders in a conflict should devise adequate reparative measures themselves, the proportionality and egalitarianism of punishment are not, in such cases, guaranteed. For example, an especially forgiving victim might not insist on any or much reparation besides an apology, while angrier victims might overestimate how much compensation they are owed. The easiest way to address this objection, if we take it to be justified, is to limit the autonomy of the stakeholders in deciding what kinds of reparation to pursue. This would still allow some room for the victims' preferences, as long as these remained within the defined boundaries of what the state considered proportional. Some restorative justice proponents might not like the idea of placing limits on stakeholders' autonomy, but said limitations help us address the proportionality worry sufficiently.

I have thus argued that reparation is sufficient for adequate penance without punishment—and that focusing on reparative sanctions can respect the proportionality worry. Finally, I wish to turn to reconciliation and to argue that reconciliation is plausibly entailed by the procedural approach to penance.

Reconciliation and the Limits of Penance as a Rationale for Criminal Law

Reconciliation is the last aspect of penance that I wish to discuss. As with the other aspects of penance, reconciliation can be understood in different ways. Garvey takes it to involve a duty of forgiveness on the part of the victim, while Radzik and Duff understand it as the restoration of the relationship among moral agents.

Whatever the details of its definition, the liberal worry lingers again: it is no business of the state to tell victims that they must forgive offenders, nor is coercing the offender to restore his or her relationship with the victim or with the community (if there was such a relationship to begin with) justified. Moreover, as already argued, most citizens are not strongly connected to each other. The most promising route for addressing objections about state overreach, I think, is to conceptualize the relevant relationship as one between moral agents, as

Radzik and Duff suggest: "The kind of reconciliation that is the goal of atonement, then, involves the restoration of a paradigmatically moral relationship. It is one wherein the parties regard one another and themselves as equally valuable moral persons" (Radzik 2009, 81).

Penance, then,

> should aim to repair any damage done to the moral relationship between [the wrongdoer] and the community and between the victim and the community. If the wrongful act gave the community any reason to doubt whether the victim or the wrongdoer is an equally valuable moral person or a reasonably trustworthy member of the moral community, then the wrongdoer should work to repair that impression. (81)

Radzik argues from the perspective of what the offender's moral obligations are. If we agree that offenders have an obligation after a wrongdoing to clarify that they (now) respect the victim as a moral equal, we are still left with the question of what the state should do in response to these wrongdoings.

One possible way to attempt reconciliation on the part of the offender—and a strategy suggested by Radzik—is to accept a legitimate sentence of punishment. "When the morally wrongful act falls under the jurisdiction of an authority who has a right to punish or penalize the wrongdoer—be it a state, an employer, or a parent—the wrongdoer's voluntary submission to that authority might also count toward the making of amends" (Radzik 2009, 103). The problem for us is that such a response presupposes the state's right to inflict punishment on offenders on the basis of penance considerations—but what we have been interested in here is whether the penance rationale gives the state the right to do so in the first place.

If the claim is that the state has a right to make offenders acknowledge the moral value of their victims, we face again the now familiar worry: offenders who have a hard time respecting a certain person might require disproportionate coercion or punishment in order to be convinced of the moral value of the victim (assuming that punishment can achieve this at all). It might therefore be infeasible, in principle, to have such an aim as a central justification for criminal law; at best, the justification might be feasible only insofar as we accepted the possibility of disproportionate punishments.

Radzik is well aware of such concerns. In her own discussion of the relevance of atonement for criminal wrongdoing, her answer to the problems at hand is ultimately similar to mine, in that many of the liberal worries can be tackled by forbidding the state or the stakeholders in the conflict from coercing atonement, instead allowing them only to arrange the circumstances such that atonement has a high chance of being achieved—via restorative justice. But this concession to the role of penance in criminal law is to accept that only the procedural thesis of penance is acceptable.

Conclusion

My aim in this chapter was to argue that penance accounts cannot offer a plausible justification for punishment but can give a plausible justification for restorative justice as a method of addressing wrongdoing. The two crucial problems that plague justifications for punishment based on the penance rationale are the liberalist objection and the proportionality demand. Penance theories fail to justify why liberal states have the right to punish offenders with the aim of making them repent their wrongdoings. Furthermore, if the aim of criminal law is indeed to make offenders undergo penance, then proportionality of punishment cannot be guaranteed.

Even though the penance rationale cannot, by itself, justify criminal punishment, I have argued that it can still inform how we should approach criminal law. Specifically, three central aspects of penance—feelings of guilt, reparation, and reconciliation—should still play a major role in considerations concerning criminal law. I argued that these aspects are best implemented by adopting restorative procedures, such as victim-offender mediation or restorative conferences. Such procedures promise to harness the productivity of all three aspects—feelings of guilt, reparation, and reconciliation—and this productivity should also be welcomed in criminal law, independently of the value of penance itself. Moreover, when we pursue these aims by implementing restorative justice, we bypass the objections to the approaches that try to justify punishment on the basis of penance considerations. As the state lacks the right to coerce penance in restorative justice settings, the liberal and the proportionality worries do not hold. In sum, the productivity of guilt can be harnessed via restorative justice without any serious moral objections.

Notes

1. Some authors use "atonement" or "repentance" instead of "penance" to refer to their approach. In this chapter, I will simply refer to "penance accounts" and slightly change and standardize the usage of the terms of other authors to simplify the discussion without altering the content of these accounts. "Account" refers to a specific version of a broader theory. Otherwise, I will differentiate between various kinds of approaches encompassed by this general rationale only when it is necessary for the discussion of the respective arguments.

2. For the sake of simplicity, I will not differentiate between feelings of guilt and remorse; I will only talk about feelings of guilt. Details of how I define and conceptualize feelings of guilt will be presented in the section "The Importance of Feelings of Guilt for Various Aims of Criminal Law."
3. Braithwaite uses the term "reintegrative shaming" but means what Baumeister and colleagues call "guilt."
4. Radzik (2009) also argues that the difference should be understood as one between pursuing a goal and guaranteeing it.

References

Baumeister, Roy F., Arlene M. Stillwell, and Todd F. Heatherton. 1994. "Guilt: An Interpersonal Approach." *Psychological Bulletin* 115, no. 2: 243–67.

Benn, Stanley I. 1958. "An Approach to the Problems of Punishment." *Philosophy* 33, no. 127: 325–41.

Boonin, David. 2008. *The Problem of Punishment*. Cambridge: Cambridge University Press.

Braithwaite, John. 2000. "Survey Article: Repentance Rituals and Restorative Justice." *Journal of Political Philosophy* 8, no. 1: 115–31.

Christie, Nils. 1977. "Conflicts as Property." *British Journal of Criminology* 17, no. 1: 1–15.

Duff, Antony. 2001. *Punishment, Communication, and Community*. Oxford: Oxford University Press.

Duff, Antony. 2003. "Penance, Punishment and the Limits of Community." *Punishment and Society* 5, no. 3: 295–312.

Feinberg, Joel. 1965. "The Expressive Function of Punishment." *Monist* 49, no. 3: 397–423.

Flew, Antony. 1954. "The Justification of Punishment." *Philosophy* 29, no. 111: 291–307.

Garvey, Stephen. 1999. "Punishment as Atonement." *Cornell Law Faculty Publications* 264: 1801–58.

Garvey, Stephen. 2003. "Restorative Justice, Punishment, and Atonement." *Cornell Law Faculty Publications* 279: 303–17.

Hampton, Jean. 1991. "Correcting Harms versus Righting Wrongs: The Goal of Retribution." *UCLA Law Review* 39, no. 6: 1659–1702.

Hart, Herbert Lionel Adolphus. 1960. "The Presidential Address: I—Prolegomenon to the Principles of Punishment." *Proceedings of the Aristotelian Society* 60, no. 1: 1–26.

Hirsch, Andrew von, and Andrew Ashworth. 2005. *Proportionate Sentencing: Exploring the Principles*. Oxford: Oxford University Press.

Hirsch, Andrew von, Julian V. Roberts, and Anthony Bottoms. 2004. *Restorative Justice and Criminal Justice: Competing or Reconcilable Paradigms?* Oxford: Hart.

Ketelaar, Timothy, and Wing Tung Au. 2003. "The Effects of Feelings of Guilt on the Behaviour of Uncooperative Individuals in Repeated Social Bargaining Games: An Affect-as-Information Interpretation of the Role of Emotion in Social Interaction." *Cognition and Emotion* 17, no. 3: 429–53.

Leith, Karen P., and Roy F. Baumeister. 1998. "Empathy, Shame, Guilt, and Narratives of Interpersonal Conflicts: Guilt-Prone People Are Better at Perspective Taking." *Journal of Personality* 66, no. 1: 1–37.

Lindsay-Hartz, Janice. 1984. "Contrasting Experiences of Shame and Guilt." *American Behavioral Scientist* 27: 689–704.

Radzik, Linda. 2009. *Making Amends: Atonement in Morality, Law, and Politics*. Oxford: Oxford University Press.

Saulnier, Alana, and Diane Sivasubramaniam. 2018. "Restorative Justice: Reflections and the Retributive Impulse." In *Advances in Psychology and Law*. Vol. 3, edited by Monica K. Miller and Brian H. Bornstein, 177–210. Berlin: Springer.

Sayre-McCord, Geoffrey. 2001. "Criminal Justice and Legal Reparations as an Alternative to Punishment." *Noûs* 35: 502–29.

Simmons, A. John. 1979. *Moral Principles and Political Obligations*. Princeton, NJ: Princeton University Press.

Stuewig, Jeff, June Price Tangney, Caron Heigel, Laura Harty, and Laura McCloskey. 2010. "Shaming, Blaming, and Maiming: Functional Links among the Moral Emotions, Externalization of Blame, and Aggression." *Journal of Research in Personality* 44, no. 1: 91–102.

Tangney, June Price. 1991. "Moral Affect: The Good, the Bad, and the Ugly." *Journal of Personality and Social Psychology* 61: 598–607.

Tangney, June Price, Rowland S. Miller, Laura Flicker, and Deborah Hill-Barlow. 1996. "Are Shame, Guilt, and Embarrassment Distinct Emotions?" *Journal of Personality and Social Psychology* 70, no. 6: 1256–69.

Tangney, June Price, Jeff Stuewig, and Logaina Hafez. 2011. "Shame, Guilt, and Remorse: Implications for Offender Populations." *Journal of Forensic Psychiatry and Psychology* 22, no. 5: 706–23.

Tangney, June Price, Jeff Stuewig, and Debra J. Mashek. 2007. "Moral Emotions and Moral Behavior." *Annual Review of Psychology* 58, no. 1: 345–72.

Wallbott, Herald G., and Klaus R. Scherer. 1995. "Cultural Determinants in Experiencing Shame and Guilt." In *Self-Conscious Emotions: The Psychology of Shame, Guilt, Embarrassment, and Pride*, edited by June Price Tangney and Kurt W. Fischer, 465–87. New York: Guilford Press.

Wicker, Frank W., Glen C. Payne, and Randall D. Morgan. 1983. "Participant Descriptions of Guilt and Shame." *Motivation and Emotion* 7: 25–39.

Zehr, Howard. 1990. *Changing Lenses: A New Focus for Crime and Justice*. Scottdale, PA: Herald Press.

7

Guilt with and without Punishment

On Moral and Legal Guilt in Contexts of Impunity

Dominik Hofmann

Introduction

In this contribution, I will locate guilt's productive force in a dialectic relationship with its destructive potential. I will show how different situations that are described with the term "impunity"—situations in which the law displays an "absent presence"—share the common feature of decoupling a wrongdoing not only from its legal punishment, but also from guilt. I will argue that guilt becomes diffuse under these circumstances and that this observation helps us to infer what a functioning legal system adds to the contours of a notion of guilt that is otherwise perceived only in moral terms. My argument will be that guilt's disruptive potential creates the demand for social institutions that regulate it and that help use its conciliatory potential in productive ways.

The Meanings of Impunity

In June 2018, when speculations began to circulate about whether Donald Trump could apply his prerogative of presidential pardon to himself, the then-president fueled them by asserting his right to do so. In other words, if he were found guilty of a misdeed, no punishment was guaranteed to ensue. Not only, then, would the core legal idea that guilt is to be followed by punishment be disrupted, but this very disruption would, as the president stressed, be provided by the same system of law. This (ultimately hypothetical) case is one of "impunity de jure," or the legally justified absence of punishment.

Even before Donald Trump's presidency, however, it was a similar practice of political actors applying constitutional rights to themselves in order to escape legally ordered punishments that led to the most recent dissemination of the term "impunity." United Nations working groups, as well as the renamed UN Sub-Commission on the Promotion and Protection of Human Rights, started using the notion in the early 1980s to denounce practices of self-amnesty

Dominik Hofmann, *Guilt with and without Punishment* In: *Guilt*. Edited by: Katharina von Kellenbach and Matthias Buschmeier, Oxford University Press. © Oxford University Press 2022. DOI: 10.1093/oso/9780197557433.003.0008 in the context of military dictatorships in Latin America's Southern Cone. When these dictatorships came to an end shortly afterward, democratic successor governments passed far-reaching amnesty laws and decrees, shielding perpetrators in the military and police ranks from being punished for crimes against humanity. The authors of these laws reasoned that the goal of appeasing deeply divided societies, and thus securing democratic transitions, ought to take precedence over retributive justice. Public discourse was quick to declare the amnesties *leyes de impunidad*—"impunity laws."

Up to this day, in many regions of the world whose pasts have been shaken by atrocities during dictatorships and ethnic conflicts, victims' organizations lead a "fight against impunity," often also referred to as a "fight against forgetting." Transitional justice processes in Latin America, Africa, and Southeast Asia have faced the dilemma of having to choose between justice and reconciliation (as this choice is perceived by the actors involved), and any social tensions that persist after political transition are often declared repercussions of impunity.

Originally,[1] the concept of impunity described an exemption from punishment, according to legal provisions and after the deed (which distinguishes it from immunity). However, this is not how the term is predominantly used today—at least not beyond the narrow scope of legal parlance. Presidential pardons and amnesties are the main tools that democracies provide for "undoing" (so to speak) a legal sentence by a political act. But not every pardon or amnesty will be publicly perceived as an act of impunity; rather, only those that are perceived as unjust or undue will be viewed this way. In the broader social discourse, therefore, rather than a legal technicality, the term "impunity" references the decoupling of a wrongdoing (along with its attendant guilt) from a punishment. This has been referred to as "de facto impunity."[2] It is not the judicial leniency itself that is perceived as scandalous in the above examples, but more precisely its invocation in response to what is perceived as an abuse of power—highlighted, of course, when executives apply these powers to themselves or to their political allies.

And yet there are further facets to de facto impunity. In Brazil's highly stratified society, for instance, impunity—*impunidade*—is seen as something that certain social classes almost naturally possess. In such a world, as long as a deed has been committed by a member of the higher strata, the nature of the crime itself does not determine the legal consequences—nor even, necessarily, does the perpetrators' willingness to pay a bribe. Instead, what counts is the perpetrator's mere social status. (Notice that here the distinction from immunity is a different one: immunity is *legally* granted.) Meanwhile, for those members of the Brazilian population living in the favelas, in their everyday struggles with police forces—struggles that long ago came to follow their own complex, internal logic—guilt has little bearing on the probability of being detained, arrested, or killed (cf., e.g., Goldstein 2013,

174–225). Impunity, in these cases, can be described as the utter separation of guilt from punishment (similarly, Fassin 2018)—and a theme similar to the recent (and not so recent) discussions about police impunity in cases of violence against Black Americans in the United States also resonates strongly. In the case of both countries, impunity refers not to the *calculated* abuse of *political* power by a small elite, but rather to the habitual use of *social* power against people from a certain milieu—the milieu, in these cases, being largely tied to ethnicity.

Impunity, ultimately, denotes a drastic discrepancy between the principle that the law should apply equally to everyone and the harsh empirical reality of a global society that is far from having completely abandoned the practice of invoking social status for legal leverage. Note also that in both the United States and Brazil, the police and other agencies that represent the law lose their status as an impartial third party that symbolizes society as a whole; they become involved in a conflict that simultaneously elevates all original conflicts to another level, one on which the separation between private and public is undermined. Where the law is not impartial, guilt can only have a contingent, subjective, and moral meaning.

The concept of impunity, however, need not be tied to power to be perceived as a social problem. In Central America, for instance, civil society organizations aim at quantifying impunity by determining the percentage of all committed crimes that remain legally unpunished (cf., e.g., the "Global Impunity Index," Le Clercq, Cháidez, and Rodríguez 2016). The numbers they arrive at sound alarming (usually lying within the scope of 95 to 98%) and include all types of crimes. The impunity that these organizations are determined to fight (they do not limit their endeavors just to gathering statistics) is not directly linked to social inequality (which, of course, nevertheless has statistical repercussions). Moreover, their mode of observation fully refrains from targeting the perpetrators; instead, it focuses solely on the authorities' inability to prosecute the deeds—an inability that is mainly attributed to a lack of resources, but often also to a culture of police corruption, powerlessness in the face of organized crime, and other factors. In any case, impunity (as a quota of legal inefficacy, according to this understanding) is seen as a deplorable fact of everyday life, mentioned in the same breath as violence, insecurity, and neopatrimonialism, and is often—much like these same phenomena—intertwined with ideas about a national or ethnic "culture" (e.g., "the Mexican culture of impunity" and the like).

Impunity and Anti-Impunity

I have so far presented different discursive understandings of the notion of impunity. Their common denominator can be found not merely, as the term would semantically suggest, in the absence of punishment, but more precisely in the absence of an effective conditional relationship between wrongdoing, guilt, and legal punishment, despite the existence of institutions tasked with securing said punishment, and despite the omnipresent claim that they should do so. The rule that conditions these elements on each other is discursively upheld, but factually broken. If this seems like a very limited common ground for different dimensions of a single term, this is because not only does the term "impunity" apply to a broad range of phenomena, but also because the discourse spreads over very diverse fields.

Nevertheless, these fields still overlap empirically, and the different notions of impunity are not as clearly separated as my way of presenting them thus far might suggest. Impunity as a means for the abuse of power, impunity as the repercussion of a violent past whose effect is to impede victims from finding closure, impunity as the most visible expression of social inequality within the legal sphere, impunity as a symptom of the state's forfeiture of the monopoly on the use of force—all these understandings of the term not only blend into each other in the everyday reality of crime and (lack of) punishment, but this blending together is also reflected in a certain ambiguity that characterizes the way in which discussion about the concept unfolds. As a matter of fact, this ambiguity is sometimes functional to the very discourse. By hinting at the semantic conglomerate into which the aforementioned meanings are blurred—without spelling out the exact reference—the International Criminal Court, for example, can justify its very existence by its founding commitment to "put an end to impunity"; the Inter-American Court of Human Rights can sentence states for violating their duty to "fight impunity" (as the "obligation to prevent, prosecute, and punish" is interpreted); political candidates can fill their agendas with promises of "eradicating impunity," thereby invoking the simultaneous semantics of law-and-order and of anti-establishment/anti-corruption policies; the media can commemorate state- and drug cartel-organized massacres by counting the "days of impunity" that have elapsed; and currents as disparate as a post-colonial human rights movement on the one hand and a new wave of the "law and development" movement on the other[3] can agree on the shared enemy of impunity—the human rights movement viewing it as an aftereffect of the colonial imposition of law, and the "law and development" movement viewing it as a persistent lag in development.

Against this backdrop, it seems more sensible to speak of "anti-impunity" (Engle 2015),[4] and to describe the ways in which this anti-impunity is championed, than to try to capture the elusive meaning of the ambiguous concept that is "impunity" itself. In the wake of the just-cited semantic developments, and especially in the Latin American discourse, impunity (called *impunidad* in Spanish) has come to be used as a political buzzword rather than a juridical term, serving as a generally accepted explanation for all kinds of perceived social shortcomings. In such a context, its character as a normatively connoted catch-all term renders further conceptual definition unnecessary. ("Impunity," in this sense, has come to join forces with the notorious rhetoric about fighting against "corruption.")

Anti-impunity, then, by way of a double negation ("anti" + "im") allows for a centrally placed indeterminacy, combined with a high level of reflexivity: crimes are scandalized only indirectly—namely, via the criminalization of their nonprosecution. Not the violation of primary norms itself, but rather the violation of secondary norms—those that regulate the *processing of* the primary norms[5]—is spotlighted. It should not come as a surprise that this is the case especially in parts of the world that are living through great social tension, since it is this variable—secondary norms being challenged—that shakes society to the very foundations, much more than do broad or extreme violations of primary norms (e.g., high rates of criminality). Where injustice is perceived to be pervasive, outrage is directed against the *secondary* injustice: the inactivity of those who are officially tasked with preventing and redressing (primary) injustice.

The conflation of these different understandings of impunity in an anti-impunity movement, coupled with the movement's pragmatic reliance on that very ambiguity, is perhaps nowhere better illustrated than in the structure and activity of the International Commission against Impunity in Guatemala (Comisión Internacional contra la Impunidad en Guatemala, or CICIG). Founded by means of a treaty between the Guatemalan government and the United Nations in 2006, this autonomously operating commission was established to fight "clandestine organizations" that commit crimes that threaten the human rights of the Guatemalan population.[6] Although created in the wake of the same transitional justice stream as the anti-impunity movement I have just described, the commission was explicitly *not* tasked with prosecuting the crimes of the country's violent past. At the time, Guatemala's military dictatorship had just come to an end; among other things, massacres against the Indigenous population (which could be called a genocide) were unprocessed. Since a mandate directed at clearing up the military's past would not have received the necessary approval of the Guatemalan Congress, the issue simply hung in the air while the commission set out to train and restructure police forces and prosecutorial agencies (cf. Anzueto 2012).

Soon after taking up its work, however, the commission modified its strategy, shifting its focus toward "criminal elite networks" entangled with the country's highest political spheres. After this strategic shift, "impunity" became a synonym for "corruption," while those who held central political and military positions during the civil war (and were not previously charged for crimes from that time) were now commonly investigated for money laundering, election fraud, and other crimes. The commission repeatedly referenced the impossibility of separating the different facets of impunity in the Guatemalan context as a justification for modifying and adapting its approach to various challenges—without changing or contradicting its original mandate of "fighting impunity."

Guilt without Punishment

Given the way in which impunity and anti-impunity have thus been delineated, it can now be asked how they relate to the forces of guilt. And given this volume's topic, at this point one might expect an argument for moral guilt as a substitute for the legal institutions whose shortcomings the anti-impunity discourse denounces. This is not the line I want to argue, though; quite the contrary.

In every case that evokes the scandalization of impunity, the connection between guilt and punishment is distorted. While common understanding holds (in accordance with Western legal tradition) that a person who has been found guilty of a wrongdoing should be punished, cases of impunity are such that, while there is no doubt about the wrongdoing or about its condemnation, punishment is nonetheless missing. Impunity and anti-impunity both point us to the fact that the joint occurrence of guilt and punishment is not a natural given. Our question then, is: What change(s) does guilt undergo under these circumstances?

In the case of presidential pardons and amnesty laws, guilt has been legally established and is not altered by the remission of punishment—but these are special cases of *legal* impunity. In all contexts of extralegal impunity, by contrast, a wrongdoing has occurred, but neither legal punishment nor a legal declaration of guilt follows. Still, guilt does not disappear entirely in contexts where the law is inactive; in fact, it is necessarily thrown back on its moral and political grounds when it is stripped of its legal dimension.[7] This structure gives us the opportunity to analyze, on the basis of such cases, what it is that the legal dimension of guilt normally adds to these moral and political dimensions, and which guilt is now missing in contexts of impunity.

On the level of theoretical reasoning, for this purpose, one can first look at practical and political philosophy. After all, these disciplines have been contemplating the relationship between guilt and punishment for centuries. For the sake of our argument (although thereby only giving a superficial account of what is, in reality, a historically complex agglomeration of multilayered debates that are often precipitously reduced to a catchy dichotomy) we can distinguish a deontological from a utilitarian tradition of theorizing about the link between guilt and punishment, and correspondingly about the question of how rules work (cf. Rawls 1955). While deontological notions define the perpetrator's guilt (understood as Kantian *Verschulden*) as the sole criterion that may factor into his or her punishment, utilitarian accounts put more weight on punishment's deterrent effects. Consequently, for the former position, impunity must appear as the deficient implementation of an ethically ordained principle; for the latter, it represents an invitation to further wrongdoing. Both strongly disapprove of impunity, but on different grounds.

The anti-impunity discourse clearly argues in line with the utilitarian account. It can, for this purpose, revert to the reasoning of the utilitarian school's Enlightenment founders: Bentham, Beccaria, and the like greatly based their theories of punishment on polemics against impunity's abhorrent (societal) effects. Meanwhile, in our era, the notion that it is inherently despicable for guilt to go without atonement is seldom the reason for opposing impunity.

Still, these theoretical reflections do not tell us much about what empirically happens to the notion of guilt in contexts where impunity is perceived to reign. In order to say more about these empirical realities, let me resort to an example from personal experience.

While on a research stay, I was taken by a friend from southern Mexico to visit his home village, a small indigenous *pueblo*. Before we arrived there, we had to pass various other communities, and whenever we stopped and revealed our destination, we encountered the same reaction: we were asked if the place we were going to was the one where the police officers had been killed. The events to which people were referring dated back 34 years. While approaching the village, I was first told about these events by my friend's elder brother, whom we had picked up on our way. I will here reproduce these reports as they were related to me, including various details and allegations; that said, I have neither independently confirmed them, nor do I believe every part of the story. My objective is not to give an account of what happened, but of what I was *told* had happened.

The village (Village A) was involved in a land conflict with a neighboring community (Village B); these conflicts are quite common in the region. On the day of the events, the neighboring village bribed a local police squad to have the disputed piece of land cleared and declared as belonging to itself (i.e., to the village making the bribe). When the women of Village A were attempting to return home from another nearby village's market (Village C) in the evening, they found their way blocked by uniformed police. I was told of the women's fearlessness, of a slap on the police commander's face, and of their eventual return to their home village, where, through the village loudspeakers (usually used for making announcements), the women offered a warning to the community in the local indigenous language regarding the coming danger. I am told that the women and children of the village were sent to the church for safety, while the village's men armed themselves with rifles. A bloodbath ensued, in which 32 policemen and residents of Village B died, as well as two inhabitants of Village A. I am told that a torrential rain fell that day (which is meteorologically almost impossible, as these events reportedly occurred in the middle of the dry season), and that the blood of the corpses was washed away within a few hours. When a prosecutor and the military were sent the next day, people told them that the corpses had only been transported to the village and that nobody knew how they had died.

Asking for details about further consequences, I am told that the surviving police commander, returning with the military and the prosecutor, described the opposing side's leader to them as a bearded, light-skinned man. The agents, the commander, and a delegation of villagers went to the church (where the women and children had hidden). The description of the bearded man fit the statue of the village's patron saint, which is exhibited there, and when the commander suddenly found himself facing this statue, he fell to his knees and asked for forgiveness. Since that day, representatives of the neighboring communities sometimes come to visit the church and offer gifts to the saint.[8] Still, the community seems to have been fairly socially isolated from its surroundings since the date of the events. This fact is presented to me as a sign of respect, earned through steadfastness (several times, I am told about the village's reputation, "*que aquí matamos gratis*" ["that here we kill for free"—which can mean "without qualms," or "without having to pay the price," although the latter interpretation is less intuitive]).

During my stay, I am repeatedly told that the events are not something people like to talk about. Still, different people bring up the topic on different occasions, even though I have neither mentioned it nor mentioned my scientific interest in the topic of impunity. On one occasion, people even played me a recording of a song that had been written about what happened.

Now, my contribution to this volume is not an ethnographic one, and I do not intend to commit the impudence of trying to give a close anthropological reading of the experience I just described. I only want to take the liberty to point to a small number of aspects that seem paradigmatic to me in the context of impunity and its relation to guilt.

The events that were described to me occurred at a time and place in which impunity, in several of the senses mentioned above, was ubiquitous—and still is. The police are expected to be corrupt, and state agencies are expected to be hostile (especially toward the indigenous population); no legal process is expected to take place, regardless of the atrocities that happen from time to time. When I asked about consequences, I had the legal system in mind, but upon hearing the answer (the anecdote about the commander and the saint), I realized that the state's reactions do not bear relevance on the negotiation of such consequences, at least not on the legal and juridical level. Questions of law and justice are not associated with the public prosecutor's office or the police (when I was told the number of victims, no distinction was made between police officers and inhabitants of the neighboring village) but instead with (divine) forgiveness and, above all, with the village's overall reputation in the surrounding region. Notice, however, that law enforcement agencies are very well involved. The law is not absent; instead, it simply does not work the way it is supposed to work. The "third party" that the police are supposed to be is not perceived as impartial; declaring before the prosecutor that no one knew what happened might be interpreted as an attempt to avoid legal consequences, but given the context it seems more plausible that what they actually hoped to avoid was extralegal retaliation). Still, people are very well aware that the state is supposed to react to violence and killings. And had the police not possessed a certain authority, they would not have been involved in the conflict in the first place. What happened was not just a clash between enemies on a lawless terrain; there is an outside world (of which I, myself, as a foreign visitor, am construed as a representative par excellence).

The law has, thus, a certain "absent presence" in such contexts, and it is exactly this absent presence that the term "impunity" describes. Justice and the law are decoupled, and guilt under such circumstances becomes an amorphous phenomenon, in this case taking the form of the village's damaged reputation in the surrounding villages. Guilt, as a social relation that is based on the expectation of others' expectations, often captures all the attention and centers all of the attendant social interactions on the unresolved topic that has birthed it; guilt causes unresolved matters to linger, and it puts everything else in abeyance until the (indeterminate) moment of resolution. As such, wherever and whenever there is no procedure for the ascription of guilt, there is also no procedure for atonement—and consequently no closure with the past.

The amorphousness of guilt is a theme shared by all contexts of impunity. The stain of having been involved in the military dictatorship stuck to many of the people prosecuted by the CICIG in Guatemala during their first years of activity, although it was procedurally not permissible to accuse them of the past atrocities. And when the commission consolidated, shifting its focus to a criminal "elite" that had "captured the state," the same names kept surfacing time and again in the contexts of different criminal schemes; in such a context, guilt lost a vast part of its significance to ascriptions of a more general and more personal form of corruption. As a result—and as is commonly the case with people who are involved in organized crime—the guilt that can be legally ascribed to them differs from the guilt that is perceived to be a characteristic feature *of their personalities* in the light of numerous unpunished crimes they committed. (Al Capone being convicted of tax evasion is the paradigmatic example that might first come to mind.)

Meanwhile, to return briefly to the example of Brazil's favelas, anthropologists report that criminality is perceived as a quasi-infectious group feature there. In this regard, the society's need to find explanations for norm violations leads to stereotyping, and this stereotyping ultimately furthers social segregation (Caldeira 2000, 19–39 and 90–101). Where impunity reins, therefore, not only does the distinction between the deed and the perpetrator's person become blurred, but ultimately so does the distinction between the perpetrator and the group to which he or she belongs.

Finally, culpability has the potential to be perceived as a characteristic of whole cultures—not in the sense of a "collective responsibility" (Arendt [1968] 1987), but rather as a pervasive latent suspicion. Where guilt is hardly ever atoned for, in the long run, everybody is assumed to be guilty of one thing or another, and the meaning of the concept as a whole begins to lose its clarity.

This loss of meaning is, to a certain degree, reflected in the way in which the law adapts to impunity. The system of international law has found a strong new paradigm in "accountability,"[9] which encompasses precisely a responsibility that is decoupled from personal guilt. Transitional justice institutions, such as truth commissions, focus on a search for truth and strive for victims to receive compensation, rather than focusing on perpetrators' guilt and striving for retaliation. Add to these factors recent developments from other fields—for instance, algorithms' growing influence on the social world (agents to which the category of "guilt" simply does not apply), or neuroscience's increasing challenges against the notion that free will forms the basis of *mens rea*—and the resulting picture looks grim for guilt as a sustainable force of social and legal action.

The Productive Force behind Guilt's Destructive Force

The picture emerging from all that has been said up to this point is that guilt is a fairly destructive force in everyday social relations and a rather atavistic concept in criminal law. How can one find a productive aspect in any of this?

First of all: *ex negativo*. Any instance of guilt losing its grip in a context of impunity can help us identify, by contrast, what guilt has accomplished where its grip has been maintained. I will therefore continue to distinguish between the two levels I have just referred to: the level of social interaction on the one hand, and the level of broader societal structures, especially legal ones, on the other.

Beginning with the former, and taking into consideration what has been said about guilt as a social mechanism, its central merit turns out to lie in the possibility that it creates for regulating the conditions under which unpleasant events can be forgotten without having to be accepted. When a norm is violated, guilt helps to restore the status quo ante. In order to achieve this, however, two things must be accomplished: (1) an *explanation* must be given for the norm violation (cf. Aubert 1982, 35), and (2) a practical means of *redemption* must be presented.

First, if the violated norm is to be upheld, it must be plausibly asserted that the violation was an exception and not a general rule. (Because, of course, a general rule that is broken most of the time is not effectively a rule.) Therefore, guilt externalizes the reason for the disappointment to the *extra*-social—to the black box that is the individual human psyche (as opposed to ascribing this disappointment to spirits or gods). The ascription that thus results from this explanatory process, then, in a second step, becomes linked to the follow-up expectation that the person who has been identified as guilty will change his or her behavior.[10] The personal ascription of guilt, in this sense, not only helps to separate the perpetrator from the deed (in the sense that the law punishes the person not for being a bad person but for the deed itself), but also frees everybody else from suspicion for and involvement with the affair. A *past* violation cannot be undone, but guilt can be atoned for in its place and within the *present*.

Using Niklas Luhmann's (1995, 75–82) distinction between three dimensions of meaning (temporal, factual, and social), we can summarize this the following way: guilt allows us to access the past from within the present (temporal), provides an explanation (factual), and ascribes the expectation of behavioral change to a particular person (social), thus creating the active role of the perpetrator as opposed to the passive one of the victim (insofar as a system based on the perpetrator's guilt—and the resulting obligation to change the behavior—substitutes for a system based on the victim's obligation to avenge). In contexts of impunity, guilt fulfills these functions only in an over- or under-generalized manner.

According to this reasoning, then, it would seem more plausible, at first glance, to speak of the "productive force" of *norm violations* rather than the productive force of guilt itself, since guilt effectively pushes back against such violations; guilt *destructs* the variation, the innovation that deviance produces.[11] For this exact reason, an overemphasis on guilt in social discourse has been denounced, the argument being that guilt's strong fixation on the past limits the possibilities for present action—thus hindering the search for appropriately complex and future-oriented solutions (cf., e.g., Esposito 2014, 66).

Under these circumstances, if it is to be argued that guilt can be a "productive force," then we must find instances where guilt instead increases the number of possibilities available, and thereby the array and complexity of solutions at our disposal. This is where the second level, the level of broader societal structures, comes into play again. I would like to argue that in the very process of balancing out disruptions, guilt yields emergent and unintended social effects—thus serving as a productive force—among which the most far-reaching can be found in the evolution of different penal institutions.

If guilt is productive insofar as it provides opportunities for atonement, this "productivity" can take the form either of forgiveness or of punishment—and in both cases, in the long run, standardized procedures for providing for them emerge, which at the same time have a regulatory function (i.e., making sure that guilt does not get out of hand). It is exactly because of guilt's inherent destructive potential that institutions for the domestication and channeling of this potential have formed within the process of social evolution; that is, mainly criminal legal institutions. Their formation is the particular product of guilt's force that I wish to highlight. This assumption—that guilt is productive in forcing into existence institutions that ultimately *rein in* its destructive potential—can be historically supported.

When the notion of guilt as a form of "pollution" emerged (cf., e.g., Arnaoutoglou 1993), the victim's obligation to avenge a wrongdoing was gradually transformed into the perpetrator's obligation to compensate for it. The precise ascription of guilt (to a specific person for a specific deed) brought about the institution of the legal trial (cf. also Gulliver 1969). Indeed, in its institutionalized form, the ascription of guilt rids everybody to whom the guilt is *not* ascribed of the duty to take a stance in the face of the conflict. In other words, as soon as guilt is personal and there are institutions tasked with legitimately arbitrating it, the rest of the community is freed from the burden of conflict resolution. Put simply, guilt that is ascribed via legal procedures helps to isolate conflicts.

Yet another legal institution that could only be established based on a notion of guilt is criminal prosecution (cf., e.g., Moore 2007; Farmer 2010). Guilt, once it was understood no longer as pollution but rather as sin, allowed societies a means for treating even those cases in which there was no apparent victim as crimes: *God* was the victim of all crimes. And since God does not want any impunity to exist ("*Deus, qui nullum peccatum impunitum dimittit*," cf. Angenendt 1994), church and state must prosecute. The institution that first performed this task systematically was the Inquisition.

There is a dialectic process under way, in which guilt's destructive potential has created the necessity for social institutions that could contain its negative effects while harvesting its merits (isolating conflicts, providing opportunities for ending conflicts). This is what I would like to call guilt's productive force. And this also brings me back to the anti-impunity movement, because it aims in the very same direction of strengthening institutions and procedures as the single most promising means of "fighting impunity." The movement does not concede much space, however, to theorizing about the fact that guilt, although absent in its legal form in contexts of impunity, still has repercussions and thus offers another reason for pursuing the same goal.

Conclusion

The exact circumstances under which different notions of guilt have led to the emergence of legal institutions like trials and criminal prosecution cannot be fully spelled out here. Much more could be said about them. Nevertheless, this analysis has illuminated the fact that legal procedures are, on the one hand, indirectly "produced" by guilt (insofar as guilt simultaneously creates a demand for them and provides a justification for why they are needed), while, on the other, legal procedures function as mechanisms for the taming of guilt, making sure that it does not "spill out" and lead to temporal, social, and factual overextensions, as is the case in contexts of impunity.

Anti-impunity is right in condemning the negative effects of inactive and ineffective legal institutions—but it does so almost exclusively by focusing on the resulting encouragement of crime (attributable to a presumed lack of deterrence) and fostering of social injustices. It should, in fact, take another factor into consideration: the failure to tame guilt's destructive forces.

Thus, when I talk about the "productive force of guilt," I do not refer to the *emotion* of guilt, nor do I claim that there is any sort of mystical "force" at work beneath guilt's surface. That is, there *is* a force, but this force is no more (and no less) than the social force of discourse and shared expectations. Neither do I aim to take a side in the controversy surrounding the debate over whether humanity should abolish the idea of guilt—guilt being an idea whose critics claim has stood in the way of human progress since its invention (Nietzsche [1887] 1999). Doubting that the fate of concepts can be decided by normative deliberation, I instead prefer to take a standpoint that is explicitly indifferent to the moral—and moralizing—aspect behind the expression "productive guilt."

This is not a plea for punitive criminal policies. If impunity were to be taken in the literal sense ("no punishment," from the Latin *impunitas*), the claim that it must be abolished would have to be read this way. But we have seen that impunity is not simply the absence of punishment; rather, impunity is the law's *absent presence*. Therefore, in the light of this assessment, if this is a plea at all, it is a plea against neglecting the importance of taming guilt with legal procedures.

Notes

1. "Originally" here refers back to ancient and even archaic Greece (cf. Velissaropoulos-Karakostas 1991).
2. The distinction stems from the influential definition of impunity in the so-called United Nations' "Joinet" and "Orentlicher Reports": "'Impunity' means the impossibility, de jure or de facto, of bringing the perpetrators of violations to account—whether in criminal, civil, administrative or disciplinary proceedings—since they are not subject to any inquiry that might lead to their being accused, arrested, tried and, if found guilty, sentenced to appropriate penalties, and to making reparations to their victims" (E/CN.4/2005/102/Add.1, 6).

3. The law and development movement, nowadays, is highly self-aware and reflexive, but still advocates a rule-of-law ideal that is strongly reminiscent of colonial notions about civilizing the "primitive" by legal means.
4. Engle's term has spread a fair bit in the past years, but exclusively within the field of international law. It seems to me that it is perfectly suited to describe the much broader coalition outlined above.
5. There is a long tradition of using the distinction between "primary" and "secondary norms" in legal theory, its most dominant point of reference being, without any doubt, Hart ([1961] 2012, 91–99). In classical German sociology of law, Hans Kelsen (1991) and Theodor Geiger (1964) are two of the main proponents of a similar theory, who developed their own understandings of the dual notion. For Geiger (1964, especially 61–72), the efficacy of the primary norm is determined by the general compliance with the secondary norm. I use the distinction in broadly the same sense.
6. The commission stopped working in 2019 after its two-year mandate had been renewed six times, in the wake of strong tensions with then-president Jimmy Morales, whom the commission had started to investigate for breaking the rules of election campaign financing. The account that can be given here of the commission is necessarily extremely short and cannot do justice to what was a unique creation of human rights advocacy (cf., however, Zamudio Gonzales 2018; Call and Hallock 2020).
7. Cf., about this point, Victor Igreja's contribution in this volume.
8. This part of the story is particularly unclear. In fact, it would be highly uncommon for neighboring communities to do that. There might be a very practical reason why this was mentioned: I visited the village during the preparations for the annual festivities with which the patron saint is celebrated. It is common and important, on the one hand, to offer flowers to the statue on this occasion, and, on the other hand, to receive guests from other communities. (However, while I was there, I did not see any other guests.) This might explain the relationship between the statement about delegations from other villages and the anecdote about the commander and the saint. In any case, I want to highlight the urge to depict the relationship with the surrounding *pueblos* as one in which one is respected (and maybe feared) instead of shunned.
9. Cf. Mulgan (2000) about the term's recent proliferation. Within international law and human rights discourse, accountability and impunity are nowadays being treated as opposites (e.g., Bassiouni 1996; Lessa et al. 2014), the "anti-impunity norm" sometimes also being called the "accountability norm" (cf. Schroeder and Thiemessen 2014).
10. This is the point at which I would place the role of guilt *feeling*. I think it can best be explained as the feeling caused by a person's awareness that something is expected from her or him.
11. I do not claim to use the terms "destructive" and "productive" in a uniform manner here: of course, destructivity in the sense of perpetuating social tensions is different from destructivity in the sense of undoing deviance's creative potential. I only want to point to the possibility of these different interpretations and, ultimately, at a way of combining them.

References

Angenendt, Arnold. 1994. "Deus, qui nullum peccatum impunitum dimittit: Ein 'Grundsatz' der mittelalterlichen Bußgeschichte." In *Und dennoch ist von Gott zu reden: Festschrift für Herbert Vorgrimler*, edited by Matthias Lutz-Bachmann, 142–56. Freiburg: Herder.

Anzueto, Marc-André. 2012. "À la croisée de la paix et de la justice: La CICIG une avancée dans la lutte contre l'impunité?" *Revue québécoise de droit international* 25, no. 2: 1–36.

Arendt, Hannah. (1968) 1987. "Collective Responsibility." In *Amor Mundi: Explorations in the Faith and Thought of Hannah Arendt*, edited by James W. Bernhauer, 43–50. Dordrecht: M. Nijhoff.

Arnaoutoglou, Ilias. 1993. "Pollution in the Athenian Homicide Law." *Revue internacionale du droit de l'antiquité* 40: 109–38.

Aubert, Vilhelm. 1982. *The Hidden Society*. New Brunswick, NJ: Transaction.

Bassiouni, M. Cherif, ed. 1996. "Accountability for International Crimes and Serious Violations of Fundamental Rights." Special issue, *Law and Contemporary Problems* 59, no. 4.

Caldeira, Teresa. 2000. *City of Walls: Crime, Segregation, and Citizenship in São Paulo*. Berkeley: University of California Press.

Call, Charles, and Jeffrey Hallock. 2020. *¿Una iniciativa demasiado exitosa? El legado y las lecciones de la Comisión Internacional contra la Impunidad en Guatemala*. CLALS Working Paper Series, no. 24. Washington, DC: American University, Center for Latin American & Latino Studies.

Engle, Karen. 2015. "Anti-Impunity and the Turn to Criminal Law in Human Rights." *Cornell Law Review* 100, no. 5: 1069–128.

Esposito, Elena. 2014. "Was binden Bonds? Das Eigentum an der Zukunft und die Verantwortung für die Gegenwart." In *Bonds: Schuld, Schulden und andere Verbindlichkeiten*, edited by Thomas Macho, 55–66. Munich: Fink.

Farmer, Lindsay. 2010. "Time and Space in Criminal Law." *New Criminal Law Review* 13, no. 2: 333–56.

Fassin, Didier. 2018. The Will to Punish. Edited by C. Kutz. The Berkeley Tanner Lectures. Oxford: Oxford University Press.

Geiger, Theodor. 1964. *Vorstudien zu einer Soziologie des Rechts*, Neuwied, Germany: Luchterhand.

Goldstein, Donna M. 2013. *Laughter Out of Place: Race, Class, Violence, and Sexuality in a Rio Shantytown*. Berkeley: University of California Press.

Gulliver, P. H. 1969. "Dispute Settlement without Courts: The Ndendeuli of Southern Tanzania." In *Law in Culture and Society*, edited by Laura Nader, 24–68. Chicago: Aldine.

Hart, H. L. A. (1961) 2012. *The Concept of Law*. Oxford: Oxford University Press.

Kelsen, Hans. 1991. *General Theory of Norms*. Oxford: Oxford University Press.

Le Clercq, Juan Antonio, Azucena Cháidez, and Gerardo Rodríguez. 2016. "Midiendo la impunidad en América Latina: retos conceptuales y metodológicos." *Íconos* 55: 69–91.

Lessa, Francesca, Tricia D. Olsen, Leigh A. Payne, Gabriel Pereira, and Andrew G. Reiter. 2014. "Overcoming Impunity: Pathways to Accountability in Latin America." *International Journal of Transitional Justice* 8: 75–98.

Luhmann, Niklas. 1995. *Social Systems*. Translated by John Bednarz Jr. with Dirk Baecker. Stanford, CA: Stanford University Press.

Moore, R. I. 2007. *The Formation of a Persecuting Society: Authority and Deviance in Western Europe 950–1250*. Malden, MA: Blackwell.

Mulgan, Robert. 2000. "'Accountability': An Ever-Expanding Concept?" *Public Administration* 78, no. 3: 555–73.

Nietzsche, Friedrich. (1887) 1999. *Zur Genealogie der Moral: Eine Streitschrift, Kritische Studienausgabe*. Berlin: De Gruyter.

Rawls, John. 1955. "Two Concepts of Rules." *Philosophical Review* 64, no. 1: 3–32.

Schroeder, Michael B., and Alana Tiemessen. 2014. "Transnational Advocacy and Accountability: From Declarations of Anti-Impunity to Implementing the Rome Statute." In *Implementation and World Politics: How International Norms Change Practice*, edited by Alexander Betts and Phil Orchard, 50–67. Oxford: Oxford University Press.

Velissaropoulos-Karakostas, Julie. 1991. "Νηποινεὶ τεθναναι." In *Symposion 1990: Papers on Greek and Hellenistic Legal History*, edited by Michael Gagarin, 93–105. Cologne: Böhlau.

Zamudio Gonzales, Laura. 2018. "La Comisión Internacional contra la Impunidad en Guatemala (CICIG): Una organización autodirigida." *Foro Internacional* 58, no. 4: 493–536.

8

Post-War Justice for the Nazi Murders of Patients in Kherson, Ukraine

Comparing German and Soviet Trials

Tanja Penter

Under German occupation in World War II, numerous crimes were committed against the Soviet civilian population in Ukraine, including the brutal murder of tens of thousands of sick and disabled people. Only a few of these crimes were later investigated by courts in the Federal Republic, and very few German perpetrators were convicted.[1] In the Soviet Union, by contrast, hundreds of thousands of Soviet citizens were sentenced to long prison terms or even death as collaborators after the war. Members of the nursing staff, including many women, often were obligated to assist in organizing these murders in hospitals and sanatoriums. The murder of more than 1,000 mentally ill people at a clinic in the Black Sea port city of Kherson serves to illustrate the (lack of) legal reprisal by German and Soviet post-war justice systems, to describe how investigative authorities and courts in Germany and Ukraine dealt with perpetrators' guilt, and to draw some preliminary conclusions about the productivity of criminal prosecutions, particularly with respect to cultures of remembrance.

Law and legal procedures make guilt visible in language; they facilitate its articulation and representation in legal rules and concepts. Guilt is enunciated and expressed in the legal process itself, which is sometimes staged and realized performatively. Court settings serve as a medium for establishing guilt recognition and guilt narratives, as well as for guilt's criticism and denial.

German and Soviet judiciaries handled perpetrators' guilt with regard to the murder of patients in significantly different ways, which will be examined in greater detail below. At the outset, however, three stark differences are immediately apparent. First, the number of convicts in each place is vastly different; overall, about 100,000 Germans and Austrians were convicted of war crimes and Nazi crimes throughout Europe after 1944 (including 21,555 German prisoners of war interned in the Soviet Union), while the number of Soviet citizens who were convicted as Nazi collaborators by Soviet courts is more than three times higher (Frei 2006, 32–5; Hilger 2006, 180–246). As far as we currently know, over

Tanja Penter, *Post-War Justice for the Nazi Murders of Patients in Kherson, Ukraine* In: *Guilt.* Edited by: Katharina von Kellenbach and Matthias Buschmeier, Oxford University Press. © Oxford University Press 2022.
DOI: 10.1093/oso/9780197557433.003.0009

320,000 Soviet citizens were arrested by the People's Commissariat for Internal Affairs (NKVD) from 1943 to 1953 as suspected collaborators and accomplices in German crimes (Mozokhin 2005; Budnitskii 2019, 447–80). In Ukraine alone, 93,590 people were prosecuted as "traitors to the fatherland and accomplices," with the majority of arrests (57%) during the war and immediately afterward in 1945 (Nikol's'kyj 2003, 206–24, 451). As in other German-occupied countries, retaliation against collaborators was not limited to courts of law, but happened extrajudicially and on the spot; not insignificant numbers of people were shot without trial by NKVD special units immediately after liberation (Epifanov 1997, 70–80).

Second, the Soviet trials happened immediately after liberation, while the Federal Republic of Germany waited 20 years before it opened criminal investigations. This was in part because German courts initially had no jurisdiction over Nazi crimes committed against Allied victims and outside German territories. These restrictions were slowly lifted beginning in 1950, after the foundation of the Federal Republic (Raim 2013, 1137–71).

Third, the severity of penalties was significantly different: Soviet citizens generally faced sentences of between 15 and 20 years of forced labor—or the death penalty—while Nazi perpetrators in the Federal Republic expected short prison sentences and quite often acquittal.

The Prosecution of "Euthanasia" Crimes in the Federal Republic of Germany and the Soviet Union

Within the framework of the National Socialist "euthanasia" program, some 250,000 disabled and mentally ill people were murdered on the territory of the German Reich between 1940 and 1945. About 70,000 of them were asphyxiated with gas; the rest were murdered by lethal injection or died of neglect and hunger (Baader 2016, 319). In addition to racial ideology and eugenics, the Reich attempted to justify the murders with economic arguments.

Jewish patients were doubly stigmatized and became the first victims of "euthanasia." In fact, scholars have established direct historical connections between the Nazi euthanasia program and the murder of European Jews: several perpetrators were first trained in the T4 program before being deployed to the murder of Jews (Friedländer 1997; Schmuhl 1999, 115–39). However, while there is considerable scholarship on euthanasia murders in the German Reich's territory, much less is known about the murder of patients in the Nazi-occupied Soviet territories. Much remains to be researched, including pinpointing the sheer number of victims in the Soviet Union. The currently assumed overall number of 17,000 to 20,000 mentally ill murder victims in Soviet territories (Fedotov 1965, 445;

Ebbinghaus and Preissler 1985, 75–107; Winkler and Hohendorf 2010, 81) is certainly too low. More recently, Paul Weindling re-estimated the total number of victims of National Socialist euthanasia in the occupied territories at about 60,000–80,000 (Weindling 2019, 67). We do know somewhat more about the murder of patients in occupied Belarus, mostly from documents used in post-war trials in the Federal Republic of Germany and the German Democratic Republic (Friedman and Hudemann 2016, 13–36; Winkler and Hohendorf 2010, 75–103). In some cases, these trials included statements by Soviet witnesses, which had been forwarded by Soviet authorities.

The murder of psychiatric patients in Kherson is unique in that it has been one of the few cases prosecuted by both Soviet and German judicial authorities. On February 3, 1966, the *Frankfurter Allgemeine Zeitung* reported on the discovery of a mass grave from World War II in the Ukrainian city of Kherson. The report was based on a *Pravda* report from the previous day, according to which the remains of 1,276 men and women and about 80 children had been uncovered during excavation work for a house. The victims had been killed by machine guns, bayonets, and blows from rifle butts. This press report prompted the chief public prosecutor of the Ludwigsburg Central Office of the State Justice Administrations for the Investigation of National Socialist Crimes, Adalbert Rückerl, to contact the Soviet judicial authorities with a request for judicial assistance.[2] The Ludwigsburg Central Office had already begun preliminary investigations in 1965 into the crimes in Kherson, including the shooting deaths of 1,000 mentally ill people at the psychiatric clinic in Kherson in October 1941, based on statements from German witnesses.

However, in November 1972, the preliminary proceedings before the regional court in Munich were discontinued. Although dozens of suspects and witnesses had been questioned, the judges concluded that "the perpetrators could not be identified and all possibilities for their investigation have been exhausted."[3] Despite the detailed testimony of numerous members of the German Air Force (who were present at the murders as spectators), the court failed to determine the identity of the perpetrators to its satisfaction.

Almost two decades earlier, on March 2, 1945, a Soviet court had handed down long prison sentences to three members of the clinic's nursing staff, including the chief physician, Konkordiia Semenovna Popova. These trial documents had long been stored in Secret Service archives but were not accessible to historians until 2015. The extensive investigative and trial materials allow a closer look at the murder of patients across various Ukrainian regions, as well as the perspectives of local communities who became involved as accomplices and witnesses (Penter 2008, 341–64; Kudryashov and Voisin 2008, 263–96; Exeler 2016, 606–29; Epifanov 2017; Dumitru 2014, 142–57). Since these criminal proceedings began immediately after liberation, witnesses were able to remember the events in great detail.

Remarkably, while comparatively rare for the overall context of Soviet trials of collaborators, it was often women who faced charges in connection with these murders, especially as nurses. This fact allows us to draw conclusions about how Soviet judicial authorities dealt with female collaboration. The trials provided a forum for witnesses and defendants to speak openly about the local contexts and conditions for these crimes, beyond the official Stalinist policies on the past. Thick investigation files testify that the investigators tried to obtain as accurate a picture as possible; they questioned witnesses and defendants extensively, including cross-examination and confrontations, as well as the opinions of medical experts. By the time these trials began in the immediate aftermath of World War II, the NKVD investigators employed violence or torture to extort confessions to a lesser extent than before and were cautious to avoid misconduct, chastened as a result of the Stalin-ordered purge of the NKVD apparatus after the Great Terror of 1937–1938.[4] Therefore, the handwritten and detailed witness testimonies serve as excellent historical documentation of crimes for which no other records exist.

By contrast, the official court documents (indictments, sentences), available in typewritten form, display the political dimension of the trials and showcase the official version of Soviet history and its semantics. It was no longer permissible to speak of "Jewish victims" there, but rather only of "victims from the peaceful Soviet civilian population." Still, these trials were more than purely political procedures controlled by Moscow, and they unfolded within the complex dynamics of their local environments. In these records, therefore, meticulously recorded facts intersect with political instrumentalization, and political control over the performance grew as the trials became larger and more public and could therefore serve the system as "show trials" (Penter 2008, 341–64; Hilger 2006, 180–246).

The Murder of the Mentally Ill in Kherson According to the Federal German Investigations

By contrast, German prosecutors depended on a few eyewitness statements to reconstruct the historical events in Kherson in October 1941. For instance, Paul Beule was assigned as field gendarme to the local commandant's office I/303 and had only recently been sent there. He had been ordered to cordon off the execution site, a few kilometers northeast of the city, near an airfield. Beule testified that airmen of the Luftwaffe dive bomber wing *Sturzkampfgeschwader* (abbreviated as "Stuka") 77 were present "as curious spectators" at the mass shootings.

Two of the airmen volunteered. According to Beule, a total of four people—two German police officers and two Luftwaffe airmen—executed the patients. He testified:

> The inmates of the mental institution were driven to the execution site by a truck. As we knew, these people were taken from an institution in the nearby village. [...] The lunatics were accompanied by guards. These guards led the victims in a group of up to 10 people in rows to the execution site, where they were then killed by gunmen who shot them in the neck. Russian prisoners then threw the victims into an empty or dry well shaft. [...] At the execution site, there was often terrible screaming from the other victims who could watch the shooting. [...] On the last truck that was brought, there were heaps of naked bodies piled up. These victims, who were barely moving, were thrown from the truck by the guards and the firing squad shot the people lying on the ground.[5]

Another member of Stuka 77 confirmed this: "The lunatics were able to watch the shootings because the vehicles were in close proximity to the pit."[6]

Another eyewitness, Erhard Biewald, an administrative sergeant with Stuka 77, testified:

> During my short stay at the execution site, I also managed to take a look into the dried-out well. One could see that not all of the sick were dead, because some of the victims were still moving. The man with the MP went there from time to time and put stray fire into the well. The sight was so horrible that I left the place and I could not eat even at noon.[7]

Some patients tried to flee from the execution site, as another witness, Paul Heinelt, a first lieutenant at the local commandant's office in Kherson, testified: "As I heard, the fugitives were supposedly chased and shot by airplanes."[8] Among the victims were women and children:

> From soldiers I spoke with there, I learned that in the morning, children were first brought to the pit. These children had presumably already been dead during the transport, and I understood it to mean that they had been injected before the transport.[9]

During his own interrogation, the local commander, Arthur Goetz, confirmed initial plans of killing by injection rather than execution. He reported that a German military doctor in uniform, who had come from out of town, told him of orders "to use a 'lethal injection' to transport the bedridden patients to the afterlife."[10] Meanwhile, other witnesses testified that an elderly lady from the nursing staff was almost shot along with the patients but managed at the last moment to convince the execution squad that she was not a patient and, thus, was saved.[11] Beule estimated the number of murdered patients in this "Aktion," which lasted from 9:00 a.m. until afternoon, to have been around 1,000.[12]

Why were the patients murdered? This question, an eyewitness recalled, was also asked of the firing squad by a desperate female victim shortly before her death. Different members of the Stuka squadron provided different answers during their interrogations: some, like the local commander Goetz, believed that the patients were murdered to clear the hospital buildings for housing purposes: "It came to my attention that the Luftwaffe wanted to keep the asylum for itself because of the laundry and accommodation available there."[13] Others believed that the reason had been that there was not enough food for all patients, or they (incorrectly) believed that only the terminally ill had been murdered, while the remaining sick had been evacuated.[14]

Witness accounts reveal the close collaboration between Wehrmacht, SS, and police forces (Pohl 2011, 271–6):[15] the Wehrmacht determined its needs for military resources (such as housing needs for military hospitals or food-supply logistics), while the SS and the police—by killing people—helped the Wehrmacht to achieve its aims. In the occupied Soviet territories, there was no systematic eugenic vision but rather economic and utilitarian motives, which led, as in this case, to the murder of patients with psychiatric and medical needs. Different agents stepped in for different reasons (Süß 2003).

The German courts implemented a particular understanding of guilt, which meant that only those whose direct involvement in the executions could be proven or who significantly abetted the crime through their actions could be considered guilty. In the 1950s and 1960s, a "subjective theory of the perpetrator" prevailed in the German justice system, according to which the objective facts of a crime were not sufficient to convict; the subjective will of the defendant was also a critical element in the determination of guilt. Hence, as a rule, judges rendered murder convictions only if the presence of an "inner affirmation" of the deeds could be proven. Unless a shooter "subjectively willed" the killings, he faced a conviction as an accomplice—and, increasingly, West German courts defined only Hitler, Himmler, and Heydrich as perpetrators and classified all lower-ranking defendants as mere accomplices, whose sentences were significantly lower (Freudiger 2002; Greve 2001).

Despite fairly precise eyewitness accounts, including from several members of the Luftwaffe who were present as spectators, the identity of the firing squad and the matter of whether they had belonged to the SS intelligence service called *Sicherheitsdienst* or the police—and to which unit—remained an open question. In the course of the investigation, all former members of Stuka 77 who could be located were interrogated. Several mentioned that two Luftwaffe airmen volunteered to take part in the executions. But their names could never be ascertained, presumably because of loyalty and silence.

Neither could it be determined which exact departments had ordered the killings. The written records were incomplete, and the court conjectured that either the supreme command of the 11th Army (*AOK11*) or the commander of the rear army area (*Korück*), who was in charge at the time, might have been responsible. Today, we know that it was the chief quartermaster at *Korück 553*, Lieutenant Colonel Doctor Benecke, who requested on September 29, 1941, that measures be taken to address the situation at the psychiatric hospital in Kherson.[16] The document proves that both *AOK11* and *Korück 553*, as well as the local commandant's office in Kherson, were involved. They considered the psychiatric patients as little more than a nuisance and a potential threat to order. At the time of the trial, however, this circumstance remained hidden to the court in Munich, which concluded:

> It cannot be determined any further which departments ordered or otherwise arranged the killing of the mentally ill. [...] Although on the basis of the interrogation of the former members of Stuka Squadron 77 it has been established that the mentally ill were transported to the place of execution by trucks of the 3rd Group of the squadron and that individual members of the 3rd Group were assigned as security guards, none of the former members of the squadron who are still alive can be convicted of an offense that is still punishable.[17]

The burden of proof for individual guilt could not be reached. The commander of Stuka Wing 77, Colonel Schönborn, had been killed in the war. His adjutant and other officers of the squadron present at trial unanimously denied responsibility for the evacuation of the hospital. The commander of the third squadron of Stuka 77, Major Helmut Bode, claimed to have learned of the massacre only afterward; no witnesses directly implicated him, and his presence in Kherson could not be proven. The sergeant who secured the trucks and drivers from the airport operations company denied any prior knowledge of their use; in the eyes of the court, his actions did not amount to collusion. The three squadron captains, at least one of whom was directly incriminated by witness testimony for providing logistical support to the firing squad, had all died in the war. Former sergeant Paul Gorzella, who participated in the execution as a security guard, was not charged because of the "insignificance of his possible contribution to the crime" and because, as a low-ranking officer, he was acting "on the basis of an alleged *Befehlsnotstand* (compulsion to obey orders)." Former sergeant Biewald claimed merely to have observed the crime as a bystander; he denied his involvement in the transport as a security guard and he was not held accountable. Lieutenant of the Field Gendarmerie Jähnig, who (despite a ban by the local commander) had demonstrably ordered several field gendarmes to cordon off the hospital grounds and was personally present at the execution site, died shortly after his interrogation, in which he denied everything, and no charges could be brought against him either.[18]

Members of the *Sonderkommando* (SK) 11a, which was temporarily stationed in Kherson and responsible for multiple killings, were also never convicted. Its leader, Paul Zapp, along with two other members, had been sentenced to prison for killing 5,000 Jewish women, men, and children—inmates of the Kherson ghetto—in September 1941 in the Munich Regional Court in 1970.[19] This mass shooting of ghetto inmates followed the same pattern: the *Sonderkommando* worked closely with the Wehrmacht, and soldiers took part in executions. Zapp served a life sentence for the murder of Jews and vehemently denied any involvement in the killings of patients. Nonetheless, there is some evidence that SK 11a and *Einsatzkommando* (EK) 12 were also involved (cf. Pohl 2011, 265, 275; Angrick 2003, 252).

The court assumed that it had exhausted all sources of knowledge and investigative possibilities. Their request for judicial assistance from the Soviet Union yielded no significant new information, as far as the files indicate; their request was forwarded to the German embassy in Moscow on April 13, 1967, and it took two years before the Ludwigsburg *Zentrale Stelle* received several protocol statements by Soviet witnesses from the Soviet public prosecutor's office. The statements did not seem to have provided any new insights. There were some cases in which judicial cooperation across the "Iron Curtain" proved more fruitful—but this was not one of them.[20]

The Soviet Prosecution of Collaborators

The Soviet prosecution of local collaborators took place in so-called military tribunals, a practice dating back to the revolution in 1917. These were special courts under the military college of the Supreme Court of the USSR, and they usually consisted of a chairman, two assessors, and a secretary. Like regular courts, they operated according to the Code of Criminal Procedure of their respective Soviet republics. However, no appeals against the verdicts of the military tribunals could be lodged. With the exception of death sentences, the punishments went into immediate effect and were carried out without delay (Strogovich 1942, 3–24.).

In Ukraine, the legal basis for the conviction of collaborators was the Ukrainian Criminal Code of 1927 and 1934. Paragraph 54-1a referred to "treason," defined as "actions carried out by Soviet citizens to the detriment of the military strength of the Soviet Union, its national independence or the integrity of its territory, such as espionage; the disclosure of military and state secrets; defection to the enemy; and flight from the country" (Maliarenka 1997, 20, translation tp/kvk). The sentences ranged from death by firing squad to 10 years imprisonment.[21]

In addition, on April 19, 1943, the Council of the Supreme Soviet of the USSR issued a decree, Ukaz 43, "on measures to punish German-fascist criminals guilty of killing and mistreating Soviet civilians and captured Red Army soldiers, as well as spies and traitors to the fatherland from the ranks of the Soviet population and their helpers." Among other things, Ukaz 43 ordered that "spies" and "traitors to the fatherland" were to be punished with the death penalty by hanging (Article 1) and that civilian accomplices (*posobniki*) from the native population be punished with exile or forced labor lasting between 15 and 20 years (Article 2). Ukaz 43 was the only Union-wide penal provision issued by the Soviet Union during the war for crimes of war and of international law. For the first time, the decree also combined the crimes of German war criminals and Soviet traitors to the fatherland under one regulation (Zeidler 1996, 16–20; Hilger et al. 2001, 177–210).

In Kherson, shortly after liberation on March 2, 1945, three members of the nursing staff of the psychiatric hospital in Kherson were tried as accomplices and convicted to long prison terms.[22] The main accused was Konkordiia Semenovna Popova, the clinic's senior physician, who was 67 years old at the time. Born into the Russian aristocracy in Rostov, she had studied medicine in Lausanne, Switzerland, before the revolution and worked as a doctor at the front during World War I. The court accused her of having participated directly in the preparation and execution of the mass killings of patients by the Germans and of having "voluntarily placed herself in the service of the Germans."[23] She was charged with having carried out the German orders without opposition and in full knowledge of the fate of the patients. Under her direction, the sick were divided into one group of "hopeless persons," some 940 patients, who were shot on October 20, 1941, and a second group of about 60 patients, considered fit for work, who were shot later, on November 25, 1941. Popova was at a meeting with German police when the decision was made, and she carried out the selection on behalf of the Germans. In the process, she was alleged to have singled out the Jewish patients. Doctor Popova also instructed the nurses to inject patients with high doses of the sleeping drug scopolamine, so that 120 patients were brought only half-conscious to the place of execution. The military tribunal found Doctor Popova guilty of treason and sentenced her to 20 years in a forced labor camp.[24]

The lack of power and agency of the accused, under conditions of an inhuman German policy of annihilation, is painfully apparent. Yet the extreme threats that Soviet defendants had faced to their lives were hardly taken into account by the courts.

Popova herself wrote a report, which presented a different version of events, later confirmed by staff testimony.[25] She explained how the entire clinic administration had been evacuated before the German invasion, leaving only herself and one colleague as the sole doctors, with 1,300 patients in the facility.[26] She immediately began to release patients back home and urged patients' relatives to come and pick up family members because of rumors of an impending liquidation of the facility by the Germans. The release of patients eventually required approval from German occupation authorities. Nonetheless, according to Popova's report, she and her colleague successfully released over 200 patients during August and September.

At the beginning of October, however, a group of German policemen[27] came to the clinic and told the doctors that the terminally ill were to be killed, claiming that every piece of bread that a patient received deprived a German soldier, and that the German command could no longer tolerate this situation. Popova (according to her statements) decisively rejected the German police officers' request to administer lethal injections and increased her efforts to release more patients over the following days: "I only knew that we had to save as many of these unfortunate patients as possible, as quickly as possible," Popova wrote in her report.[28]

A few days later, a group of German doctors and psychiatrists visited the clinic and reviewed patients' medical histories and diagnoses. On the morning of October 20, 1941, field gendarmes arrived at the clinic in trucks and took the patients to be murdered. In consultation with the German doctors, 40 child and adolescent patients and 70–80 bedridden patients were sedated with scopolamine for easier transport. Popova testified at the court hearing that "[t]he injections were given to the children so that they were not conscious at the time of the shooting."[29] According to Popova, she and her colleague successfully negotiated with the German police authorities to spare 150 patients who had a good prognosis for recovery. (Originally, the doctors requested that 250 patients be spared, but this request was refused.) The doctors aimed to place the 150 patients in surrounding villages but managed to find housing for only 90; the remaining 60 were picked up later, on November 25, 1941, and shot. Popova cited in her defense that through her own and the other doctor's efforts, about 300 of the clinic's 1,300 patients survived: 90 were saved as a result of later negotiations in addition to the 200 patients who had been saved months earlier (i.e., back in August and September).

Charged with selecting Jewish patients, Popova claimed in her interrogation that a German police order prohibited the release of such patients. She testified:

> After we received such an order, we—myself and Doctor Bitner—went to the departments together and reviewed all medical records. According to our estimates, about 20–25% of the patients in the psychiatric clinic at that time were

> Jews. We had been asked to compile lists of these sick Jews, but we did not compile them. I did not provide the Gestapo with any information about them, nor did the Gestapo ask me to do so again.[30]

Other witnesses confirmed that Jewish patients were shot alongside the others and not separated from other patients.[31]

In their sentencing, the Soviet Military Tribunal rejected Popova's defense of having saved 300 patients, of having refused to administer poison, and of having refused to select Jewish patients. The fact that she sedated underage patients with scopolamine, which impaired their awareness during the executions, was taken by the court as complicity rather than as an act of mercy. Moreover, her decision to remain in the clinic with the patients while most of the clinic's administrators and medical staff were evacuated to the Soviet hinterland was not considered favorably as a sign of courage or willingness to sacrifice, but rather as an indication of willingness to collaborate. The decisions of those who fled the German invasion and left the patients behind were not questioned; in fact, according to Wehrmacht records, it was precisely the shortage of and abandonment by nursing and medical staff that caused the military administration to fear chaos and disorder among the patients—a fear that prompted these murders. As the remaining two doctors at the clinic then confronted the inevitable lack of resources amid war and occupation, they were faced with the realization that many terminally ill patients would not be able to receive adequate care. Their choices fall into the category of a "choiceless choice" (Langer 1982, 72).

Food shortages were part of the considerations of the occupying Germans. Hermann Opitz, a member of Stuka 77, testified that "[t]he German doctor also justified the shootings by citing the complete lack of food, medicine and bandages. She also mentioned that the idiots to be shot were chosen because they could not be cured."[32] A similar argument was made by Doctor Otto Andresen, a staff doctor of the Stuka squadron: "At that time, it was said that the Russian doctors sorted out the sick according to their ability to work, since the food, which was in any case scarce, could only be given to those who were able to work."[33] These statements by German Wehrmacht members shift the blame to the Soviet doctors but are evidence of a critical food shortage and lack of medical supplies. Mass starvation occurred across the occupied Soviet territories in institutions for the sick and the disabled, as supplies were cut off. This strategy is known as "hunger euthanasia" and was practiced widely in the German Reich. Given the circumstances, the Soviet doctors may have considered a quick death by firing squad a more humane alternative to the agonies of slow starvation.

The scope for ethical action had become radically reduced for Soviet actors under the conditions of the German occupation. Italian Holocaust survivor Primo Levi addressed this issue in the context of the Holocaust in his paradigmatic essay on the "gray zone," in which he argued against casting moral judgments on victims who chose survival by cooperating with their persecutors (Levi 1988). Judged by today's moral standards, Doctor Popova appears as a courageous heroine who made great efforts to use her diminished agency to maneuver and negotiate on behalf of her patients, and who resisted ruthless instructions as much as possible. Nonetheless, the Soviet military tribunal convicted her as a traitor and henchwoman of the Germans.

This trial reflects both the Soviet state's broad definition of collaboration, as well as its stereotypical division of people into categories of collaborator, resistance fighter, and victim, which left no room for "gray zones" and failed to do justice to the complexity of biographies in wartime (Penter 2011, 335–48). A completely decontextualized understanding of guilt emerges on the part of the Soviet judiciary, which viewed all work for the enemy in the occupied territory as treason and misjudged the real predicament and everyday threat of death that people faced under the everyday realities of the German extermination policy. While the German courts assumed that the lower ranks of the security forces followed orders and lacked authority and command over the situation, the Soviet Union applied no such distinctions even for cleaners, cooks, and interpreters who faced long prison sentences for working with the German occupiers.

The criminal case against Doctor Popova went through two revisions: first, after Stalin's death in 1953 when Khrushchev assumed power, and then again in 1993 after the collapse of the Soviet Union in newly independent Ukraine (Rebitschek 2017, 262–81). In both cases, the appeals commissions concluded that Popova's guilt had been proven beyond doubt and that her sentence was justified. The already aged doctor presumably did not survive her imprisonment in a camp for long, but there is no further information about her death.

There are two more defendants whose Soviet trials are pertinent because they later appealed their judgments and were eventually rehabilitated. First, there was the Ukrainian Ivan Semenovich Leleko, a native of the Kherson region (b. 1898), who was charged with assisting Doctor Popova in selecting Jewish patients, as well as Communist and Komsomol members. Leleko participated in the logistical preparations and helped load patients onto trucks. He was charged with helping German policemen track down a female patient who fled and was hiding; she was subsequently shot. Like Popova, Leleko was sentenced to 20 years in a camp with forced labor.[34] However, Leleko insisted on his innocence during the trial and appealed his sentence several times. After the death of Stalin, he sent a direct appeal to the USSR's Supreme Military Prosecutor on November 20, 1956, claiming that the evidence against him was slanderous:

> Under the influence of threats and other methods of compulsion that were practiced at that time, I was forced to sign a freely invented, fabricated accusation, on the basis of which I was taken to court without proper examination or fact-finding, and above all in the knowledge that the investigation material was fabricated, where I received a severe punishment.[35]

Leleko complained that two prosecution witnesses had been pressured by investigators and deliberately slandered him in their statements. He insisted that he was not a traitor against his fatherland but saw no alternative under German occupation but to continue working at his place of employment; otherwise, he would have been shot for sabotage. Furious, his complaint asked pointed questions:

> What circumstances proved my counterrevolutionary action? What is my betrayal of the homeland? The fact that I could not be evacuated and was in the territory temporarily occupied by the Germans? Where is the justice of the local organs of Soviet law? Surely it cannot be that in this case the guilt of the Germans is simply transferred to a Soviet man?[36]

But this seemed to be *precisely* the logic of the Soviet military tribunal: since the actual German perpetrators were not available for conviction, Soviet citizens would be put in the dock and sentenced. Although those who were tried could be considered victims of the occupation themselves, they fulfilled the function of substitute culprits for the state and local communities.

Leleko's efforts to have his sentence reviewed were unsuccessful, and his appeal was rejected by the military prosecutor. The reason given was that he was proven to have participated in loading patients who were executed onto trucks; hence, he was implicated in the German crimes. Four years later, in 1959, during the general amnesty of the Khrushchev period, Leleko was released from prison, having served 15 years in a labor camp.[37] It was only in 1993, three decades later and after his death, that his conviction was overturned. According to a new law that was passed to rehabilitate victims of political repression in the newly independent Ukraine, it was concluded that the evidence against Leleko was inconclusive and that he did not collaborate.[38] He was fully rehabilitated—a decision that could only be communicated to his son.[39]

The third case involves a 44-year-old Ukrainian nurse, Ustiniia Fedorovna Dudko, a widow and mother of two minor children. She was charged with injecting patients with the sedative scopolamine and sentenced to 10 years' forced labor in the cold northwestern region of the Komi Autonomous Soviet Republic.[40] In 1954, having almost completed her sentence, she submitted a petition for a pardon because she feared not being allowed to return home. She pleaded:

> My fault lies solely in the fact that I continued working in the hospital. I did not know that the Germans were killing the patients. The Germans picked up the sick and took them away. Only then was it revealed that these people had been shot. [...] My guilt is only indirect. I should have given up my work, but the material hardship forced me to stay.[41]

The "material hardship" she refers to was real and claimed the lives of hundreds of thousands of people who fell victim to the starvation policy of German occupation. In Ukrainian cities and industrial regions, this hunger strategy had not only ensured food supplies for German troops but also transferred enormous quantities of food back to the Reich. The number of starvation victims under German occupation can only be roughly estimated today, due to a lack of sources and the difficulty of clearly attributing causes of death to hunger. Estimates range from hundreds of thousands to millions, including those who fell victim to hunger only indirectly (Berkhoff 2014, 54–75; Gerlach 1998; 1999, 44–80, 265–319; 2002, 287–360, Dieckmann 2011, 178–209, 555–85; Gerhard 2009, 45–65; Kay 2006).

Nurse Dudko's two children had grown up in orphanages after their mother's arrest, and they supported her petition to the Presidium of the Supreme Soviet of the Soviet Union. Her daughter wrote:

> I, Komsomol member Tamara Vasil'evna Fokina, address the Supreme Soviet of the USSR and ask you to take pity on us orphans who have not seen their mother for a long time and to give her the opportunity to return to the Kherson region. We have been looking forward to this for many years.[42]

Her pardon application was rejected by the audit commission in Kherson.[43] As in Leleko's case, it took until 1993 for Ustiniia Dudko to be rehabilitated by a Ukrainian court, on the grounds of risks to her own life. The court in 1993 exonerated her of the charges of treason against the fatherland. It is noteworthy that the elements of coercion and threat to survival only became relevant to post-Soviet courts in the newly independent Ukraine in determining responsibility and guilt (cf. Penter 2019, 353–76). Neither she nor her children could be located for this notice of rehabilitation, however; she may never have seen them again.[44]

Soviet trials made no distinction between male and female perpetrators; women were treated with equal severity from the very beginning. In fact, in some cases, women who were not guilty of any crime more serious than simply working for the Germans as cooks, cleaners, or translators—or even individual

Jewish women who held functions in the ghetto—were sentenced to long prison terms of 10 to 25 years (cf. Penter 2008). In the West, male and female perpetrators were often coded differently, and women often received milder sentences because of their gender (Kretzer 2009; Braun 2003, 250; Edwards 1985, 133). It was not until the 1950s and 1960s that female Nazi perpetrators became more visible in public perception and new images of the female "beast" and "concentration camp megalomaniac" emerged that were in no way perceived as inferior to the male perpetrators in terms of aggressiveness (Eschebach 2003, 104). The "euthanasia" proceedings, though, seem to be a certain exception because, at least in the very early German proceedings before the founding of the Federal Republic of Germany, female doctors and nurses were sometimes also sentenced to very long prison terms or death (cf. Freudiger 2002, 360–3.).

Conclusion: The Social Impact of Criminal Verdicts of Guilt

What do we make of the different aftermaths of the crimes in Kherson in the German and Soviet criminal systems? Did these differences affect the recognition of the crimes and the status of the victims in these post-war societies? Nazi "euthanasia," in particular, reflects general social (in)sensitivities and biases, as Götz Aly has shown in his provocative study. The mass murder of mentally ill and disabled people was not treated like other Nazi crimes by courts in the Federal Republic and remained taboo for a long time (Aly 2013; Rotzoll et al. 2010).

The first euthanasia charges were brought in the "Doctors' Trial" (Nürnberger Ärzteprozess) before the American subsequent military tribunal in 1947, which convicted 16 defendants to imprisonment and/or death. West German courts also tried 105 defendants from 1946 to 1974 in a total of 30 euthanasia trials, all of which involved German medical staff working in German medical institutions. Murder convictions remained the exception in the prosecution of euthanasia crimes, and were limited to the period before the founding of the Federal Republic. Before the founding of the Federal Republic of Germany in 1949, two-thirds of perpetrators of euthanasia murders were convicted for murder, but after 1950, more than 80% were sentenced as "accomplices." The decisive criterion of the "subjective will" of the agent served to exonerate the accused beginning in the 1960s, and responsibility shifted up the chain of command until only Hitler and his closest team were left. Hence, doctors who had been actively involved in children's "euthanasia" in medical facilities were acquitted by German courts on the grounds of having acted in good faith in the legality of their actions (Freudiger 2002, 119, 344). Some verdicts even reveal judges' (tacit) approval for the killing of disabled children, as seen in the light sentence given by the Munich Regional Court to Doctor Pfannmüller in 1951 (Freudiger 2002,

272–8). Widespread approval, among the general population and even relatives, made it difficult to press charges (Aly 2013). It took a long time before the victims of "euthanasia" were recognized and remembered, a shift that occurred only in the 1980s. A memorial and information center for the victims of the National Socialist "euthanasia" murders at the site of the planning center, Tiergartenstraße 4 in Berlin, was not inaugurated until 2014 (Westermann et al. 2011; Hohendorf et al 2014; Rotzoll et al. 2010).

Did the failure of the German judiciary to charge and convict perpetrators contribute to this lack of recognition for these victims? The type of language with which psychiatric patients were described even decades after the crimes reveals ongoing dehumanization. Defendants and prosecutors routinely referred to the sick and disabled in degrading terms, such as "insane" (*Geisteskranke*), "lunatics" (*Irre*), or "imbeciles" (*Schwachsinnige*). Until the 1970s, there was virtually no semantic difference or discontinuity from the language of the Nazi era. The court's inability to deliver justice in the form of guilty verdicts certainly contributed to the lack of transformation in German society with regard to the dehumanizing ideology and practices of National Socialism.

Of course, the guilty verdicts in post-war Soviet society did not markedly improve the status of the sick or the disabled, either. Under Stalin, guilt was framed alongside the charges of collaboration and "betrayal of the fatherland," especially if this "betrayal" was presumed to have arisen from a "hostile attitude toward Soviet power" (Cadiot and Penter 2013, 161–71; Penter 2008, 341–64; Exeler 2019, 189–219).

For the Soviet state and post-war society, such trials served complex functions. They were an instrument for the re-Stalinization of the population, yet at the same time they also fulfilled other functions: while the regime condemned collaborators for their "betrayal of the fatherland" and their lack of loyalty to Soviet power, local communities found an outlet for retribution and the processing of individual war traumas. The trials thus reveal a complex guilt interaction between local and state actors, which facilitated the re-Stalinization of society precisely because the trials served to release local "pressure from below" (Fitzpatrick 1993, 299–320). Moreover, in contrast to the show trials of the 1930s, which primarily sentenced completely innocent people on purely political grounds, the trials of German war criminals and Soviet collaborators sometimes put actual killers in the dock. In that way, the Stalinist regime was able to succeed in legitimizing its post-war purge policy before Soviet society by bringing legal practice into greater harmony with the population's sense of justice.

In these post-war trials, the Soviet state pursued a demonstrative legal formality that created an appearance of the rule of law, both internally and externally, and situated itself in a symbolically political manner against the Western culture of the rule of law. In this way, the Stalinist leadership sought to distance itself from the condemnation practices of the Great Purges, when hundreds of thousands were sentenced in summary trials without ever having seen a prosecutor or judge. The post-war trials, on the other hand, were ostensibly based on the rule of law and served the Soviet state to establish its own legitimacy.

On the local level, these trials provided a podium to process complicity and collaboration, and also (to some extent) an alternative discourse of memory that differed from the official version. While, in the post-war years, the Soviet government attempted to establish an official culture of remembrance of the Second World War that integrated the heterogeneous Soviet population into one identity—simultaneously concealing conflicts and obscuring the actual experiences of people under German occupation—the trials exposed the dilemmas and entanglements of war and occupation. Nevertheless, in general, the Soviet citizens' collaboration in Nazi crimes could not be openly addressed until the end of the Soviet Union itself (cf., in detail, Penter 2010, 394–411; 2008, 341–64).

There is no memorial in Kherson. There is no place or date to remember the more than 1,000 patients of the psychiatric hospital or, for that matter, the more than 5,000 Jewish victims in the city, even though the exact location of the mass graves was known to Soviet and post-Soviet judicial authorities. However, when one of these mass graves was uncovered accidentally during construction in 1966, the city hastily erected a memorial stone for "1,276 peaceful Soviet citizens—victims of fascism." The patients, like the Jews in general, remained nameless and hidden as generic "victims of fascism" until the end of the Soviet Union.

The guilty verdicts in the Soviet trials did not ultimately lead to greater recognition of the victims in post-war society. The murders of patients represent a difficult legacy for the Soviet state and the local communities alike. The Soviet state had failed to evacuate the sick and disabled before the German invasion and instead abandoned them to fend for themselves after the nursing staff had fled. The families failed in not taking their sick relatives home from the hospital, whatever their reasons; some families even sent their relatives *back* to the hospital after the hospital had released them. (Note that, after the German invasion, Ukrainian families faced a dwindling availability of life-sustaining resources, which might have made them fear that they could not provide for their ill relatives.) In cases where families' actions resulted in their relatives' deaths, there might (in some respect) be a burden of failure comparable to the one that Aly sees in German families' failure to condemn euthanasia policies—although this burden was undoubtedly heavier in the German case. Guilty verdicts for the doctors and nurses thus functioned here as a sort of moral relief for the community; once the guilt was legally established and punished harshly, there was no more need for broader public debate over what had happened (cf. the introduction to this volume), even if the containment of guilt in criminal prosecution did not go so far as to *vindicate* the victims.

Paradoxically, the opposite obtains for West Germany, where the unsatisfying and deficient legal containment of guilt *increased* the pressure for public debate (cf. Buschmeier 2018, 38–50)—hence the cultural adaptation of criminal trials such as the Auschwitz trials, which became occasions for theater and literature, and films such as *Spiegelgrund* (1999) and *Nebel im August* (2016) that dramatize euthanasia in the Nazi era. The lack of a voice among the victims and their relatives has contributed to the fact that, in Ukraine, a cloak of silence lies over the sites of the crimes to this day, and that no systematic historical reappraisal of the Nazi murders of the sick or disabled has yet taken place. In the memory of the local communities who witnessed the horrible deeds, however—and who live next to the mass graves—the crimes are still present today.

Notes

1. For the verdicts of the 42 German criminal trials (30 West and 12 East German) for National Socialist "euthanasia" crimes in the Reich, see Mildt (1996, 2009).
2. See Rückerl's letters of January 3 and February 23, 1967. In Zentrale Stelle Ludwigsburg. B 162/27709, 9–13. Archival materials are referenced in Endnotes hereinafter, and have been translated by Katharina von Kellenbach.
3. "Order of discontinuance of the preliminary proceedings against unknown persons on suspicion of murder or aiding and abetting it (shooting of about 1,000 mentally ill persons near Kherson)." In Zentrale Stelle Ludwigsburg. B 162/27950, pp. 1691–1704, here p. 1692.
4. See, for this "purge of the purgers," Viola (2017).
5. Paul Beule. Interrogation, October 20, 1965. In Zentrale Stelle Ludwigsburg. B 162/27950, 30–31.
6. Paul Gorzella. Interrogation, February 5, 1970. In Zentrale Stelle Ludwigsburg. B 162/27948, 1170.
7. Erhard Biewald. Interrogation, November 19, 1968. In Zentrale Stelle Ludwigsburg. B 162/25762, 446.
8. Paul Heinelt. Interrogation, November 9, 1965. In Zentrale Stelle Ludwigsburg. B 162/25761, 68.
9. Willi Fischer. Interrogation, February 5, 1970. In Zentrale Stelle Ludwigsburg. B 162/27948, 1167.
10. Arthur Goetz. Interrogation, January 13, 1967. In Zentrale Stelle Ludwigsburg. B 162/27950, 102. Another witness reported that the patients were originally intended to be killed by a serum that was injected by a Russian doctor but that the serum did not have the desired effect. Cf. Heinelt (1965, 68–69).
11. Richard Geisler. Interrogation, October 1, 1968. In Zentrale Stelle Ludwigsburg. B 162/25762, 364.

12. Beule 1965, 31–2.
13. Goetz 1967, 102.
14. "Disclosure Order." In Zentrale Stelle Ludwigsburg. B 162/27950, 1696.
15. On the murder of the inmates of a psychiatric clinic in the Leningrad district, see Hürter (2001, 377–440).
16. Bundesarchiv-Militärarchiv (BA-MA). RH 23/69, 122. I would like to thank Dieter Pohl for providing me with a copy of this document. (See also Pohl 2011, 275).
17. "Disclosure Order." In Zentrale Stelle Ludwigsburg. B 162/27950, 1697.
18. "Disclosure Order." In Zentrale Stelle Ludwigsburg. B 162/27950, 1693–1704.
19. Judgment of the Munich I Regional Court of 26 February 1970, IV 9/69.
20. See Söhner (2020); Tytarenko and Penter (2020).
21. By a decree of May 26, 1947, the Presidium of the Supreme Soviet of the USSR had abolished the death penalty for all crimes. But on January 12, 1950, the death penalty for "traitors, spies, subversive elements, and saboteurs" was reintroduced.
22. Archive of Ukrainian security service in Kherson region (ASBUChO). F. 5(r), Op. 1, D. 22-o. A fourth defendant was acquitted at the trial.
23. "Judgment." In ASBUChO. F. 5(r), Op. 1, D. 22-o, 144.
24. "Judgment." In ASBUChO. F. 5(r), Op. 1, D. 22-o, 144–6.
25. Report of the physician Popova. In ASBUChO. F. 5(r), Op. 1, D. 22-o, 9–10. See witness account of the hospital's chief feldsher, Dmitrii Shramko, December 18, 1945, in a different trial. USHMM Archives, RG-06.025 [Central Archives of the Federal Security Services (former KGB) of the Russian Federation records relating to war crimes trials in the Soviet Union, Nikolaev Trial, H-18755, volume 1], Reel 22, 216.
26. Ethnic-German Russian physician Tina Genrichovna Bitner.
27. Popova uses the term "Gestapo" here, which Soviet witnesses often used to describe members of the Security Police and the SD.
28. Report of the physician Popova, November 5, 1944. In ASBUChO. F. 5(r), Op. 1, D. 22-o, 9.
29. Minutes of the court hearing. In ASBUChO. F. 5(r), Op. 1, D. 22-o, 133.
30. Popova. Interrogation protocol. In ASBUChO. F. 5(r), Op. 1, D. 22-o, 11.
31. Prusanova. Interrogation protocol. In ASBUChO. F. 5(r), Op. 1, D. 22-o, 73.
32. Hermann Opitz. Statement, May 21, 1971. In Zentrale Stelle Ludwigsburg. B 162/27950, 1507–8.
33. Otto Andresen. Statement, June 4, 1969. In Zentrale Stelle Ludwigsburg. B 162/25762, 524.
34. Judgment. In ASBUChO. F. 5(r), Op. 1, D. 22-o, 170–2.
35. Leleko. 1956. Letter to the Supreme Military Prosecutor of the USSR, November 20. In ASBUChO. F. 5(r), Op. 1, D. 22-o, 182–3.
36. Ibid., 183.
37. ASBUChO. F. 5(r), Op. 1, D. 22-o, 197–198.
38. Zakon Ukrainy, "O reabilitacii zhertv politicheskikh repressii na Ukraine," April 17, 1991.
39. ASBUChO. F. 5(r), Op. 1, D. 22-o, 207. In the Soviet context the word transliterates as "reabilitatsiya," it was akin in practice to "acquittal."

40. Judgment. In ASBUChO. F. 5(r), Op. 1, D. 22-o, 170–2.
41. Request for pardon by Dudko. In ASBUChO. F. 5(r), Op. 1, D. 22-o, 165–6.
42. Letter of request from the adult daughter Tamara Vasil'evna Fokina. In ASBUChO. F. 5(r), Op. 1, D. 22-o, 167–8.
43. Idib., 173–4, 180.
44. ASBUChO. F. 5(r), Op. 1, D. 22-o, 208.

References

Aly, Götz. 2013. *Die Belasteten: "Euthanasie" 1939–1945; Eine Gesellschaftsgeschichte.* Frankfurt am Main: S. Fischer.

Aly, Götz, and Susanne Heim. 1993. *Vordenker der Vernichtung: Auschwitz und die deutschen Pläne für eine neue europäische Ordnung.* 2nd ed. Frankfurt am Main: S. Fischer.

Angrick, Andrej. 2003. *Besatzungspolitik und Massenmord: Die Einsatzgruppe D in der südlichen Sowjetunion 1941–1943.* Hamburg: Hamburger Edition.

Baader, Gerhard. 2016. "Eugenik, Rassenhygiene und 'Euthanasie.'" In *Diskriminiert—vernichtet—vergessen: Behinderte in der Sowjetunion, unter nationalsozialistischer Besatzung und im Ostblock 1917–1991*, edited by Alexander Friedmann and Rainer Hudemann, 311–20. Stuttgart: Franz Steiner.

Bazyler, Michael J., and Frank M. Turekheimer. 2014. *Forgotten Trials of the Holocaust.* New York: New York University Press.

Berkhoff, Karel C. 2004. *Harvest of Despair: Life and Death in Ukraine under Nazi Rule.* Cambridge, MA: Belknap Press of Harvard University Press.

Berkhoff, Karel C. 2014. "'Wir sollen verhungern, damit Platz für die Deutschen geschaffen wird': Hungersnöte in ukrainischen Städten im Zweiten Weltkrieg." In *Deutsche Besatzung in der Sowjetunion 1941–1944: Vernichtungskrieg, Reaktionen, Erinnerung*, edited by Babette Quinkert and Jörg Morré, 54–75. Paderborn: Ferdinand Schöningh.

Braun, Christina von. 2013. "Die unterschiedlichen Geschlechtercodierungen bei NS-Tätern und -Täterinnen unter medienhistorischer Perspektive." In *"Bestien" und "Befehlsempfänger": Frauen und Männer in NS-Prozessen nach 1945*, edited by Ulrike Weckel and Edgar Wolfrum, 250–65. Göttingen: Vandenhoeck & Ruprecht.

Budnitskii, Oleg. 2019. "The Great Terror of 1941: Toward a History of Wartime Stalinist Criminal Justice." *Kritika: Explorations in Russian and Eurasian History* 20, no. 3 (Summer): 447–80.

Burds, Jeffrey. 2018. "'Turncoats, Traitors, and Provocateurs': Communist Collaborators, the German Occupation, and Stalin's NKVD, 1941–1943." *East European Politics and Societies and Cultures* 32, no. 3: 606–38.

Buschmeier, Matthias. 2018. "Felix Culpa? —Zur kulturellen Produktivkraft der Schuld." *Communio* 47: 38–50.

Cadiot, Juliette, and Tanja Penter 2013. "Law and Justice in Wartime and Postwar Stalinism." *Jahrbücher für Geschichte Osteuropas* 61: 161–71.

Dieckmann, Christoph. 2011. *Deutsche Besatzungspolitik in Litauen 1941–1944.* Göttingen: Wallstein.

Dumitru, Diana. 2014. "An Analysis of Soviet Postwar Investigation and Trial Documents and Their Relevance for Holocaust Studies." In *The Holocaust in the East: Local*

Perpetrators and Soviet Responses, edited by Michael David-Fox, Peter Holquist, and Alexander Martin, 142–57. Pittsburgh, PA: University of Pittsburgh Press.

Ebbinghaus, Angelika, and Gerd Preissler. 1985. "Die Ermordung psychisch kranker Menschen in der Sowjetunion—Dokumentation." In *Aussonderung und Tod—Die klinische Hinrichtung der Unbrauchbaren*, edited by Aly Götz, Angelika Ebbinghaus, Matthias Hammen, Friedemann Pfäfflin, and Gerd Preissler, 75–107.Beiträge zur nationalsozialistischen Gesundheits- und Sozialpolitik 1. Berlin: Rotbuch.

Edwards, Susan. 1985. "Gender 'Justice'? Defendants and Mitigating Sentence." In *Gender, Sex and the Law*, edited by Susan Edwards, 129–158. London: Croom Helm.

Epifanov, Aleksandr E. 1997. *Otvetstvennost' gitlerovskich voennych prestupnikov i ich posobnikov v SSSR (istoriko-pravovoj aspect)*. Volgograd: VA MVD RF.

Epifanov, Aleksandr E. 2017. "Institutional and Legal Framework for the Commissions' Activities for Establishing and Investigating Nazi Crimes." *Legal Concept* 16, no. 4 (October): 54–61.

Eschebach, Insa. 2003. "Gespaltene Frauenbilder: Geschlecht-Dramaturgien im juristischen Diskurs ostdeutscher Gerichte." In *"Bestien" und "Befehlsempfänger": Frauen und Männer in NS-Prozessen nach 1945*, edited by Ulrike Weckel and Edgar Wolfrum, 95–116. Göttingen: Vandenhoeck & Ruprecht.

Exeler, Franziska. 2016. "The Ambivalent State: Determining Guilt in the Post-World War II Soviet Union." *Slavic Review* 75, no. 3: 606–29.

Exeler, Franziska. 2019. "Nazi Atrocities, International Criminal Law, and Soviet War Crimes Trials: The Soviet Union and the Global Moment of Post–Second World War Justice." In *The New Histories of International Criminal Law: Retrials*, edited by Immi Tallgren and Thomas Skouteris, 189–219. Oxford: Oxford University Press.

Fedotov, Dmitrii. D. 1965. "O gibeli dushevnobol'nych na territorii SSSR, vremenno okkupirovannoi fashistskimi zachvatchikami v gody Velikoi Otechestvennoi voiny." *Voprosy social'noi i klinicheskoi psichonevrologii* 12: 443– 59.

Fitzpatrick, Sheila. 1993. "How the Mice Buried the Cat: Scenes from the Great Purges of 1937 in the Russian Provinces." *Russian Review* 52: 299–320.

Frei, Norbert, ed. 2006. *Transnationale Vergangenheitspolitik: Der Umgang mit deutschen Kriegsverbrechern in Europa nach dem Zweiten Weltkrieg*. Göttingen: Wallstein.

Freudiger, Kerstin. 2002. *Die juristische Aufarbeitung von NS-Verbrechen*.Beiträge zur Rechtsgeschichte des 20. Jahrhunderts 33. Tübingen: Mohr Siebeck.

Friedländer, Saul. 1997. *Nazi Germany and the Jews*. Vol. 1, *The Years of Persecution 1933–1939*. London: Weidenfeld & Nicolson.

Friedman, Alexander, and Rainer Hudemann, eds. 2016. *Diskriminiert—vernichtet—vergessen: Behinderte in der Sowjetunion unter nationalsozialistischer Besatzung und im Ostblock 1917–1991*. Stuttgart: Steiner.

Ganzenmüller, Jörg. 2005. *Das belagerte Leningrad 1941–1942: Die Stadt in den Strategien von Angreifern und Verteidigern*. Paderborn: Ferdinand Schöningh.

Gerhard, Gesine. 2009. "Food and Genocide: Nazi Agrarian Politics in the Occupied Territories of the Soviet Union." *Contemporary European History* 18, no. 1: 45–65.

Gerlach, Christian. 1998. *Krieg, Ernährung, Völkermord: Forschungen zur deutschen Vernichtungspolitik im Zweiten Weltkrieg*. Hamburg: Hamburger Edition.

Gerlach, Christian. 1999. *Kalkulierte Morde: Die deutsche Wirtschafts- und Vernichtungspolitik in Weißrussland 1941 bis 1944*. Hamburg: Hamburger Edition.

Greve, Michael. 2001. *Der justitielle und rechtspolitische Umgang mit den NS-Gewaltverbrechen in den sechziger Jahren*. Frankfurt am Main: Peter Lang.

Gross, Jan T. 2002. *Neighbors: The Destruction of the Jewish Community in Jedwabne, Poland*. London: Penguin Books.

Hamburger Institut für Sozialforschung, ed. 2002. *Verbrechen der Wehrmacht: Dimensionen des Vernichtungskriegs 1941–1944. Ausstellungskatalog*. 2nd extended ed. Hamburg: Hamburger Edition.

Hilger, Andreas. 2006. "'Die Gerechtigkeit nehme ihren Lauf?' Die Bestrafung deutscher Kriegs- und Gewaltverbrecher in der Sowjetunion und der SBZ/ DDR." In *Transnationale Vergangenheitspolitik: Der Umgang mit deutschen Kriegsverbrechern in Europa nach dem Zweiten Weltkrieg*, edited by Norbert Frei, 180–246. Göttingen: Wallstein.

Hilger, Andreas, Nikita Petrov, and Günther Wagenlehner. 2001. "Der 'Ukaz 43': Entstehung und Problematik des Dekrets des Präsidiums des Obersten Sowjets vom 19. April 1943." In *Sowjetische Militärtribunale*. Vol. 1, *Die Verurteilungen deutscher Kriegsgefangener 1941–1953*, edited by Andreas Hilger, Nikita Petrov, and Günther Wagenlehner, 177–210. Cologne: Böhlau.

Hohendorf, Gerrit, Stefan Raueiser, Michael V. Cranach, and Sibylle V. Tiedemann, eds. 2014. *Die "Euthanasie"-Opfer zwischen Stigmatisierung und Anerkennung: Forschungs- und Ausstellungsprojekte zu den Verbrechen an psychisch Kranken und die Frage der Namensnennung der Münchner "Euthanasie"-Opfer*. Berichte des Arbeitskreises zur Erforschung der nationalsozialistischen Euthanasie und Zwangssterilisation 10. Münster: Kontur.

Hürter, Johannes. 2001. "Die Wehrmacht vor Leningrad: Krieg und Besatzungspolitik der 18. Armee im Herbst und Winter 1941/42." *Vierteljahrshefte für Zeitgeschichte* 49: 377–440.

Kaiser, Claire P. 2014. "Betraying their Motherland: Soviet Military Tribunals of *Izmenniki Rodiny* in Kazakhstan and Uzbekistan, 1941–1953." *Soviet and Post-Soviet Review* 41, no. 1 (Winter): 57–83.

Kay, Alex J. 2006. *Exploitation, Resettlement, Mass Murder: Political and Economic Planning for German Occupation Policy in the Soviet Union, 1940–1941*. New York: Berghahn.

Kretzer, Anette. 2009. *NS-Täterschaft und Geschlecht: Der erste britische Ravensbrück-Prozess 1946/47 in Hamburg*. Berlin: Metropol.

Krin'ko, E. F. 2004. "Kollaboratsionizm v SSR v gody Velikoi Otechestvennnoi voiny I ego izuchenie v rossiiskoi istoriografii." *Voprosy istorii* 11: 153–64.

Kudryashov, Sergey, and Vanessa Voisin. 2008. "The Early Stages of 'Legal Purges' in Soviet Russia (1941–1945)." *Cahiers du monde russe* 49, no. 2–3 (May): 263–96.

Langer, Lawrence L. 1982. *Versions of Survival: The Holocaust and the Human Spirit*. Albany: State University of New York Press.

Levi, Primo. 1988. *The Drowned and the Saved*. London: Michael Joseph. Originally published as *I sommersi e i salvati* (Turin: Giulio Einaudi, 1986).

Maliarenka, V., ed. 1997. *Reabilitatsia represovanykh: Verkhovnyi sud Ukraïny: Zakonodavstvo ta sudova praktyka*. Kiev: Jurinkom.

Melnyk, Oleksandr. 2013. "Stalinist Justice as a Site of Memory: Anti-Jewish Violence in Kyiv's Podil District in September 1941 through the Prism of Soviet Investigative Documents." *Jahrbücher für Geschichte Osteuropas* 61, no. 2: 223–48.

Mildt, Dick de. 1996. *In the Name of the People: Perpetrators of Genocide in the Reflection of Their Post-war Prosecution in West Germany: The "Euthanasia" and "Aktion Reinhard" Trial Cases*. The Hague: Martinus Nijhoff.

Mildt, Dick de, ed. 2009. *Tatkomplex: NS-Euthanasie. Die ost- und westdeutschen Strafurteile seit 1945*. 2 vols. Amsterdam: University Press.

Mozokhin, Oleg Borisovič. 2018. *Repressii v cifrakh i dokumentakh: Dejatel'nost' organov VChK—OGPU—NKVD—MGB (1918–1953gg.)*. Moscow: Veche.

Moskoff, William. 1990. *The Bread of Affliction: The Food Supply in the USSR during World War II*. Cambridge: Cambridge University Press.

Nikol's'kyj, V. M. 2003. *Represyvna diyal'nist' orhaniv deržhavnoï bezpeki SRSR v Ukraïni (kinec' 1920-ch-1950-ti rr.)*. Istoryko-statystyčne doslidžhennja. Donec'k: n.p.

Penter, Tanja. 2005. "Collaboration on Trial: New Source Material on Soviet Postwar Trials against Collaborators." *Slavic Review* 64: 780–90.

Penter, Tanja. 2008. "Local Collaborators on Trial: Soviet War Crimes Trials under Stalin (1943–1953)." *Cahiers du Monde russe* 49, no. 2–3: 341–64.

Penter, Tanja. 2010. "'Das Urteil des Volkes': Der Kriegsverbrecherprozesse von Krasnodar 1943." *Osteuropa* 60, no. 12: 117–31.

Penter, Tanja. 2010. *Kohle für Stalin und Hitler: Arbeiten und Leben im Donbass 1929 bis 1953*. Veröffentlichungen des Instituts für soziale Bewegungen, series C, vol 8. Arbeitseinsatz und Zwangsarbeit im Bergbau. Essen: Klartext.

Penter, Tanja. 2011. "Die Ukrainer und der 'Große Vaterländische Krieg': Die Komplexität der Kriegsbiographien." In *Die Ukraine: Prozesse der Nationsbildung*, edited by Andreas Kappeler, 335–48. Cologne: Böhlau.

Penter, Tanja. 2019. "Vergessene Opfer von Mord und Missbrauch: Behindertenmorde unter deutscher Besatzungsherrschaft in der Ukraine (1941–1943) und ihre juristische Aufarbeitung in der Sowjetunion." *Journal of Modern European History* 17, no. 3: 353–76.

Pohl, Dieter. 2011. *Die Herrschaft der Wehrmacht: Deutsche Militärbesatzung und einheimische Bevölkerung in der Sowjetunion 1941–1944*. Frankfurt am Main: Samuel Fischer.

Prusin, Alexander V. 2003. "'Fascist Criminals to the Gallows!' The Holocaust and Soviet War Crimes Trials, December 1945—February 1946." *Holocaust and Genocide Studies* 17, no. 1: 1–30.

Raim, Edith. 2013. *Justiz zwischen Diktatur und Demokratie: Wiederaufbau und Ahndung von NS-Verbrechen in Westdeutschland 1945–1949*. Quellen und Darstellungen zur Zeitgeschichte 96. Munich: De Gruyter Oldenbourg.

Rebitschek, Immo. 2017. "Feindbilder auf dem Prüfstand: Sowjetische Kollaborateure im Fokus der Revisionskommissionen, 1954 und 1955." *Jahrbücher für Geschichte Osteuropas* 65, no. 2: 262–81.

Rotzoll, Maike, Gerrit Hohendorf, Petra Fuchs, Paul Richter, Christoph Mundt, and Wolfgang U. Eckart, eds. 2010. *Die nationalsozialistische 'Aktion T4' und ihre Opfer: Historische Bedingungen und ethische Konsequenzen für die Gegenwart*. Paderborn: F. Schoeningh.

Schmuhl, Hans-Walter. 1999. "Vergessene Opfer: Die Wehrmacht und die Massenmorde an psychisch Kranken, geistig Behinderten und 'Zigeunern.'" In *Wehrmacht und Vernichtungskrieg: Militär im nationalsozialistischen System*, edited by Karl Heinrich Pohl, 115–39. Göttingen: Vandenhoeck & Ruprecht.

Schneider, Wolfgang. 2019. "From the Ghetto to the Gulag, from the Ghetto to Israel: Soviet Collaboration Trials against the Shargorod Ghetto's Jewish Council." *Journal of Modern European History* 17, no.1: 83–97.

Skorobohatov, Anatolij V. 2004. *Kharkiv u chasy nimec'koï okupaciï (1941–1943)*. Kharkiv: Prapor.

Söhner, Jasmin. 2020. "Politicization and Practice of German-Soviet Judicial Cooperation in the Cold War." PhD diss. project, University of Heidelberg.

Strogovich, M.S. 1942. *Voennye tribunaly sovetskogo gosudarstva* [Military tribunals in the USSR]. Moscow: n.p.

Süß, Winfried. 2003. *Der "Volkskörper" im Krieg: Gesundheitspolitik, medizinische Versorgung und Krankenmord im nationalsozialistischen Deutschland 1939–1945.*Studien zur Zeitgeschichte 65. Munich: Oldenbourg.

Tytarenko, Dmytro. 2016. "Medizinische Betreuung und nationalsozialistische Krankenmorde in der Ukraine unter der deutschen Okkupation." In *Diskriminiert—vernichtet—vergessen: Behinderte in der Sowjetunion, unter nationalsozialistischer Besatzung und im Ostblock 1917–1991*, edited by Alexander Friedman and Rainer Hudemann, 355–72. Stuttgart: Steiner.

Tytarenko, Dmytro, and Tanja Penter. 2021. "Die Ermordung von Psychiatrie-Patienten in Poltava/Ukraine unter nationalsozialistischer Besatzungsherrschaft im Spiegel deutscher und sowjetischer Ermittlungsakten und Justizkooperation im Kalten Krieg." *Jahrbücher für Geschichte Osteuropas* 68, no. 3.

Vasylyev, Valeryi, Natalija Kashevarova, Olena Lysenko, Marina Panova, and Roman Podkur, eds. 2018. *Nasyl'stvo nad cyvil'nym naselenniam Ukraïny: Dokumenty specsluzhb 1941–1944*. Kyiv: Vydavec' Zakharenko V.O.

Viola, Lynne. 2017. *Stalinist Perpetrators on Trial: Scenes from the Great Terror in Soviet Ukraine*. Oxford: Oxford University Press.

Weindling, Paul Julian, 2019. "The Need to Name: The Victims of Nazi 'Euthanasia' of the Mentally and Physically Disabled and Ill 1939–1945." In *Mass Murder of People with Disabilities and the Holocaust*, edited by Brigitte Bailer and Juliane Wetzel, 49–85. Berlin: Metropol, 2019.

Westermann, Stefanie, Richard Kühl, and Tim Ohnhäuser, eds. 2011. *NS-"Euthanasie" und Erinnerung: Vergangenheitsaufarbeitung—Gedenkformen—Betroffenenperspektiven*. Medizin und Nationalsozialismus 3. Münster: LIT.

Winkler, Ulrike, and Gerrit Hohendorf. 2010. "Nun ist Mogiljow frei von Verrückten: Die Ermordung der PsychiatriepatienInnen in Mogilew 1941/42." In *Krieg und Psychiatrie 1914–1950*, edited by Babette Quinkert, Philipp Rauh, and Ulrike Winkler, 75–103. Beiträge zur Geschichte des Nationalsozialismus 26. Göttingen: Wallstein.

Zeidler, Manfred. 1996. *Stalinjustiz kontra NS-Verbrechen—Die Kriegsverbrecherprozesse gegen deutsche Kriegsgefangene in der UdSSR in den Jahren 1943–1952. Kenntnisstand und Forschungsprobleme*. Hannah-Arendt-Institut für Totalitarismusforschung e. V., Berichte und Studien, vol. 9. Dresden: T.U. Dresden.

Zeidler, Manfred. 2004. "Der Minsker Kriegsverbrecherprozess vom Januar 1946: Kritische Anmerkungen zu einem sowjetischen Schauprozeß gegen deutsche Kriegsgefangene." *Vierteljahreshefte für Zeitgeschichte* 2: 211–45.

PART III

GUILT AS CREATIVE IRRITATION

9

Rituals of Repentance

Joshua Oppenheimer's *The Act of Killing*

Katharina von Kellenbach

In her essay "Of Monsters and Men: Perpetrator Trauma and Mass Atrocity," the legal scholar Saira Mohamed (2015) argues that the terminology of trauma should be used to describe the effects that committing atrocities have on perpetrators. She is concerned that the extremity of war crimes brought before the International Criminal Court in The Hague creates the temptation to exclude perpetrators from the circle of humanity by declaring them *hostis generis humani*, "enemies of humankind." In her argument to treat perpetrators as humans and not as monsters, Mohamed is rightly concerned that declarations of guilt justify retributive violence and tend to exclude people from the circle of care. People who are guilty of severe crimes no longer "merit empathy and deserve to be heard" (1173). Therefore, she pleads for an extension of the discourse of trauma, because a person who suffers psychological and moral trauma is a "subject worthy of attention and respect" (1174). "Greater attention to perpetrator trauma," she contends, "can transform our thinking on the role of rehabilitation in postconflict reconciliation, about the breadth of trauma itself, and about the humanity of perpetrators" (1201).

In conversation with her analysis of Joshua Oppenheimer's 2012 docudrama *The Act of Killing*, I contend that we should reclaim the language of guilt in order to affirm the humanity of perpetrators. Guilt does not disqualify people from humanity. On the contrary, the guilty person has moral standing and is neither to be silenced nor expelled from human rights and responsibilities. Guilt requires and deserves remedy. Across the different religious and cultural landscapes, rituals of repentance make guilt productive of change and provide performative pathways for atonement and expiation. While the specifics diverge, penitential rituals involve the cultivation of feelings of remorse and contrition, confessional modes of telling the truth about the past, and symbolic gestures of penal suffering that vindicate the victims and restore the political and cosmic order.

Katharina von Kellenbach, *Rituals of Repentance* In: *Guilt*. Edited by: Katharina von Kellenbach and Matthias Buschmeier, Oxford University Press. © Oxford University Press 2022. DOI: 10.1093/oso/9780197557433.003.0010

Guilt, Trauma, and Moral Injury

Experiencing guilt is quite different from suffering trauma, a term that should remain reserved for the wounds that afflict victims of violence. This distinction is important for moral and political reasons, but also because the consequences of violence affect victims and perpetrators differently: guilt presupposes agency and the power of choice, which has been taken from the victim. Ironically, it is more often the victim who *feels* guilty, ashamed, and blameworthy, rather than the wrongdoer. For instance, a victim of rape feels sullied and guilt-ridden (what did she wear, why was she in this place at that time, how could she have trusted this person?), while the rapist feels a sense of power and satisfaction. The rape of innocents is pleasurable to those who commit rape, or it would not happen; the violation of victims feels empowering and gratifying, because it is *their* dignity that is degraded and their reputation that is tarnished. As a general rule, perpetrators do not *feel* guilty, tainted, or ashamed.

Guilt manifests as blindness to the suffering of others, an indifference to their dehumanization. It is the cause and the effect of doing harm, a deliberate dismissal of empathy. Every act of violence hardens the heart, to use a biblical concept, and entails a refusal to see, feel, and respond to the suffering of the victim. Perpetrators must avert their eyes and block their ears to avoid perceiving the pain inflicted on victims. The psychiatrist Robert Jay Lifton (1986), who studied the psychological condition of Nazi medical doctors, called this phenomenon "doubling" and "numbing." Doubling refers to the psychological construction of an internal wall that separates the inhumane professional from the humane personal self of a person, while "numbing" speaks to the willful desensitization that is necessary to inflict harm. Unlike those with dissociative identity disorder, people who engage in doubling are perfectly functional members of society, albeit capable of extreme cruelty. Compartmentalizing one's "lives" by keeping segments of one's professional and private, personal and public, intimate or religious lives separate is a particular feature of modern existence. At some point, doubling is no longer subject to conscious control, as certain emotions and memories remain closed off.

Guilt, understood through the prism of psychological doubling, is not a character trait that defines the essence of a person. It is a situational category: people make choices that require the obstruction of empathy in particular situations. This turns one person into a perpetrator and another into a victim. Moreover, a victim of violence can become a perpetrator, and vice versa. For instance, the rapist may himself have been the victim of child abuse or have suffered the traumatic effects of neglect, poverty, and violence. In turn, the victim of rape may act out violently against her children and other people under her control. Victimhood is not a moral category that defines character or identity. The oppressed become the oppressors, as revolutions devour their own children. Victimization is not morally uplifting or spiritually purifying. It is traumatizing.

Trauma occurs when victims find themselves in situations of extreme powerlessness that flood the system with stress hormones and trigger fight-or-flight responses. The experience of victimization is defined by a lack of agency and options. Perpetrators, by contrast and by definition, suffer less stress because they retain some control over the situation. Trauma is a symptom of powerlessness, the condition of victimization. The National Institute of Mental Health identifies distorted feelings of guilt and self-blame as one of four symptoms that define post-traumatic stress disorder (PTSD) (NIMH 2019). Victims and survivors of violence are haunted by guilt feelings, which are now known as "survivor guilt" (Leys 2009, 2010). But those who control the situation *are* guilty, a state that presents symptomatically as desensitization and blocked empathy for the suffering of others.

Already in 1948, Martin Buber tried to distinguish between "Guilt and Guilt Feeling" at the London International Conference of Medical Psychotherapy, shortly after World War II. He feared that psychotherapy's focus on guilt as an internal feeling arising from conflicts between ego and superego risked ignoring the actuality of objective, culpable wrongdoing. The psychotherapeutic paradigm, he maintained, marginalizes the "factual occurrences of guilt in the lives of 'patients,' suffering men,. . . because in the theory and in the practice of this science only the psychic 'projection' of guilt, but not the actual events of guilt, is afforded room" (Buber 1957, 114). But the actions and omissions of people cause real harm in the world, and people incur ontic responsibility for healing the suffering and repairing the damage they inflicted. Guilt, in his view, occurs in the real world and should not be reduced to a phenomenon that transpires purely within the interiority of individuals.

Buber also argued that guilt transcends particular rule systems, be they legal or familial, political or cultural. While right- and wrongdoing are inculcated by parental discipline, religious commandments, and educational programs, atrocities leave wounds in the cosmic fabric that must be healed on multiple levels. He pleaded against the fracturing of human existence into specific areas of competence, where the law, therapy, and religion treat different spheres of life. The consciousness of guilt, he maintained, is grounded in fundamental human reflexivity in the face of the irreversibility of time and the irreparability of our actions. Psychotherapy, he wrote, plays an important role in helping people gain the necessary "strengths and elements of a unified being that guards against conflict and contradiction" (Buber [1958] 2008, 141, my translation). But any process of repentance and atonement must integrate the internal and the external, the personal and the public, the unconscious and the relational. As a Jewish philosopher, Buber speaks of the soul as code for the psychological, moral, and spiritual center of a person within an interconnected and interdependent matrix. We do not live as atomized individuals, disconnected from each other, but rather in a social and cosmic matrix, within which we are (to play on the term) response-able to ourselves and each other.

The concept of "moral injury" is another concept emerging in the United States to address the predicament of military veterans who return from war and face mental health crisis (cf. Shay 1994, 2002; Brock and Lettini 2012). In this volume, Susan Derwin discusses this concept in the context of the literature of U.S. veterans, who face moral distress and alienation when re-entering civilian life. U.S. military veterans have exceptionally high rates of suicide and are "disproportionately homeless, unemployed, poor, divorced, and imprisoned" (Brock and Lettini 2012, 53). In trying to diagnose what ails military veterans, the concept of moral injury has gained popularity. Killing in war has always been considered exceptional and different from other forms of killing. While all of the world religions prohibit killing and murder, all of the sacred scriptures and legal codes provide exclusions and exceptions that range from war and self-defense to honor killings and executions, not to mention sacrifices (cf. Meagher 2014, 42). Approved killings cause no guilt. Warriors, executioners, and butchers traditionally bear no moral taint. Their violence is carefully circumscribed and ritually contained. As Paul Ricoeur pointed out in *Symbolism of Evil*:

> [T]he maleficent power of which the murderer is the bearer is not a taint that exists absolutely without reference to a field of human presence, to words that express defilement. A man is defiled in the sight of certain men, in the language of certain men. Only he is defiled who is regarded as defiled; a law is required to say it; the interdict is itself a defining utterance. [...] This "education" of the feeling of impurity by the language which defines and legislates is of capital importance. (Ricoeur 1968, 36)

U.S. soldiers returning from combat zones suffer moral injury, as Susan Derwin argues in this volume, because they encounter communities that are unwilling or incapable of accepting their wartime experiences. Moral injury grows out of emotional and cognitive dissonance that takes root in the crevices of geographical distance and political alienation. While in combat, soldiers accept the accidental shooting of children, the errant missile strike on a hospital, the destruction of a speeding car as reasonable and inevitable. But back in the civilian world, they face disbelief and questions of legitimacy, which undermines codes of conduct that shield against empathy.

What is considered approved killing is subject to community norms and sanctions. For instance, members of the South African security forces did not feel guilty for brutalizing and killing enemies of the apartheid state until the political regime collapsed, at which point they were forced to confront victim testimony before Truth and Reconciliation Commissions (cf. Gobodo-Madikizela 2003). Similarly, CIA interrogators have not expressed feeling guilty for using "enhanced interrogation methods" despite the cruelty of their actions. The U.S. government has not (yet) legally declared and sanctioned these practices as torture. In sum, guilt feelings are unreliable indicators of morality, because they are conditioned by community validation and codes of professional conduct that authorize the infliction of suffering on others.

Contrition and Remorse

Since guilt feelings are capricious, I suggest distinguishing them from feelings of contrition and remorse, which are morally loaded emotions. Although sometimes described as spontaneous and overpowering, contrition emerges in the course of ethical deliberation and communal negotiation. Contrition and remorse are cultivated emotions that play a central role in the process of repentance in which people aim for personal change and communal transformation. In contrition, guilt becomes productive as it is directed toward conversion and fundamental change.

Martin Luther, for instance, described contrition in the Augsburg Confession as a thunderbolt of God that induces terror and despair, a hammer that breaks rock into pieces, and an agony that annihilates the self (cf. Melton 1991, 58). Contrition is painful because it causes self-loathing and self-hatred that increase with the gravity of harm and degree of complicity in evil. It is one thing to feel contrition for the mistreatment of a colleague or a minor theft, and quite another to come to the recognition that mass killings and torture constitute "a mistake" (Kellenbach 2013). While Martin Luther was right to compare contrition to a thunderbolt of excruciating destructive force, his metaphor is misleading to the extent that it suggests miraculous insight and instantaneous comprehension of culpable wrongdoing. Contrition often, and regrettably, does not strike wrongdoers like a thunderbolt from heaven. Rather, contrition must be cultivated in rituals that repudiate and expose the wrongfulness of past action.

The Christian and Jewish traditions have elaborated complex and distinct theories and practices of repentance to facilitate recovery from culpable wrongdoing. According to the medieval Jewish philosopher Maimonides (1135–1204), intellectual and moral recognition of one's sins (*hakarát ha-chét'*), as well as remorse and heartfelt self-loathing (*charatá*) are obligations mandated by *teshuvah*, (repentance). Contrition is as much an intellectual as a moral and emotional recognition of wrongdoing. The Christian church speaks of *contritio cordis* (heartfelt contrition), similarly mandating an integrated intellectual and emotional recognition of wrongdoing. But how does a penitent prove that he or she feels genuine remorse, which literally means, etymologically, "bites of conscience," a crushing experience that undermines a person's self-confidence and self-worth? Long-standing polemics among the religious traditions and within religious communities have debated the impossibility of proving the sincerity of contrition. Similar suspicions attend the second mandate, namely the truthful confession (*confessio oris* or *vidúi*) in the Jewish tradition. What if contrition is fake and confession is false?

In the following, I want to analyze Joshua Oppenheimer's docudrama *The Act of Killing* (2012) as a participatory ritual event that transformed inchoate guilt into contrition by exposing the absurdity of reigning codes of legitimacy. This film has been both criticized and celebrated for showcasing a parable of repentance, which duped Indonesian death squad leaders, most prominently Anwar Congo, into a psychodrama of contrition and remorse. The film became an international spectacle that involved an interreligious production team: Joshua Oppenheimer is an American Jew, his Indonesian film crew and protagonists are predominantly Muslim, and the film's co-producer Werner Herzog brings a German Christian sensitivity to the topic.

Oppenheimer is the son of German Jews who fled Nazi Germany. In Indonesia, he found a country where the agents of mass killing remained in power and got to shape historical memory (cf. Oppenheimer 2015). Oppenheimer first interviewed surviving family members of the mass killings, which targeted alleged communists and ethnic Chinese during 1965–1966. He found them too fearful of speak on camera. They sent him to the perpetrators, who they knew would feel no such constraints. In fact, the perpetrators boasted about their deeds in full awareness that their side had won and they were owed gratitude for conducting the purge. In other words, 50 years after the events, the victims lived in fear and ignominy, under surveillance, in silence and abjection, while the perpetrators enjoyed the spoils.

For the film, Oppenheimer recruited a group of paramilitary killers and strongmen under the ruse of making a musical that celebrates their glorious past. The men are prideful braggarts who willingly recreated scenes of torture, replaying the burning and pillage of an entire village, and reenacted murder for the camera. One man, in particular, became the focus of the film. Anwar Congo is introduced as a local gangster who was recruited by the Pancasila Youth movement, a fascist, nationalist, and fiercely anti-communist political organization. This movement has been allied with those in governmental power through the reigns of Sukarno (1947–1967) and Suharto (1967–1998). The film does not provide much historical context and leaves the politics of the killings, their administrative structure, and military organization in the background. The film makes clear that Congo received orders from superiors who transferred suspects to him for torture and murder. He claims to have killed around a thousand people, although there is no proof given in the film. Congo begins the film as a strutting gangster-hero who is proud of his skills and his discovery of the most efficient killing technique: a steel cable that strangles people without leaving a bloody mess. There is no evidence of remorse or contrition at the beginning of the film.

But there are two scenes, which have become the center of discussion, in which Congo seems to develop recognition of culpable wrongdoing and feelings of contrition. In the first scene, he re-enacts the role of a victim during an interrogation and is himself strangled with a steel cable. He seems to feel the pain and asks to stop the filming. He walks off to the side, his face twisted in tears. When he speaks of feeling pain and terror, Oppenheimer gently rebukes him, pointing out that his victims knew they were actually going to be murdered. The physical pain of the steel cable against his throat seems to initiate cognitive insight, and he appears to empathize with his victims' fear—possibly for the first time. His tears, his inability to continue, and his verbal exchange with Oppenheimer signal a sort of moral and intellectual reckoning with the violence of his acts. In the second scene, which is placed at the end of the film, Congo returns to the rooftop that was the site of many of his murders. At the beginning of the film, Congo recounts these murders laughingly and dances the cha-cha-cha. But at the end of the film, when he revisits the site at night, he appears haunted and overcome by remorse; he looks visibly aged and dry-heaves helplessly. As he retches wordlessly for what seems like a long time, his pride has vanished and his memories turn into dis-ease.

These two scenes have become the focus of debate and criticism. For instance, Daniella Mina Dadras comments in a blog:

> Read in this light, Anwar's moments of remorse and authentic humanity–such as his "breakdown" while performing the role of victim and his subsequent epiphany of empathy for the people he killed—start to look like very lucky coincidences [...]. Using Oppenheimer's belief in the healing power of performance, Anwar creates a masterpiece of guilt and repentance that his director buys hook, line and sinker [. . .]. Oppenheimer is both Anwar's author (or editor, at least) and his willing dupe, the ideal audience for his continued self-evasion. (Dadras 2014)

The suspicion of hypocrisy is echoed by other reviewers, such as Robert Cribb, a professor of Asian history and politics, who considers the film a "manipulation," "misleading," and "staged," especially the "rooftop scene of the murders, [where] Congo seems to experience remorse. [. . .] The incident seems staged" (Cribb 2013).

Does it matter if these scenes (and frankly the entire film) are staged as a parable of repentance? Questions over the sincerity of contrition have been a matter of fierce theological debate for centuries. It is the classic theological flashpoint between Roman Catholicism and Protestantism: Martin Luther rejected contrition as a precondition for the grace of forgiveness, noting that it was impossible to ascertain whether someone felt genuine contrition or was merely faking it. Is Congo really experiencing such a life-shattering emotional insight, or is he merely play-acting? This is the conundrum viewers face in these scenes in *The Act of Killing*. For the "Protestant," the likelihood of hypocrisy disqualifies the performance; contrition must be truthful and sincere, or it must be rejected as phony, as Maria-Sibylla Lotter maintains in this volume. However, in the Roman Catholic and Jewish traditions, contrition and confession are performed ritually and externally, and their emotional sincerity is not mandatory. In ritual, performance counts. As Edith Wyshogrod points out, "for Maimonides, praxes, bodily actions, can constitute acts of contrition. On specific ritual occasions, repentance can be fulfilled through performance: [. . .] Practice gives proof of obedience. Doing before hearing, as Levinas often points out" (Wyshogrod 2006, 161). The alternative between sincerity and performativity is a false choice, as Adam Seligman et al. (2008) show in their groundbreaking essay *Ritual and Its Consequences: An Essay On the Limits of Sincerity*. They point to rituals of politeness that require saying "please" and "thank you" at the dinner table, thereby creating an "as if" reality of gratitude and harmony despite the underlying reality of emotional tension and disorder. Saying one is sorry, or performing remorse, creates an "as if" reality, which shifts external relationships and affects internal and emotional conditions.

Congo's life has changed. Having played a contrite character on camera, he is pushed into a new role in public that is radically different from his previous character as a hero deserving of adulation. Whether he faked his contrition or not, his social standing has changed dramatically. Calculated speech acts are communicative events that are as capable of destabilizing worldviews and power arrangements as are pretend performances. In contrition, faked or sincere, guilt assumes a language. The film creates moral and cognitive dissonance in audiences, both national and international, by forcing viewers to observe scenes of revolting cruelty and appalling behavior that shatter norms of decency and morality. There is no obvious resolution, and this very lack of closure in itself bypasses the question of sincerity and accounts for the film's ritual power.

The Sincerity of Contrition

Not only has Congo's sincerity been attacked for hypocrisy, but the entire project and its director have been criticized for their ostensibly insincere intentions. In questioning the reasons and motivations of Joshua Oppenheimer, who initiated and arranged this film project, as well as the penitential performance of Anwar Congo, critics miss the ritual dimension of this film to repudiate and expose guilt percolating through Indonesian politics. Spectators and commentators, along with the protagonists and film crew, are drawn into a ritual (re-)enactment of history and compelled to engage in emotional and intellectual responses to a bizarre spectacle of cruelty and contrition. Through the techniques of theater, the past is rewritten in the present in order to cast a different future—the very definition of repentance.

There is a lively scholarly debate, which cannot be fully reviewed here, in which criticism runs the gamut from laudation and praise to rejection and condemnation. Here, I want to focus on the potential for theater to create a communal ritual of repentance that transposes guilt into contrition and fosters personal and collective change; instead of viewers interrogating the motivations of Oppenheimer or the sincerity of Congo's personal contrition, confession, and conversion, the film should be evaluated on its ritual efficacy.

In the widening gulf between the secular and the religious, the arts may increasingly be called upon to serve the functions of religion: to provide the time and create the space for people to gather together and to engage painful subjects through stylized performances, symbolic speech acts, and visualizations. Artistic performances can create intentional communities that turn spectators into participants and break through the indifference that shrouds the unbearable and the unspeakable. As vicarious spectacles, they construct a ritual process that creatively disrupts personal and communal denial and allows for mourning and repentance. *The Act of Killing* sets out to create a ritual of mourning and repentance at the local, national, and international levels. Oppenheimer recruits protagonists into confessional reenactments that risk these individuals' future criminal prosecution as a result of their exposure.

Some commentators have accused Oppenheimer of Western-centric and personally selfish motivations for the fact that he went to Indonesia to shoot a film about a murderous past not his own. Oppenheimer gained fame and fortune and built his career as a brilliant and innovative filmmaker on the suffering of Indonesians. Upon completion, he retreated into the safety and comfort of the Western world, leaving his (anonymous) collaborators and exposed subjects to fend for themselves in Indonesia. He has not returned to the country, but he has become an international spokesperson for its history and politics, thereby possibly further silencing Indigenous voices. Moreover, he brought a particular gaze and agenda as the son of Holocaust survivors, thus projecting his Jewish sensibilities of memory and repentance onto a Muslim, Buddhist, Hindu, and Christian Indonesian canvas. He has been accused of casting a preconceived storyline that universalizes his personal faith in humanity and the existence of a moral order. Here is Dutch critic Lucien van Liere:

> What Oppenheimer expects to see and hear (his amazement about what happened in Indonesia), what he wishes former killers to express (regret, confrontation) and how he understands the link between a violent past and an adjured present expressed in the gestures, rituals and routines of his protagonists form a soteriological perspective on humanity. (van Liere 2018, 16)

Soteriology is the theological subdiscipline that spells out the path to salvation and the promise of redemption. In his role as priest-confessor, Oppenheimer's ethics of recruitment and the way he cultivates the trust of his subjects is also seen as questionable. He coaxes aging former members of the paramilitary forces into re-enacting scenes of violence and revealing their dreams and fears to him in intimate settings, albeit before running cameras. While the filmmaker's presence is never visible on the set, he is guiding the performance and answering questions in his role as presiding priest. In several scenes, Congo turns to ask "Joshua" for reassurance, advice, and opinion. Is Oppenheimer exploiting his trust, especially in scenes of somatic distress, when Congo is overcome by tears and nausea (cf. van Liere 2018, 28)? Using grotesque masks and garish costumes, the film creates fantastic dream sequences in order to visualize the ghosts that begin to haunt Congo's dreams. Oppenheimer allows Congo to watch previously shot scenes on a small TV monitor and asks him to comment on his performance as mediated on screen. This suggests that Congo is given a measure of control over his performance, but the charge remains that Congo is more object of manipulation than subject of repentance. As the film proceeds, the strongman Congo transforms into a vulnerable, elderly grandfather, who is visibly overcome by contrition and remorse—for the benefit of a viewing public that is both spectator and participant in a transformative process of historical revision.

There are also questions about the film's intended audiences and their appetite for redemptive story lines. Hollywood films are notorious for happy endings that must continuously prove the ultimate goodness and coherence of the universe. *The Act of Killing* seems to cater to audiences that harbor what Lynne Arnault called "the will to redemption." In her critique of Hollywood Holocaust films, Arnault excoriated the "essentially optimistic view of the potential of humanity to do good" and "the idea that good eventually triumphs over evil, thereby restoring meaning and purpose to our lives" (Arnault 2003, 157). Her point is especially pertinent for the scene in which Congo returns to the rooftop to retch helplessly, which seemingly satisfies the desires of "redemptive logic": as a contrite sinner, Congo confirms the moral order of the universe, in which the evildoer is punished and the righteous are vindicated. We do not live in this universe.

Arguably, there is a strongly ethical mandate to face and portray the world in all of its immoral, nihilistic absurdity. But does this obviate the need to strive for a different, moral, and liberating vision of and practice in the world?

While some respond to *The Act of Killing* as a narrative of redemption, others watch the same scenes of Congo's contrition and come to the opposite conclusion. For instance, Saira Mohamed deduces that "Congo has not taken full responsibility for his crimes, and he has not atoned for his sins. As he retches on that rooftop where he killed so many people, we see his attempt at catharsis fail: There is nothing there" (Mohamed 2015, 1199). If two viewers watch the same scene and come to opposite conclusions, something else is going on: remorse and redemption are never "objective" but rather participatory events. The meaning of a ritual depends on the disposition of its participant(s). The transformative power of ritual cannot be empirically verified or measured. Whether spectator-participants identify with Anwar Congo, on whom the narrative arc centers, or they focus on his unrepentant comrades depends on their subject position. As Lucien van Liere points out, the story of denial and refusal is in many respects the more compelling message of the film:

> The real issue in the movie, however, is not Congo, but his accomplice Zulkadry, who shows no remorse, who has learned to master his ghosts through therapy (00:48:23–00:48:25) and who advises Congo to do the same (00:48:46). Zulkadry points to the natural way of things and bounces the question of responsibility back to the audience: war crimes are defined by the winners, he argues (01:07:45) [. . .]. These complex men show no visible repentance and thus do not satisfy a democratic audience. But the very complexity of these men evokes the ghosts of Oppenheimer ("Oh my God") and they can live happily ever after with their banality of ghosts. These men do not vomit to save the director's idea of humanity. (van Liere 2018, 32)

Oppenheimer leaves their stories in the film. It is up to the viewers to fill in the gaps and to wrestle with issues of guilt and contrition. And this wrestling will take different forms, depending on location and political context. Outside and inside Indonesia, audiences arrive with their own experiences and perspectives on what constitutes heroism and villainy as they watch Oppenheimer's stylized world of bizarre cruelty. Camilla Møhring Reestorff (2015, 3) speaks of a "participatory documentary ecosystem" to frame these multiple levels of authorship, agency, and viewing experiences. She points out, for instance, that Western critics who focus on Oppenheimer's foreign status and American gaze often overlook the role of the anonymous co-director as well as of the 49 anonymous crew members. They were part of the creative team and shaped the narrative. The fact that their names have to be concealed for security reasons is part of the "documentary ecosystem." They should not be ignored and relegated to a secondary, silent, and passive status. They were as much part of the creative process as the perpetrators whose fantasies populate the scenery of the film within the film. Reestorff argues that this film should be seen as

> an assemblage of various agencies, participants and conflicting takes on the genocide. The analysis must therefore account for at least four agencies: the gangsters, the director, the anonymous Indonesian codirector and the 49 other anonymous crewmembers, the victims and the extended audience that partake in the distribution and circulation of the film. These four agencies all influence the ecosystem that constitutes the film and they work across platforms and distribution systems. (Møhring Reestorff 2015, 3)

These multiple interpretative levels create a ritual space, in which observers become participants engaging multiple subject positions.

Within this ecosystem, Congo's performance of contrition, whether fake or real, takes on political significance. For Indonesian audiences, the ritual staging of repentance poses the question of what "ought" to be the correct response and memory. The film was shown across Indonesia in private screenings and was made available free of charge for downloading and streaming within Indonesia. It was also presumably discussed across Indonesia. Adam Tyson makes an attempt at empirical analysis by interviewing dozens of audience members, conceding that "a claim about political impact based on the volume of downloads and number of screenings is unhelpful" (Tyson 2015, 188). Although critical of the film, Tyson concludes from his qualitative interviews "that there was a general consensus regarding the merits of historical debate and an appreciation that *The Act of Killing* presents Indonesians with the opportunity to rethink their traumatic past" (189).

The Truth of Confession

If the sincerity of contrition is elusive, so is the veracity of confession. Nevertheless, most religious rituals of repentance require some form of oral confession and verbal articulation. Speech externalizes internal truths, where they can be contested, proven and disproven. In confession, personal stories can be verified and challenged. This is clearly spelled out in the Jewish tradition, which has no priests to receive confessions or to pronounce absolution. Still, an oral confession is mandated before reconciliation and atonement are granted. Confession (*vidúi*) is necessary, according to the Talmud, because "unspoken matters that remain in the heart are not significant matters" (Kiddushin 49b).

It is in the public realm that truth is adjudicated. It is futile to expect that the truth, the full truth, and nothing but the truth is revealed without conflict and contestation.

Oppenheimer stages the truth in the most surreal imagery and on the basis of the boastful tales of the posse of perpetrators that he assembles around Congo. He creates a confessional event against the intentions of the men with whom he is in conversation and who create the re-enactments. Their celebratory mood turns to horror in the hands of the director. Oppenheimer exposes the underside of the terror. In turn, Oppenheimer's ethics have been challenged as deceptive and manipulative:

> The participants appear to show hospitality to Oppenheimer that he perhaps was not expecting, especially from men openly confessing past acts of murder they had participated in. In part this shows how perverse Indonesian social reality is: political rhetoric that celebrates *preman* (thugs), mass murderers unpunished, and acts of killing discussed without remorse or guilt. (Elyda 2016)

"Boasting," Oppenheimer claimed in an interview on *kunstundfilm.de*, as cited by Lucien van Liere, "is a means of hiding," by which a person is "desperately running away from the guilt" (van Liere 2018, 25). Boasting and confession are both speech acts, and Oppenheimer transmogrifies one into the other. It is in the surrealist amplification and bizarre symbolization that Oppenheimer mediates and facilitates the confessional confrontation with different truths about the past. Once the spoken words assume external reality on screen, their meaning transforms and the rules of what may be said change:

> This ending, in which Anwar both seems to somatically realize his actions and the humanity of his victims, has been rejected with reference to Anwar not being truthful. However, if we maintain the focus on the participatory documentary ecology it does not necessarily matter if Anwar's performance is truthful. Because the mere fact that he deems it necessary to perform somatic affects and grief is a testimony to the importance of the participatory assemblage. It has become impossible for him to continue the repetition of the rhythms of the past and he must acknowledge the existence of the competing rhythms, i.e. acknowledge the genocide as difficult heritage. Therefore, it is through competing affective rhythms that the victims become grievable. (Møhring Reestorff 2015, 15–6)

The film aims at a transgressive truth that transcends the political and therapeutic, legal and religious frameworks of confession. Its garish imagery violates the rules of truth-telling in law, religion, politics, and therapy. As an "artivist" project, the film creates a forum for truth-telling that engages in playful and experimental exploration that exceeds political objectives. The Cambodian American filmmaker Kalyanee Mam cautions that "political documentary has always existed on the border between art and actuality, and [. . .] the art of film is being compromised by impact agendas that position films politically as means to an end" (quoted in Tyson 2015, 187). Møhring Reestorff invents a new term to characterize the transformative potential of art. She describes Oppenheimer's film as an "art activist—artivist—documentary practice" (Møhring Reestorff 2015, 6). As an artivist project, the film aspires to change current conditions in Indonesia (and the world), which is arguably idealist, optimistic, and naïve. Such aspirations are easily ridiculed. But the anonymous Indonesian co-director and Indonesian crew members were willing to risk their safety and welfare for this aspiration, hoping that their activist art might contribute to changing community conventions. Their sacrifice will not come to fullest fruition until they are allowed to step out of the shadow of anonymity. This has not (yet) happened.

But the publicity and impact of the film may one day be compared to the significance of the 1978 broadcast in Germany of the NBC miniseries *Holocaust*. The nationwide broadcast of this fictional historical drama is widely credited with changing public and personal perceptions of German history (cf. Hayden and Markovits 1980). *The Act of Killing*, in many respects, is a much more disturbing film than the U.S.-made NBC miniseries; its crass portrayal of violence and surreal confessions serves as a powerful and disturbing counternarrative to the official government history of the massacres as having been a heroic purge and a necessary police measure against the threat of communism.

The Pain in Penance

Oppenheimer maintains that Congo "has escaped justice, but he has not escaped punishment" (Oppenheimer 2016, 10:50–11:07). Anwar Congo has not been convicted in a court of law, but his public exposure and participation in the film caused hard treatment and harm. While it is impossible to measure or verify his penitential suffering, his participation disrupted his life, unsettled his family, and impacted his social standing. Indeed, Oppenheimer's ruse unleashed a dynamic that brought a certain amount of suffering on the protagonists of his film, beyond simply Anwar Congo. For instance, the Indonesian filmmaker Shalahuddin Siregar notes that "the speakers were forced to bear the adverse consequences after the public watched the film while the director was able to waltz away, back to their comfortable life in big cities. Or maybe the supporters of *The Act of Killing* will think, why bother thinking about the impact of the film to a vicious killer like Anwar Congo?" (Siregar 2015). Congo himself spoke of having been put "in an awkward position" in an interview with Al Jazeera, according to Siregar. He faced not only national and international condemnation, but also disapproval from his former comrades, who presumably expressed contempt for the repentant traitor and contrite defector.

Anwar Congo, for his part, is said to "(insist) that he is satisfied with his portrayal in *The Act of Killing* even though he has suffered from anxiety and somatic malfunction since the film debuted at the Toronto Film Festival in 2012" (Tyson 2015, 193). He feels "to an extent betrayed," according to Adam Tyson, but "resigned to the fact that journalists and human rights campaigners from around Indonesia have his mobile number and residential address, and that calls for retributive justice are only growing louder" (193). Even if he was initially seduced into making this film, expecting glory and celebration, his acceptance of public humiliation and exposure is ennobling. Congo continued the film project after others realized the implications and withdrew their participation. His willingness to continue and to endorse the film constitute a form of penitential suffering.

The Latin word *poena* means "pain, punishment, penalty" and is at the root of the English word "pain." It is derived from the ancient Greek term *poiné* (ποινή) that meant "blood money, fine, penalty and punishment." Valerij Zisman in this volume questions the need for punishment as an unethical infliction of suffering, while Dominik Hoffman wrestles with the implications of impunity. Both doubt the function and need for suffering in the productive transformation of guilt. Their arguments are well taken, and there is great urgency to call for more reservations and misgivings before penal suffering is inflicted, usually on someone other than oneself. But human suffering is inevitable, as Buddhism teaches, and, under the right circumstances, it can become meaningful and instructive. Suffering, while unpleasant, is not always destructive. Indeed, suffering is an integral element in personal and communal transitions and transformations, and it is used deliberately in rites of passage. Noting the prevalence of violent imagery and sacrificial practices across a wide range of rituals, Margo Kitts points out in *Elements of Ritual and Violence* that "experiences of pain, violence and trauma are held to resonate more deeply in personal development and social relations than are experiences of pleasure and joy" (Kitts 2018, 21). Pain concentrates the mind and is strongly inscribed in memory, which makes it useful in rituals that mark transitions. The meaning and context of pain changes its experience dramatically. To give a simple analogy, consider the difference between the physical pain of torture and the physical pain of extreme sports. Both involve physically painful sensations, but one is voluntary while the other is involuntary. Gasping for air under conditions of torture is vastly different from struggling to breathe while climbing Mount Everest. Torture is traumatic, while athletic agonies are empowering. Penitential suffering is undertaken voluntarily and more akin to hiking the Himalayas than to torture. Religious disciplines involve the denial of pleasure, comfort, and privilege as spiritual exercises in order to increase empathy and sympathy. "Com-passion," in the sense of suffering with others, requires emotional and mental preparedness that is practiced in penitential austerities. It is in this sense that the legal scholar Antony Duff defends penance as a necessary harsh treatment for structured exercises that aim to focus a "sinner's attention on his sin and its implication" (Duff 2003, 299). At its best, the pain in penance serves to cultivate mindfulness, compassion, and respect for self and others. At its worst, the harsh treatment of penal suffering victimizes, disempowers, and dehumanizes—and the difference between the one and the other is often not clearly marked or readily apparent.

Congo's suffering, to the extent that he endured loss of status, comfort, and privilege, is not only necessary for his personal redemption and transformation; it also serves the vindication of his victims. And survivors and their communities not only desire to see their tormenters suffer; they have a right to. Any genuine "satisfaction" requires not only monetary compensation and administrative restitution, but also affective redress in the form of shared grief and affliction. In punishment, perpetrators begin to empathize with the experience and anguish of victims and survivors. The debt of guilt is paid in the coin of suffering. The measure of guilt is calculated in the degree of penal affliction. It is this "expressive function" of punishment that makes it not only ethically legitimate but even *obligatory* for the vindication of victims (cf. Feinberg 2004). As an act of communication, penal suffering is expressed symbolically and restrained ritually. Already in biblical times, the *ius tallionis* of an "eye for an eye, tooth for a tooth, life for life" was transformed into a principle to guide monetary—that is, symbolic—compensation. It aimed to reduce vengeful cycles of violence. Penitential suffering can be channeled into constructive sacrifices of time and resources that restitute victims, serve the community, and repair the social fabric. It thereby transforms the guilt of the past into genuine new beginnings that ground the openness of the future, as Hannah Arendt argued in *The Human Condition* (Arendt [1985] 1998, 247).

Conclusion

The concept of guilt, with its rich history of repentance, is a better term to safeguard the humanity of perpetrators than is the concept of trauma. While it is true, as Saira Mohamed argues in the following quote, that perpetrators and victims share suffering as members of humanity, their conditions are not comparable:

> Understanding the perpetrator as a suffering person need not undermine the goal of respecting victims and giving voice to their experiences. As elucidated in *The Act of Killing*, to recognize perpetrators as suffering trauma does not entail a reconfiguration of the perpetrators as a victim. It does not require a denial of Anwar Congo's crimes. It does not require sympathy or forgiveness. It is painful to watch Anwar Congo's failed catharsis on that rooftop, but one does not feel sorry for him and certainly does not forgive him. Recognizing trauma does, however, require recognition of the humanity of perpetrators, and it does therefore allow them, in one sense, to share the same space with victims. Both perpetrator and victim are thinking, feeling beings. (Mohamed 2015, 1157)

Perpetrators suffer from distorted visions, deceptive impulses, and self-centered perspectives. Religious traditions variously describe this condition as brokenness and alienation (sin), suffering (*dukka*), the evil impulse (*yetzer hara*), and karmic captivity in the wheel of becoming (*samsara*) propelled by greed, hatred, and ignorance. Guilt humanizes perpetrators because monsters cannot make false choices, become guilty, feel remorse, and engage in repentance. While the religious pathways differ, the release from captivity-in-guilt intensifies perpetrators' suffering (in the form of self-loathing) in contrition, the scandal of confessional exposure, and the imposition of penitential sacrifices. Given the right contexts, such suffering is humanizing and not dehumanizing. Rituals of repentance channel unspeakable feelings into stylized movements and formal speech acts. Such rituals "are not just ways of evidencing feelings that were already fully formed; they are ways of adding depth and structure to those feelings themselves" (Duff 2003, 299). With the decline of religious rituals in modern secular societies, artistic projects such as *The Act of Killing* create ritual space-times that metabolize guilt into repentance and thereby make guilt productive of personal transformation and political change.

References

Arendt, Hannah. (1958) 1998. *The Human Condition*. Chicago: University of Chicago Press.

Arnault, Lynne S. 2003. "Cruelty, Horror, and the Will to Redemption." *Hypatia* 18, no. 2: 155–88.

Brock, Rita Nakashima, and Gabriella Lettini. 2012. *Soul Repair: Recovering from Moral Injury after War*. Boston: Beacon Press.

Buber, Martin. 1957. "Guilt and Guilt Feelings." *Psychiatry* 20, no. 2 (May): 114–29.

Buber, Martin. (1958) 2008. "Schuld und Schuldgefühle." In *Schriften zu Psychologie und Psychotherapie*, edited by Judith Buber Agassi, 127–52. Martin Buber Werkausgabe (MBW) 10, edited by Paul Mendes-Flohr and Peter Schäfer. Gütersloh: Gütersloher Verlagshaus.

Cribb, Robert. 2013. "Review: An Act of Manipulation?" *Inside Indonesia*, April 20. Last modified March 3, 2019. https://www.insideindonesia.org/review-an-act-of-manipulation.

Dadras, Daniella Mina. 2014. "'The Act of Killing' and How Not to Get Conned by a Charming Madman" *Popmatters*, January 15. Last modified September 3, 2019. https://www.popmatters.com/176219-the-act-of-killing-and-how-not-to-get-conned-by-a-charming-madman3-2495711244.html.

Duff, Robin Anthony. 2003. "Penance, Punishment, and the Limits of Community." *Punishment & Society* 5, no. 3: 295–312.

Elyda, Corry. 2016. "Can We Defend *The Act of Killing* and *The Look of Silence*?: A Response to Shalahuddin Siregar." *Cinemapoetica*, March 31. Last modified September 3, 2019. https://www.cinemapoetica.com/can-defend-act-killing-look-silence/.

Feinberg, Joel. 2004. "The Expressive Function of Punishment." In *Philosophy of Law*, edited by Joel Feinberg and Jules Coleman, 761–71. Belmont, CA: Thomson, Wadsworth.

Gobodo-Madikizela, Pumla. *A Human Being Died That Night: A South African Woman Confronts the Legacy of Apartheid*. New York: Harcourt, 2003.

Hayden, Rebecca S., and Andrei S. Markovits. 1980. "'Holocaust' before and after the Event: Reactions in West Germany and Austria." *New German Critique* 19: 53–80.

Kellenbach, Katharina von. 2013. "'The Truth about the Mistake.' Lessons of a Nazi Perpetrator to His Son and the Intergenerational Transmission of Guilt." *Historical Reflections/Reflexions Historiques* 39, no. 2: 14–30.

Kitts, Margo. 2018. *Elements of Ritual Violence*. Cambridge: Cambridge University Press.

Leys, Ruth. 2009. *From Guilt to Shame: Auschwitz and After*. Princeton, NJ: Princeton University Press.

Leys, Ruth. 2010. *Trauma: A Genealogy*. Chicago: University of Chicago Press.

Lifton, Robert J. 1986. *The Nazi Doctors: Medical Killing and the Psychology of Genocide*. New York: Basic Books.

Meagher, Robert Emmet. 2014. *Killing from the Inside Out: Moral Injury and Just War*. Eugene, OR: Cascade Books.

Melton, J. G., ed. 1991. *American Religious Creeds*. Vol. 1. New York: Triumph Books.

Mohamed, Saira. 2015. "Of Monsters and Men: Perpetrator Trauma and Mass Atrocity." *Columbia Law Review* 115, no. 5: 1157–216.

Møhring Reestorff, Camilla. 2015. "Unruly Artivism and the Participatory Documentary Ecology of *The Act of Killing*." *Studies in Documentary Film* 9 no. 1: 1–18.

NIMH (National Institute of Mental Health). 2019. "Post-traumatic Stress Disorder." Bethesda, MD: NIMH. https://www.nimh.nih.gov/health/topics/post-traumatic-stress-disorder-ptsd/index.shtml.

Oppenheimer, Joshua, dir. 2012. *The Act of Killing*. http://theactofkilling.com.

Oppenheimer, Joshua. 2015. "As if the Nazis Were Still in Power: Interview with Joshua Oppenheimer about *The Look of Silence*." *Kunst und Film*, January 10. https://kunstundfilm.de/2015/10/interview-oppenheimer-silence/.

Oppenheimer, Joshua. 2016. "Interview: Making the Invisible Visible." By Roxanne Bagheshirin Lærksen, *Louisiana Channel*, April 2016. Video, 26:30. https://channel.louisiana.dk/video/joshua-oppenheimer-making-invisible-visible.

Peli, Pinchas H. 1984. *On Repentance: The Thought and Oral Discourses of Rabbi Joseph Dov Soloveitchik*. Ramsey, NJ: Paulist Press.

Ricoeur, Paul. 1968. *The Symbolism of Evil*. Boston: Beacon Press.

Sambath, Thet, and Rob Lemkin, dirs. 2009. *Enemies of the People*. Cambodia/United Kingdom: Old Street Films.
Seligman, Adam B., Robert P. Weller, Michael J. Puett, and Bennett Simon. 2008. *Ritual and Its Consequences: An Essay on the Limits of Sincerity*. New York: Oxford University Press.
Shay, Jonathan. 1994. *Achilles in Vietnam: Combat Trauma and the Undoing of Character*. New York: Simon & Schuster.
Shay, Jonathan. 2002. *Odysseus in America: Combat Trauma and the Trials of Homecoming*. New York: Scribner.
Siregar, Shalahuddin. 2015. "Ethics behind *The Look of Silence*." *Cinema Poetica*, November 29. https://cinemapoetica.com/ethics-behind-the-look-of-silence/.
Tyson, Adam. 2015. "Genocide Documentary as Intervention." *Journal of Genocide Research* 17, no. 2: 177–99.
van Liere, Lucien. 2018. "The Banality of Ghosts: Searching for Humanity with Joshua Oppenheimer in *The Act of Killing*." *Journal for Religion, Film and Media* 4, no. 1: 15–34.
Wyshogrod, Edith. 2006. "Repentance and Forgiveness: The Undoing of Time." *International Journal for Philosophy of Religion* 60, no. 1/3: 157–68.

10

Performing Guilt

How the Theater of the 1960s Challenged German Memory Culture

Saskia Fischer

Peter Weiss's play *The Investigation* (1965) marked a caesura in the theater and memory discourse of the 1960s. For the first time, the eyewitnesses and victims of the Holocaust themselves had the chance to be heard on stage. Although an increasing amount of Holocaust testimonial literature was published in Germany between 1945 and 1949, its response remained fairly limited. Not only did the authors frequently find themselves confronted with ignorance and a defensive attitude by many Germans; in the years after 1949 in the Federal Republic of Germany the so-called *Schlussstrichdebatte*—the desire to literally "draw a line" under the Nazi past—also suppressed an intensive engagement with these texts (see Feuchert 2012). It was not until *The Investigation* by Jewish playwright Peter Weiss that the testimonies of Holocaust survivors strongly demanded unprecedented recognition in the public eye.

The perspective of *The Investigation*, which brought the Frankfurt Auschwitz Trials (1963–1965) to the stage and gave vast room to the cruelest experiences of suffering and murder, deliberately challenged the way German society dealt with its past. While going far beyond just documenting the legal convictions of the Nazi perpetrators, Weiss's piece also fiercely denied the self-perception among many Germans that they, too, were allegedly victims of "the" Nazis and that they themselves had neither supported nor actively participated in the National Socialist regime. In contrast to such a falsified narrative of German history, *The Investigation* did not depict the concentration camps (and thus the organization of the systematic extermination of humans) in isolation from society and thereby delegate guilt to a Nazi elite alone. Rather, the play strove to unmask as self-delusion the supposed innocence and the belief of many Germans that they were not responsible at all. Hence the play not only focused on the complicity of German society during National Socialism, but also, as Weiss himself put it, on a "description of the present" (Schumacher 1986, 83)—yet with particular emphasis on the capitalist system in the Federal Republic of Germany— in which the guilt of the past seemed to continue. Given the broad rejection of an alleged

Saskia Fischer, *Performing Guilt* In: *Guilt*. Edited by: Katharina von Kellenbach and Matthias Buschmeier, Oxford University Press. © Oxford University Press 2022. DOI: 10.1093/oso/9780197557433.003.0011 collective guilt in Germany's politics, general public, and official memory culture (see Buschmeier's contribution in this volume), the controversies that *The Investigation* caused were hardly a surprise. On the contrary, the play and its performance were indeed *intended* to "disturb" post-war society and to cause vast "indignation" (Weiss [1965] 1971, 101). Weiss's play hit the very core of a society in which former Nazi perpetrators and NSDAP members continued to hold high offices, even in politics, and which had, up until then, successfully concealed its guilt and made itself comfortable in its supposed victim role.

In this chapter, I argue from a binary point of view that highlights the discrepancy between "the Germans" and "the Jewish Holocaust survivors." This is not to say that Jews in the Federal Republic of Germany were not part of German society after 1945. Neither do I aim to mislabel them *or* the members of any other groups persecuted during National Socialism (such as Sinti, Roma, etc.) as not being part of German post-war society. On the contrary, *The Investigation* brings attention to the ways in which Germany after the war continued to marginalize Jewish victims, who in the 1960s were still struggling for public acceptance. Nevertheless, the problem of guilt at stake here, as well as this striving among Holocaust survivors for public recognition, becomes apparent when we consider the fact that the majority of Germans—who were not Jewish—attempted to conceal their own complicity and responsibility, sometimes even by silencing the voices of Holocaust survivors in public. Therefore, this general juxtaposition between "the Germans" and "the Holocaust survivors" will be maintained in this chapter in order to highlight the distinctions that Weiss's play drew (and exposed) between the victims on the one hand and former perpetrators, bystanders, and accomplices on the other.

Weiss's *The Investigation* serves as an impressive example of 1960s German theater's attempt in particular, and art's attempt in general, to rebel against a complacent, distorted, and one-sided self-image of German history by rendering transparent the guilt-defense mechanisms of the society it portrays. And yet Weiss's understanding of theater is of astonishing contemporary relevance. His approach, which revealed the Germans' guilt entanglement and called on them to face up to their responsibility, has recently led to a renewed interest in the power and potential of theater and film to help societies come to terms with genocide and war crimes. The Swiss theater director Milo Rau, for instance, whose highly acclaimed play and documentary *Congo Tribunal* (2015 and 2017) symbolically brought to trial the atrocities committed during the civil war in Eastern Congo—a conflict that has been going on in the Kivu and Ituri areas for over 20 years without legal prosecution hitherto—explicitly referred to the aesthetic concept of *The Investigation* (Thurn, forthcoming). Therefore, my observations on the cultural productivity of publicly presented and negotiated guilt, as evident through Weiss's play and its reception, have implications that extend to more recent forms of political art and performances as well.

The productive power of *The Investigation* in coming to terms with guilt is threefold, as I will outline in this chapter: (1) On the one hand, it owes in part to the vehemence with which the play's premiere in 1965 attempted to create a critical public space for addressing the "question of German guilt." (2) On the other hand, it owes to the way in which the play depicted collective guilt, responsibility, and complicity; essentially, both the text itself *and* its staging can be seen as an attempt to transform the mentality and morals of German society, aiming toward a (self-)critical engagement with the past. (3) In its staging, *The Investigation* was not a mere re-enactment. In fact, it demonstrated how allegedly "past" guilt stretches into the present, as well as how this guilt is retrospectively denied, concealed, or reinterpreted. Thus, Weiss's piece illustrated a processual understanding of guilt, showing guilt as a temporal phenomenon that has lasting and even devastating consequences for the present so long as atrocities remain unresolved—and by presenting guilt as a highly contested topic, it highlighted the fact that guilt itself is shaped and illustrated in and through narratives.

Performing Guilt: Disturbing and Provoking Post-War Society

Based on the testimonies of the Frankfurt Auschwitz Trials, and thus on the grounds of authenticated oral documents, *The Investigation* vigorously raised the question of German guilt; that is, of German society's complicity in and responsibility for the Holocaust. With this work, Weiss sought to expose a "veiling" and "concealment of the past," and thereby to force the "sleeping" population of a divided Germany in the 1960s (Weiss [1965] 1971, 92, my translation) to engage in a relentless and self-critical confrontation with the atrocities committed under National Socialism. Even rhetorically, in his quotes, the enlightening and educational impetus of Weiss's play is already obvious. The aim of *The Investigation* was nothing less than to exert a moral and political influence "on the consciousness of as many Germans as possible," as the Suhrkamp Verlag, Weiss's publisher, summarized the play's intention (Weiß 2000, 208, my translation). Already, these prior public statements by Weiss and his publishers, and even more so the premiere of *The Investigation* in 1965, provoked controversial and heated debates, which were carried out in the feuilleton, in letters to the editors of daily newspapers, on radio and TV, and in numerous podium discussions (Weiß 1998). Nevertheless, the question remains as to how the performance of one single piece of art could have led to such sweeping reactions.

Convinced of the play's central importance for a shift in German memory discourse, Weiss and his publisher agreed on an "open premiere." On October 19, 1965, as a major media event unparalleled to this day, Weiss's play premiered simultaneously on 15 stages across East and West Germany. Among the directors and actors were leading figures in the political theater of the former Weimar Republic, such as Erwin Piscator, Paul Dessau, and Helene Weigel. Roughly a week after the premiere, almost all of the West and East German broadcasters produced radio adaptations, which, like the television adaptation in March 1966, reached large audiences (Wannemann 2004, 215). Stagings in London and Stockholm by two of the most famous directors of the time, Peter Brook and Ingmar Bergmann, respectively, followed. The widespread presence of the play in the cultural sphere and its interpretation by outstanding and well-known artists ultimately realized Peter Weiss's and Suhrkamp's far-reaching educational ambitions for society.

Moreover, staging the play throughout Germany was meant, simultaneously, as a symbolic act in itself: "theater," as a somewhat universal artistic institution, assigned itself the role of educating the German public by prompting that public toward a self-critical examination of the Holocaust—and theater at that time tried to accomplish this by using all of its influence, and thereby to prove its transformative cultural power. Indeed, one must keep in mind that, in the 1960s, the theater in Germany was still one of the leading mass media; it had not yet been replaced by television to the extent that it increasingly did shortly thereafter (in the following decade). In this sense, the play's vast reach and the fact that the theater became a highly influential place for debating social change is perhaps even more impressive but also understandable.

Apart from the support by leading actors, directors, and intellectuals, the provocative rethinking of collective guilt that was ignited as Weiss's play and its performances spread across the country would have been inconceivable without the beginning of the legal reappraisal of the National Socialist atrocities; legal processes such as the Eichmann Trial in Jerusalem (1961) and the Frankfurt Auschwitz Trials (1963–1965) were both widely reported in the media (Weiß 2000; Wenzel 2009). There is widespread agreement in historical research about the significance that the Auschwitz Trials, in particular, had on the perception of the Holocaust in Germany. The historian Norbert Frei, for example, emphasizes that the trials were "undoubtedly the historically and politically most significant attempt" to "deal with the criminal events in the largest of the National Socialist concentration and extermination camps by means of criminal law" (Frei 1996, 123, my translation). However, it was also Frei who admitted that the structure of the judicial process decisively limited any broader, collective reflection on the systematic genocide (Frei 1992, 105). To put it more pointedly, the inevitable reliance on the individualistic focus of criminal law prevented the public from perceiving the crimes of National Socialism in a society-wide dimension. Hence, Mirjam Wenzel highlights in her study that it was mainly the reflections, interpretations, and commentaries on the court proceedings by literature and philosophy scholars that led to a "paradigm shift" regarding awareness about the Holocaust—an awareness, she concludes, that *was enhanced by* the deliberate invitation of authors and intellectuals, such as Hannah Arendt, to write about the trial (Wenzel 2009, 9–10).

In other words, by the very first time that Weiss's play reached the German public, there was already a certain climate of willingness to promote social and critical discourse about the Holocaust— at least in some parts of society. All the same, in this gradually transforming discourse about the Holocaust, *The Investigation*, both along with and as a part of theater and literature more broadly, proved to be a special medium for shaping the discussions on guilt in a decisive, differentiated, and influential way.

The "Jurisdiction" of the Stage in the Face of the Holocaust

Historical studies have shown that the accusation of collective guilt, such as that found on posters distributed by the Allies with the words "Diese Schandtaten: Eure Schuld!" (These atrocities: your guilt!), was not at all based on an official statement by the military government (Dutt 2010, 8). The Allies did not, in fact, seek to criminalize all Germans; they did not wish to condemn them in terms of a legal understanding of guilt, nor to impose *collective* punishment. Rather, the Allies, in the name of a so-called re-education, wanted to appeal to Germans' *sense of guilt* and thus Germans' conscience by forcing them to acknowledge their complicity in (as bystanders, denouncers, profiteers, etc.) and their support of the Nazi regime. Dutt quotes C.G. Jung who explained in a widely noted press interview in 1945: "The question of collective guilt [...] is a fact of life for the psychologist, and it will be one of the most important tasks of therapy to get the Germans to acknowledge this guilt" (8, my translation). The Allies believed, as Jung also had in mind, that the growth of a sense of guilt among the Germans would help them rediscover and redevelop their humanity. This is why the Allies forced the Germans to visit the concentration camps, to look at the piles of corpses and thus (literally) face up to the crimes committed there. This conviction can be related to more recent psychological studies, which insist that feelings of guilt may be able to foster social relationships by leading to an idea of collective responsibility (see the chapter by Lisa Spanierman in this volume; see further Baumeister, Stillwell, and Heatherton 1994).

Nevertheless, the failure of such beliefs belongs to the history of the so-called *Vergangenheitsbewältigung* (coming to terms with the past). The Allies' intention to hold all Germans morally responsible ultimately led to the opposite and a harsh rejection of guilt in large parts of German society. In German politics and across the country's intellectual elite since the end of the Second World War, the rhetorical strategy of limiting collective guilt to a *legal* understanding of guilt served mainly as a means of refuting the alleged accusation that all Germans were criminals, thereby reducing the accusation to an absurdity (see Kämper 2010; Frei 1996). In this way, the Germans (especially in the Federal Republic of Germany) succeeded in distancing themselves from the perpetrators and asserting for themselves, as Matthias Buschmeier stresses in this volume, a "neutral position under the law," which they equated with not being responsible at all, and even, in fact, contributed to their tendency to see themselves as *victims* of "the" Nazis. In turn, this strategy—namely, the impulse to reject a (false equivalency of) near-universal *criminal* guilt—prevented the general public from developing *any* meaningful sense of guilt and therefore becoming intensively involved with the collective dimension of guilt that exceeded the realm of law.

In the face of these guilt-rejecting maneuvers, it becomes clear what an enormous provocation *The Investigation* posed to post-war society, particularly as the stage, as Weiss phrased it, "took the form of a tribunal" (Weiss [1965] 1971, 100, my translation) and, metaphorically speaking, the German audience was placed in the dock. But what exactly does it mean for the play's aesthetic concept when Weiss views the theater as a "tribunal," and therefore—one could conclude—as a means of bringing about some sort of justice on stage?

It was none other than Fritz Bauer, the State Attorney General responsible for initiating the Frankfurt Auschwitz Trials, who addressed this in a panel discussion with Peter Weiss in 1965:

> There should be a division of labor, dear Peter Weiss, between the Auschwitz judge and the Auschwitz poet. The Auschwitz judge punishes; the Auschwitz poet should educate. This division of labor is necessary, and I, as a lawyer, tell you that we lawyers in Frankfurt have cried out in fright, cried out with all our soul for the poet who gives voice to that which the trial is unable to pronounce. (Weiß 1998, 75, my translation)

According to Bauer, the poet should step in for the judge, so to speak, where the court—due to the criminal procedure code—loses its power. In other words, the poet should act where the law has no authority. In saying this, Bauer effectively stressed two points at once regarding guilt: he indicated that there are dimensions of guilt that cannot be grasped with the instrument of law, and he simultaneously highlighted the significant contribution that the arts could *and should* bring into the engagement with it—namely, education. If one remembers the high aspirations for education and enlightenment that Weiss and his contemporaries aimed to effect in their staging of *The Investigation*, it seems as if Bauer clearly summed up Weiss's original intent. Yet the question remains as to how education, or as Weiss formulated it, the awakening of the "sleeping" German population (Weiss [1965] 1971, 92), ought to be realized.

Bauer's demands of the poet can arguably be traced all the way back to Friedrich Schiller, who most influentially coined the concept of the "jurisdiction of the stage." He proclaimed the special significance of literature as an independent but highly relevant medium of social and individual self-reflection and moral education. In his famous lecture *Was kann eine gute stehende Schaubühne eigentlich wirken* (What a Good Standing Stage Can Actually Do), Schiller stressed that the "jurisdiction of the stage" begins "where the territory of the secular courts end" (Schiller [1784] 1992, 187, my translation). One sees immediately how strongly Fritz Bauer's phrases ultimately referred to these high humanistic ideals of art and education from the late eighteenth century, and therefore how Bauer implicitly related Peter Weiss's work to this line of tradition. Schiller further explained, emphatically: "Bold criminals, who have long since moldered into dust, are now summoned by the almighty call of poetry and repeat a shameful life to the horrifying instruction of posterity" (190). That is, their lives and actions, as can be concluded from Schiller's words, form a kind of "didactic play" for the audience that exposes the "bold criminals" as morally reprehensible. By declaring that the stage is not only akin to a "trial," but also that it constitutes a "moral institution," Schiller viewed the theater as a supplement both to moral-philosophical categories that are all-too-narrow to capture the complexity of guilt in real life and to a legal understanding of guilt.

Not only did Schiller and his concept of theater focus on a notion of guilt that concentrates on the individual ("bold criminals"), but he also emphasized the purifying effect that the theatrical staging of guilt—and thus an emotional experience of and engagement with it—would cause in the audience:

> Healing shivers will seize mankind, and in the silence everyone will praise his good conscience, when *Lady Makbeth* [*sic*], a terrible sleepwalker, washes her hands and summons all the perfumes of Arabia to exterminate the foul smell of murder (191).

These words highlight theater's capacity for presenting and condemning guilt in a deeply impactful way, while they also—and perhaps more importantly—point to the potential for theatrical stagings of guilt stories to offer moral instruction through the intense, emotional, aesthetic experience ("healing shivers") they offer to their audiences. In this sense, Schiller's understanding of the "jurisdiction of the stage" was closely related to Gotthold Ephraim Lessing's *Mitleidspoetik* (poetics of compassion), in which he reinterpreted the Aristotelian theory of tragedy and declared the theater to be *the* place for the development and fostering of humanity.

Lessing's and Schiller's concepts of the impact of theater ultimately followed the idea that through emotional engagement with culprits' stories, the audience will develop what in the so-called history of "Western" thought has been described as "conscience"—"everyone will praise his good conscience," as Schiller put it (190). Their notion of theater's effectiveness was deeply rooted in the Christian tradition, which fostered the assumption that, when developing a sense of guilt, man also recognizes himself as a morally responsible being with a free will. Therefore, in theological discussions and in Christian ethics, feelings of guilt are directly related to virtue development and responsible behavior (Herdt 2016.). Nonetheless, such a theater focusing on an individual and his deeds, and the attempt to educate the audience for the better through a cathartic effect, had, according to Weiss, become impossible when dealing with the Holocaust on stage. That is why, in the notebook he kept while attending the Auschwitz Trials as an observer, and in which he already outlined his initial thoughts on *The Investigation*, he jotted: "This has nothing to do with Lessing" (Weiss [1965] 1982, 328).

While, on the one hand, Weiss referred directly to this central literary trope of the "jurisdiction of the stage," he rejected, on the other, the established notion of how guilt ought to be performed that had been associated with theater since Schiller and Lessing. The Holocaust—which, by exploiting all means developed in modernity, aimed to oppress and murder people systematically in the cruelest possible way, and which entangled not just all the henchmen involved in it but, even more so, also the inmates of the camps in the bureaucratically organized processes of mass murder—exceeds the leading conventions of representation. Genocide of this sort surpasses an idea of theater that focuses on "tragic heroes," on an individual and causal concept of guilt, and it defies theater's aims of purifying the audience through emotional experiences like "healing shivers." Weiss thus brought the failure of the humanistic ideals of art and education—a failure so cruelly exposed by National Socialism—into crystal-clarity when a witness in his play reported that an SS man "was engaged in a discussion" with concentration camp inmates "on the humanism of Goethe" (Weiss [1965] 2015, 108), only for the same SS man, shortly thereafter, to shoot a woman and her two children without hesitation or even the slightest emotion. It is a central ambivalence surrounding literature in general after 1945 that unfolded in Weiss's play: on the one hand, the effort to maintain the "jurisdiction of the stage," and thus the continuity of cultural tradition in the face of Auschwitz—and on the other, the simultaneous knowledge of literature's possible inadequacy and futility in the face of the task.

Yet Weiss did not fundamentally dissociate his play completely from Lessing and Schiller's theater concepts. In his play, too, the victims and survivors of the Holocaust were shown with deep compassion. Thus, Peter Weiss did not fully abandon the focus on a main character when he dedicated an entire canto to "The End of Lili Tofler," who in his play did not betray her friend but instead sacrificed herself for him, thereby trying to uphold moral integrity and loyalty in the midst of the reprehensibility and cruelty of this extermination camp. Moreover, with regard to the witnesses' intense and illustrative narratives about the conditions in Auschwitz, it can be assumed that Weiss wanted to stimulate a moral stance of empathy toward Holocaust survivors in the audience as well, and that he even relied on the idea that this emotional involvement, while watching the performance, would provoke a feeling of guilt and responsibility in the spectators.

Nonetheless, for Weiss, a theater that focuses *merely* on provoking an emotional "catharsis" ran the risk that audiences would leave the theater "purified" only in the sense of an emotional relief, but without further reflection on what they heard and saw. In Weiss's view, such a theater was more likely to exonerate from guilt, rashly aiming to overcome the unpleasant feelings that guilt happens to carry, than to encourage continued engagement with it. This is why Weiss—in close relation to Brecht's epic theater—divided his play into single cantos; why he emphasized an anti-illusionistic mode of representation by refraining from any action on stage, apart from the testimonies of the witnesses and defendants; and why he focused on the Auschwitz Trial as the process of investigating and negotiating guilt. After all, the "question of German guilt," as one could say in reference to Karl Jaspers, is too complex to be dealt with conclusively, much less can it be fully resolved within the short span of time allotted to a theater performance. Hence, the stage is no longer in a position to speak justice. What remains for theater is the process, the "investigation" of the guilt itself, as an open task. The title of Weiss's play can be read as an aesthetic statement, as well as an appeal to the audience to ask the question of guilt and to raise awareness about it—continuously.

One must stress, however, that Schiller and Lessing were not concerned solely with theater's emotional cleansing effect; they, too, always strove for moral purification and, thus, a reflective process and engagement with what was performed on stage. Yet it was the dangers outlined above—the dangers associated with a theater that focuses strongly on emotion—that Weiss wanted to avoid. The aesthetic radicalization of Weiss's piece, by vehemently discarding tragedy, even surpassing Brecht's epic theater, and already heralding the so-called "post-dramatic theater," was the result of a conception of theater that sought to encourage a wide-ranging social process of reflecting on guilt. It broke with all established theater traditions that threatened to prevent this approach.

Although, for Weiss, the problem of guilt in the face of the Holocaust could not be conclusively clarified and was difficult to perform, the aesthetic concept of his play underlined that it is literature—if anything at all—that can most adequately meet this challenge. A theater that strictly relied on narration remained a place that could engage with guilt productively. For Weiss, as for Schiller, guilt seemed to need a special place on stage, since art and literature in particular are capable of illuminating those areas—including the complexity of guilt dynamics—that are central to human life, which are not covered fully by rigid legal customs, nor by fixed moral-philosophical notions. It is in art above all that this reflection of moral questions and responsibilities takes place in secular societies, as Maria-Sibylla Lotter has pointed out (Lotter 2012, 320; see further Buschmeier 2018). Relatedly, the literary scholar Fritz Breithaupt has emphasized that unclear, elusive guilt practically calls for narration in order to make its ambiguity culturally manageable (Breithaupt 2012, 8). However, while Breithaupt sees in this the reason for and the realization of a "culture of excuse," whereby the lack of clarity is quite often seized to free oneself from guilt and attribute it to others with the help of narratives, Lotter stresses—more in line with Schiller—"the moral dimension of perception" and thus a self-critical approach to guilt that is expressed in guilt stories (Lotter 2012, 320, my translation).

Nonetheless, literature can ultimately serve either intention. As a discursive, but also emotionally influential, medium, it can intensively express both the rational investigation of what we call guilt and the interpersonally elusive dimensions of guilt and feelings of guilt in relationships, which often exceed or undermine the basis of moral principles. Moreover, literature is capable of self-reflective examination as part of rhetorical and performative strategies of guilt avoidance, as well as a means of strengthening a sincere and critical perception of guilt. It is because of literature's nature, being such a complex art, that it need not dissolve the ambiguities inherent to dealing with guilt, but rather can depict them as such. Thus, in *The Investigation*, the trial situation, and thereby the focus on the witnesses' and defendants' narrations that solely constituted the action on stage, demonstrated in a very literal way the communicative effort of the remembrance process and the discussion of guilt. Only in this complex and (self-)reflective way might the theater, for Weiss, still be able to "educate"—as Fritz Bauer demanded of the "Auschwitz poet."

Weiss's Processual Understanding of Guilt: Continuity, Comparability, Repeatability

With its aesthetic concept, *The Investigation* went far beyond a mere portrayal of the Auschwitz Trials. Weiss held more than just the individual perpetrators accountable; through the testimonies, the defendants' evasions, the strategies of the Defence Counsel, and the far-reaching historical reflections that the Prosecutor and some of the witnesses made, the play attempted to focus not just on the particular but also on the broader social, moral, and economic conditions that had enabled the Holocaust in the first place. Ultimately, the play revealed the genocide's potential to repeat on account of persistent mental, personal, and systemic similarities in the structure of society—especially in a capitalist system like the Federal Republic of Germany. As I will outline further in the following, it was precisely this "system-mediating character" (Söllner 1988, 172, my translation) of *The Investigation* that followed a processual and dynamic concept of guilt and demonstrated the culturally productive and transformative dimension of Weiss's play.

In Weiss' play, Witness 3 and the Prosecutor were the two characters who most rigorously sought to eliminate the supposed distinction between past and present. Their statements, more than anyone's, underscored the comparability of the (then) present-day political and social conditions to those during the Third Reich. This attempt at exposure did not concern itself only with the sadistic murders committed by *individuals*; rather, the statements of Witness 3 and the Prosecutor—as well as the entire structure of the play, in the way that it traces the path from "The Loading Ramp" to "The Fire-Ovens"—represented Auschwitz as a vast "machinery of killing." By virtue, thus, of the fact that the play focused on Auschwitz as a *system* in which the technical efficiency of murder lay at its core, *The Investigation* characterized the Holocaust as modernity's first bureaucratically and industrially organized genocide.

In the fourth canto about "The Possibility of Survival," Weiss had Witness 3 say:

Witness 3: We must get rid of our exalted attitude
that this camp world
is beyond our comprehension
We all knew the society
which had produced the regime
that could bring about such a camp
we were familiar with this order
from its very beginnings
and so we could still find our way
even in its final consequences
which allowed the exploiter
to develop his power
to a hitherto unknown degree
and the exploited
had to deliver up his own guts. (Weiss [1965] 2015, 88)

Witness 3 already referred here to a system of exploitation outside the camps, which, at Auschwitz, reached the extreme of human self-sacrifice ("the exploited / had to deliver up his own guts"). Similarly, the Prosecutor further specified this system as coming from a society "[w]e all knew" by explicitly mentioning "the beneficial friendship / between the camp administration" and industry: "You Herr Witness / and the other directors / of your great company / achieved through unlimited human sacrifices / yearly turnovers of unprecedented sums" (Weiss [1965] 2015, 103–4). Together with the statements of Witness 3, a distinct critique of modernity and capitalism became evident in Weiss's piece, in the form of a critique of purely instrumental reason and capitalist profit maximization.

The Investigation, at this point, very much in the spirit of Horkheimer and Adorno's *Dialektik der Aufklärung* (1944) (Dialectic of Enlightenment), drew a connection from the conditions of capitalist society to National Socialism and the systematic genocide in the concentration camps. In *Dialektik der Aufklärung*, Horkheimer and Adorno had emphasized the extent to which the strictly rational calculation of profit maximization, when coupled with a striving for ruthless self-preservation, surpasses the cruelty of archaic violence (Horkheimer and Adorno 2003, 67–103). For Weiss, following the reflections of the Frankfurt School, it was not only this brutally rational form of capitalism *but also* the rejection of an enlightened, Jewish as well as Christian concept of the individual, which values the dignity of humans independently from the principles of their functional effectiveness and utility, that led German industry not only to approve of Auschwitz, but to support it actively as well. When the Prosecutor continues: "Let us consider once again / that the successors of this company / amassed glittering fortunes / and that they are now about to enter / what is called a period of expansion" (Weiss [1965] 2015, 103–4), then it is indicated that the prevailing capitalist conditions continue to bear a risk of again (or still) supporting radical forms of exploitation or even genocide. Above all, however, it was explicitly emphasized in the play not only that industry passed through National Socialism with impunity and even managed to increase its profits, but also that the defendants themselves were able to switch from one system to another without harm: "They live without shame / They enjoy high office / They multiply their possessions / and continue those works / for which the prisoners were formerly employed" (199–200). In this respect, Weiss's play tended to encourage the conclusion that every capitalist social system is potentially prone to tolerating fascism.

As such, there has been much discussion among scholars about the extent to which *The Investigation* expressed a socialist position by supposedly agreeing with the "Dimitrov thesis" (Breuer 2004, 301; see also Young 1988), according to which capitalism not only supports mass murder, but actually makes it possible in the first place. However, Weiss did not follow such a simplified explanatory model that attributes the blame solely to a political system. Rather, the Prosecutor summed up damningly at the end of the play: "That represents a conscious / and deliberate disregard and offence / against those who died in the camp / and also against those who have survived [. . .] Such behavior on the part of the Defence / obviously demonstrates the persistence / of that very sentiment / which inspired those actions / for which the Defendants / are arraigned" (Weiss [1965] 2015, 201). Yet, the terms "conscious" and "sentiment" do not quite match the German original, in which the more complex term, *Gesinnung*, transcended simple reference to an individual belief or feeling. *Gesinnung* also describes a person's stance and way of thinking, implicitly influenced and informed by values and morals related to the society's norms. By concentrating on the *Gesinnung*, guilt and complicity in *The Investigation* proved to illuminate a complex, dynamic interaction between social structures and ethics, as well as the individual's responsibility for these currents within society.

When Weiss had the defendant Boger speak of cruel torture methods as though they were an allegedly necessary disciplinary tactic, his choice of words did more than just downplay the atrocities committed in the concentration camps; they revealed that Boger still considered torture to be appropriate, still devalued and discriminated against the victims, and still regarded any rebellion against the National Socialist regime, and the order within the concentration camps, as treason (see further Breuer 2004, 227). Boger had not developed any sense of guilt at all, as he testified within *The Investigation*: "According to orders / I had this responsibility / moreover I am of the opinion / that even today / if flogging were brought in / for example in case of juvenile delinquency / you might be able to control / all these outbreaks of violence" (Weiss [1965] 2015, 73). And, before that, he stated, "In the interests of camp security / strong measures had to be taken / against traitors and other vermin" (67). In this way, the defendant openly continued to humiliate camp inmates and Holocaust survivors in court and degraded them to "vermin" or even objects, thereby revealing that the cruelty of his *Gesinnung*, his beliefs and moral stance, had remained—and (as one can conclude) still influenced society at the time the play was staged. Furthermore, the reference Boger made to the appropriateness of "flogging" in relation to "juvenile delinquency" is of particular importance here. The terms suggest a direct connection between the torture methods used in concentration camps and the penal system in the Federal Republic of Germany, as well as the education and corporal punishment of children and adolescents. In this way, his language opened up a whole network of social spheres in which, even in the 1960s, this mentality—which was dominant in the concentration camps—had survived.

On the whole, the defendants and Defence Counsel's behavior and comments were based on three central patterns of argument: first, *denial and evasion*; second, *concealment and softening*; third, *gross contempt*. The *evasions and denials* ranged from reversing the roles of victim and perpetrator (Defendant 5: "I only came to the camp / under compulsion [. . .] / I told them again and again / I am a doctor in order to save human life"; Weiss [1965] 2015, 28), to claiming a mere obedience to orders, to attempts at justifying the cruelty by invoking the higher goals of the state or nation. As the Defence Counsel stressed in Weiss's piece:

> Defence Counsel: Our clients acted in the best faith
> and according to the principle of absolute obedience
> With their oath of allegiance until death
> they all bowed to the objectives
> of the state leadership of the time
> as the administration of justice and the Wehrmacht
> did. (102)

The defendants' language in Weiss's play was interspersed with administrative terminology that betrayed a rhetoric of *concealment and softening*. Defendants talked of "segregation" and "transfer," of "aggravated interrogation" and "natural departures," where, point blank, torture or murder were ultimately the issue. For example, Weiss had one of the defendants say, "I had no time / to look at the content of the trains" (98–102), when the topic was actually not "content" but rather people being ruthlessly deported to death camps.

However, it was especially the openly expressed *gross contempt* that the defendants held for the victims—evident in the diabolical laughter into which the defendants break from time to time—that ultimately exposed the extent to which German society continued enabling and supporting the perpetrators' mentality into the present. This interpretation, though, requires a more detailed explanation: "The Defendants laugh in agreement" (201) mainly appeared in the play whenever the discrepancy between the severity of a crime alleged by the prosecution or witnesses and the proof of corresponding evidence became obvious (Söllner 1988, 188). With the certainty of having the legal principle of evidence-based judgment on his side, the Defence Counsel rejected all of the Prosecutor's remarks—for example, on the personnel and institutional continuities from the Third Reich to the present—by referring to the individual guilt that was to be negotiated. The Defence Counsel stated, "Our discussion is only concerned / with what can be attributed to our clients / on the basis of evidence / This kind of general reproach / is trivial / particularly reproaches / against an entire nation" (Weiss [1965] 2015, 199). In this and other ways, repeatedly, *The Investigation* revealed the limits of the criminal process, which could not grasp the actual extent of the guilt and certainly could not condemn it. Accordingly, the defendants in Weiss's piece felt safe to protest loudly and to laugh cruelly at the victims, as well as about the judicial system's attempt to convict them. It was because of the perpetrators' trust that their society—which previously accepted them as blameless and respectable—would keep the punishment within reasonable limits (so to speak) that they laughed and felt confident about publicly ridiculing their victims' suffering.

The play further accentuated this problem when the Defence Counsel considered any general statements by the Prosecutor about German society's guilt entanglement and its responsibility for the Holocaust as a (socialist) political attack against the economy in general: "We protest against this question / which has no other purpose / than to undermine confidence in our industries" (200). Yet it was precisely such remarks ("to undermine confidence in *our* industries") that revealed the fact that the Defence Counsel's attempts to ward off any recognition of continuity served the purpose of defending not only the defendants but simultaneously also the "system" of the Federal Republic of Germany at large—"just as if the legitimacy of the FRG depends on the former Nazis being able to lead their lives undisturbed. And this is exactly what should be conveyed" in the play, as Ingo Breuer has emphasized (Breuer 2004, 225, my translation). If one considers how strongly—as Matthias Buschmeier demonstrates in his chapter — the defense against collective guilt among the Germans was based on the supposed necessity of protecting their "reputation" before the world, then it becomes clear why Weiss pushed this very conviction to the obvious extreme: by revealing how it does not hesitate to protect even the worst criminals for the "sake" of the nation.

Consequently, Weiss had the main defendant, Camp Adjutant Mulka, of all people, using the collective "we," speak the last lines:

> Defendant 1: Today
> now that our nation
> has once again worked its way up
> to a leading position
> we should be concerned with other things
> than with recriminations
> These should long ago
> have been banished from the lawbooks
> by the Statute of Limitations.
> *Loud agreement from the side of the Defendants*
> (Weiss [1965] 2015, 203)

While Mulka's speech, which closed the play, directly referred to a widespread opinion within the German population, the person *giving* this speech demonstrated who actually benefits most from such a *Gesinnung*: the perpetrators. Guilt and responsibility were shifted here onto the audience, as the play's final monologue underlined the extent to which it is the Germans' repressive attitude toward their own history that allowed a perpetrator to bluntly demand the ending of any further "investigation." Confronted with this cunning and challenging finesse at the end of the play, the audience itself was performatively forced into the role of complicity: should they, for their part, lapse into "loud approval" with the defendants and "applaud" Mulka's closing remarks, as theater conventions usually demand at the end of a play? And if they do so, would their applause not expose them as agreeing with Mulka and thus sharing responsibility with him for having the audacity to call the Auschwitz Trial an unnecessary attempt at "recriminations?" Theater here came very close to a ritual, in the sense that it tried to eliminate the difference between actors and audience, and thus already heralded the form of "political" performance that became more and more influential for the theater of the 1960s (Fischer 2019). This type of ending for the play connects closely to the expectations of Weiss's "documentary theater," as he himself had formulated them: such a theater should, according to Weiss, "involve the audience in the performance in a way that is not possible in the real trial room, it can put the audience on an equal footing with the defendants or the prosecutors, it can make them participants in a commission of inquiry" (Schumacher 1986, 83, my translation). Thus, the play performatively widened the circle of guilt by assigning the role of silent consent to the spectators, who might not have been active perpetrators but who benefited from the persecution and murder of the Jews nevertheless, and who, as they knew about the injustices, were morally complicit (see further Maria-Sibylla Lotter's chapter). It is precisely for this reason that Weiss refrained from a guilty verdict at the end of his play. He did not want to mark any "scapegoats" and thereby absolve the audience, when the play's entire objective was to encourage audiences to confront and reflect on their responsibility for the Holocaust. What the play centered, and what lingered once the last line was spoken, was the process itself, the "investigation" of this guilt, in which Weiss's piece performatively entangled the audience.

Weiss's commitment to socialism in the 1960s certainly led to a rather one-sided presentation, which, as Nike Thurn precisely illustrates, is particularly difficult in view of Weiss's distinction between victim groups, since the play tends to favor the rebellious political camp inmates over the Jewish, predominantly "passive," victims (Thurn, forthcoming). Yet it is the self-reflective aesthetic form through which Weiss tried to grasp, by way of the theater itself, the complexity of guilt and the consequences that derive from it that continues to fascinate. It is the narratives and symbolic performances in a society that contribute significantly to the extent to which a certain idea of guilt gains influence; at the same time, it is also the narratives and symbolic performances, especially as a complex, critical piece of art, that can serve to *expose* the strategies of guilt defense.

In this sense, inherent to Weiss's *Investigation* was also a critique of the forms and cultural practices of guilt representation, interpretation, and negotiation in German society, as expressed, for instance, in the memory culture at that time, as well as in the Frankfurt Auschwitz Trials. A theater that wishes not to become "guilty" itself, so to speak, by reassuring or merely entertaining the audience is, according to Weiss, forced to render society's strategies, cultural practices, and narratives surrounding guilt transparent. More than that, however, Weiss's work also reminds us that, when performing guilt, such a theater should actively reflect on and critique its own rhetoric and modes of expression, so as to avoid becoming part of a "culture of excuse" or a participant in the guilt's repression. Only in this way can theater relentlessly awaken the audience, as Weiss put it, from their slumber of guilt denial and of numbness—and this is what can be described, using Weiss's *Investigation* as an example, as the productive quality of theater when coming to terms with guilt.

References

Baumeister, Roy F., Arlene M. Stillwell, and Todd F. Heatherton. 1994. "Guilt: An Interpersonal Approach." *Psychological Bulletin* 115: 243–67.

Breithaupt, Fritz. 2012. *Kultur der Ausrede*. Frankfurt am Main: Suhrkamp.

Breuer, Ingo. 2004. *Theatralität und Gedächtnis: Deutschsprachiges Geschichtsdrama seit Brecht*. Cologne: Böhlau.

Buschmeier, Matthias. 2018. "Felix Culpa? Zur kulturellen Produktivkraft der Schuld." *Communio* 47: 38–50.

Dutt, Carsten. "Vorwort." In *Die Schuldfrage: Untersuchungen zur geistigen Situation der Nachkriegszeit*, edited by Carsten Dutt, 7–16, Heidelberg: Manutiu.

Feuchert, Sascha. 2012. "Fundstücke: Bemerkungen zu Darstellungskonventionen und paratextuellen Präsentationsformen früher Texte deutschsprachiger Holocaustliteratur." In *Berührungen: Komparatistische Perspektiven auf die frühe deutsche Nachkriegsliteratur*, edited by Günther Butzer and Joachim Jacob, 217–30, Munich: Wilhelm Fink.

Fischer, Saskia. 2019. *Ritual und Ritualität im Drama nach 1945: Brecht, Frisch, Dürrenmatt, Sachs, Weiss, Hochhuth, Handke*. Paderborn: Wilhelm Fink.

Fischer, Saskia, Mareike Gronich, and Joanna Bednarska-Kociołek, eds. 2021. *Lagerliteratur. Schreibweisen—Zeugnisse—Didaktik*. Bern: Peter Lang.

Frei, Norbert. 1992. "Auschwitz und Holocaust: Begriff und Historiographie." In *Holocaust—Grenzen des Verstehens: Eine Debatte über die Besetzung der Geschichte*, edited by Hanno Loewy, 101–9. Reinbek: Rowohlt.

Frei, Norbert. 1996. "Der Frankfurter Auschwitz-Prozeß und die deutsche Zeitgeschichtsforschung." In *Auschwitz: Geschichte, Rezeption und Wirkung; Jahrbuch 1996 zur Geschichte und Wirkung des Holocaust*, edited by Fritz Bauer Institut, 123–38, Frankfurt am Main: Campus.

Herdt, Jennifer. 2016. "Guilt and Shame in Virtue Development." In *Developing the Virtues: Integrating Perspectives*, edited by Julia Annas, Darcia Narvaez, and Nancy E. Snow, 235–54. New York: Oxford University Press.

Horkheimer, Max, and Theodor W. Adorno. (1944) 2003. *Dialektik der Aufklärung: Philosophische Fragmente*. Frankfurt am Main: S. Fischer.

Kämper, Heidrun. 2010. "Kollektivschuld—Die diskursive Instrumentalisierung eines gesellschaftlichen Konstrukts." In *Die Schuldfrage: Untersuchungen zur geistigen Situation der Nachkriegszeit*, edited by Carsten Dutt, 17–44. Heidelberg: Manutiu.

Lotter, Maria-Sibylla. 2012. *Scham, Schuld, Verantwortung. Über die kulturellen Grundlagen der Moral*. Frankfurt am Main: Suhrkamp.

Schiller, Friedrich. (1784) 1992. "Was kann eine gute stehende Schaubühne eigentlich wirken?" In *Schiller: Theoretische Schriften*, edited by Rolf-Peter Janz, 185–200. Friedrich Schiller, Werke und Briefe in zwölf Bänden 8, edited by Otto Dann, Axel Gellhaus, et al. Frankfurt am Main: Deutscher Klassiker Verlag.

Schumacher, Ernst. 1986. "Engagement im Historischen: Ernst Schumacher unterhielt sich mit Peter Weiss, August 1965." In *Peter Weiss im Gespräch*, edited by Rainer Gerlach and Matthias Richter, 82–93. Frankfurt am Main: Suhrkamp.

Söllner, Alfons. 1988. *Peter Weiss und die Deutschen: Die Entstehung einer politischen Ästhetik wider die Verdrängung*. Berlin: Westdeutscher Verlag.

Thurn, Nike. Forthcoming. "Recht und Radio: Theatrale Bearbeitungen von Völkermord-Prozessen und -Propaganda in Peter Weiss' *Die Ermittlung* und Milo Raus *Hate Radio*." In *Peter Weiss 1916–2016: Experiment und Engagement heute*, edited by Matteo Galli and Marco Castellari. St. Ingbert: Röhrig Verlag.

Wannemann, Klaus. 2004. *Erwin Piscators Theater gegen das Schweigen: Politisches Theater zwischen den Fronten des Kalten Krieges (1951–1966)*. Tübingen: De Gruyter.

Weiß, Christoph. 1998. "... eine gesamtdeutsche Angelegenheit im äußersten Sinne... Zur Diskussion um Peter Weiss' *Ermittlung* im Jahre 1965." In *Deutsche Nachkriegsliteratur und der Holocaust*, edited by Stephan Braese, 53–69. Frankfurt am Main: Campus.

Weiß, Christoph. 2000. *Auschwitz in der geteilten Welt: Peter Weiss und Die Ermittlung im Kalten Krieg*. Sankt Ingbert: Röhrig.

Weiss, Peter. (1965) 1971. *Rapporte 2*. Frankfurt am Main: Suhrkamp.

Weiss, Peter. (1965) 1982. *Notizbücher 1960–1971*. Frankfurt am Main: Suhrkamp.

Weiss, Peter. (1965) 2015. *The Investigation: Oratorio in 11 Cantos*. Translated by Alexander Gross. London: Marion Boyars.

Wenzel, Mirjam. 2009. *Gericht und Gedächtnis: Der deutschsprachige Holocaust-Diskurs der sechziger Jahre*. Göttingen: Wallstein.

Young, James E. 1988. *Writing and Rewriting the Holocaust: Narrative and the Consequences of Interpretation*. Bloomington: Indiana University Press.

11

Guilty Dreams

Culpability and Reactionary Violence in Gujarat

Parvis Ghassem-Fachandi

In late February and early March of 2002, the city of Ahmedabad in Gujarat, India, descended into a pogrom. On an overcrowded train, Hindu activists were returning from the temple town of Ayodhya, a popular pilgrimage site outside the state. They had traveled there to support the building of a Hindu temple to the god Ram on the site of a former Muslim mosque, which had been destroyed by activists ten years earlier. The train stopped briefly in Godhra. After an altercation between Muslim station vendors and Hindu activists, stones were thrown onto the train, which stopped again outside the station. Then four coaches of the train caught fire. Many passengers were killed. In the following days and weeks, the entire Muslim minority community of Gujarat became a target in a statewide pogrom. In cities like Ahmedabad, Muslims faced economic boycotts, attacks on their residential neighborhoods, the destruction of their property, and the indifference or complicity of the police in these acts. Hundreds of Muslim shrines and mosques were attacked, burned, and razed. Mass rape, arson, and deadly, violent attacks by large organized crowds and gangs armed with swords took place in front of a gaping, knowing, and partly approving or even participating public (Varadarajan 2002).

These events are usually referred to as the "Gujarat riots." The passage from *pogrom* to *riot* constitutes an act of reduction that does two things at once. It integrates a particular event into a series of preceding events, eliminating its specificity.[1] Furthermore, the term *riot* complicates the assertion of culpability because it invokes two equal communities mutually attacking one another. Pogroms, by contrast, are organized events following a planned objective, characterized by a psychological mobilization that far exceeds the immediate group of actors involved in a riot. There is no concept in Gujarati that can be assimilated semantically to the term *pogrom*, though there are many words that allow for the rendering of *riot* (Ghassem-Fachandi 2012, 60).

When the state had barely begun recovering in May, Uma Bharati, the Union Minister for Youth and Sports at the Centre in Delhi, responded to a query by a journalist about why the "riots" had been so exceptional. She responded:

Parvis Ghassem-Fachandi, *Guilty Dreams* In: *Guilt.* Edited by: Katharina von Kellenbach and Matthias Buschmeier, Oxford University Press. © Oxford University Press 2022. DOI: 10.1093/oso/9780197557433.003.0012

> The rise of intolerance among Hindu youth and the fact that they are unapologetic about such inhuman acts is a cause of great concern. That such elements—even if small in number—exist among Hindus is terrible. They must be destroyed from the roots.[2]

A few weeks later, the commentator Mahesh Daga, in the same newspaper, commented on the bewilderment of the well-known violence expert and police consultant K. P. S. Gill. Similar to Bharati, Gill expressed consternation at "the apparent lack of remorse" shown by ordinary Hindus and the extreme level of brutality unleashed against the Muslim minority in the state. Confirming this understanding, Daga asked what might explain "this cynical disregard for the sanctity of human life, especially among a people who otherwise pride themselves on their non-violence, pacifism, and, need one add, vegetarianism."[3]

The oddity in these platitudinous exchanges is what they bring into play without becoming explicit: the suggestion that, in past communal violence, violent actors were somehow more apologetic about the inhuman acts they committed. Gujarat had witnessed extreme forms of communal violence in the late 1960s, mid-1980s, and at both ends of the 1990s. It is rather absurd that a "riot expert" like K. P. S. Gill would be astonished about perpetrators' lack of remorse; Daga points out that such a lack is exactly what characterizes instances of ethnic violence in general. In ethnic violence worldwide, victims are held responsible for becoming targets, and perpetrators fall short of either developing or acknowledging retroactive feelings of guilt. Bharati and Gill seem to suggest that there exists a "normative" case in which violent actors are ostensibly somewhat less violent, or that they at least feel less bad about their actions afterward. Significantly, however, Muslims are never mentioned, nor is the nature of the violence unleashed against them; in this, the statements by Bharati and Gill succeed in passively confirming a stable stereotype; namely, if it was unusual for Hindus to engage in such extreme behaviors, by implication it was not so for Muslims, who in Gujarat are regularly closely associated with violence, and for whom there exist a store of applicable stereotypes (Ghassem-Fachandi, 2012).

This chapter describes three non-Muslim responses to the pogrom and to its aftermath that are segmented in time and context. It seeks to understand whether and how shame, remorse, or regret were expressed or curtailed in public discourse and private interactions, leaving the state under the shadow of an adumbrated guilt, with all its productive and unproductive possibilities.[4] While it remains unclear to what extent empathy might lead or contribute to reparative action in each instance, I will nonetheless turn to contexts in which scholars might not expect such expressions. The violence of the Gujarat pogrom was exceptional, but not because "ordinary Hindus" or "Hindu youth" felt or acted differently than they did in years prior; rather, it was exceptional because violent unanimity was officially tolerated and even encouraged. Residents felt emboldened because parts of the state administrative apparatus, political leaders, and civil society were complicit in the pogrom. I address guilt here in two ways: as external attribution where it is explicitly rejected, and as reactive feeling that manifests in social relationships in indirect ways.

Flower Petals over Asphalt

Shortly after the pogrom, in late March 2002, Bharat, a Hindu friend and my roommate, drove on his Hero Honda motorcycle past the site of a former Muslim shrine in the city of Ahmedabad. When Bharat approached the site, he saw a middle-aged man, barefoot, straddling the middle of the road unprotected from the two-way traffic. The man wore a typical white Muslim skullcap (*topi*) and a light, long, flowing *kurta* (Indian chemise) tailored in the Pathani style. Puzzled, Bharat stopped his motorcycle on the side of the road, and, taking off his helmet, observed the scene for a while. Negotiating the busy traffic, the man threw flowers from a basket wrapped in green cloth on freshly poured asphalt, making sure only to use his right hand in a pious gesture. He had left his leather sandals neatly on the side of the road, out of the way of traffic, next to his two-wheeler.

His movements traced the contours of an invisible structure that until a few weeks ago had been situated in the middle of the road: the shrine of Vali Gujarati. Vali Gujarati is a historical figure (often called *Wali Muhammad Wali*), a Muslim poet-saint who is alternately referred to as "Wali Aurangabadi," "Wali Dakhani," or "Wali Gujarati," depending on the geographic origin of the speaker. A 17th-century poet, the saint is known for his many *ghazals*, lyrical stanzas that emotively invoke the experience of love and the impasses of a longing heart. His verses were interspersed with appreciation and affection for life in urban centers.[5] Born in Aurangabad (which today lies in the state of Maharashtra), he is often described as the grandfather of Urdu poetry and is believed to have died in Ahmedabad, Gujarat's largest city, in 1707. Regarding his name, "*Vali*" denotes the title of a guardian or warden of an institution, as well as an exalted personage; the geographic qualifier "*Gujarati*" stresses a particularly intimate relation to the state.

Muslim shrines in India are called *dargah* (tombs) and the city of Ahmedabad, founded by Sultan Ahmed Shah in the early 15th century, has many of them. Unlike mosques, Muslim shrines are frequently visited by Hindus and members of other non-Muslim communities. Some are large sandstone structures of architectural beauty and significant historical age, while others are the size of a single human grave—a mere cement rectangle hidden in plain sight at a street corner and covered under a fine film of urban dust.

Occupying the center of a busy road in the Shahibaug area, the shrine of Vali Gujarati in particular was a symbol of religious harmony and mutual cultural overlap. Scholars of South Asian Islam have described the unique blend of Middle Eastern and Central Asian traditions that have seeped into the soil of India over many centuries as "Islamicate India," a term initially coined by the historian Marshall G. S. Hodgson (1974, 57–60; see also Frembgen 1993). Today, however, these cultural formations are increasingly separating and redefining what is considered "Hinduism" and what is considered "Islam," with pernicious consequences for Hindus and Muslims on the subcontinent. What are left are myriad mushrooming temples that loudly announce Hindu identity and freshly poured cement mosques, painted in gory greens, into which Hindus never enter.

On the evening of February 28, 2002, the shrine of Vali Gujarati was bulldozed in the early days of the Gujarat pogrom. This nocturnal attack came as a shock for many local residents, as, unlike in other such instances, no remnants of the shrine survived. Attacks on religious structures are common in communal violence, and evidence of the destruction is usually visible for weeks and months, humiliating the injured community publicly, but also allowing for the articulation of indignation from various sides. In this case, however, the miscreants ferried away all debris to the last speck, leaving behind only the quickly refurbished, smooth, flat surface baking in the sun. No one could point to any ruins and express sorrow, regret, or remorse. It was as if the shrine of Vali Gujarati had never existed. Yet passing drivers knew what had stood there, and some avoided the spot assiduously, dodging oncoming traffic as they veered dangerously toward the left or the right. They raised their hands to their foreheads while driving at full speed, nodding to the enduring presence of a Muslim saint's grave on the empty black asphalt. Perhaps some also feared that driving over the saint's spot would bring his wrath.

My friend Bharat immediately understood the rationale of the barefoot man's pious gestures and the reverence he displayed. Throwing pink and purple flower petals on asphalt, where they wilt quickly, the man made visible for a brief moment the contours of what had been the resting place of the "Guardian of Gujarat." In local understanding, shrines grow out of the soil on which they stand, underscoring the unique relationship to their place of emergence. It is not uncommon in India for divine forces to impede traffic, compelling a reorganization of the surrounding urban space. The man moved slowly and with rapt attention, despite the fact that hot asphalt is no friend to bare feet. Many structures in India have white marble floors to stave off the sun's unrelenting rays, which, if captured instead by dark surfaces, mercilessly singe naked skin or delicate flower petals. The disappeared shrine of the poet-saint, too, had been made of white stone.

When Bharat described the scene to me late that afternoon, he was disturbed. His emotional participation in the barefoot man's travails surprised me. A staunch Hindu nationalist, Bharat had supported the violence throughout the preceding weeks, during which hundreds of Muslim shrines, large and small, historical and insignificant, had been destroyed or maimed all over the city. These acts of vandalism were considered "revenge" (*badlo*), designed to discipline the Muslim community collectively after 59 passengers on the train from Ayodhya, many of them pilgrims, including Hindu *karsevaks* (national volunteers) had been burned alive in the town of Godhra. The tragedy is referred to as the "Godhra incident" (*godhraa ghatnaa*), but the chief minister referred to it as the "Godhra massacre" (*godhraa hatyakaand*).[6]

At the outset of the pogrom, I had many tense discussions with Bharat, as I disapproved strongly with the political position he had taken. The nightly bulldozing had been reported in vernacular and English-language papers as major news. Bharat seemed indifferent at the time, both to the destruction and to the large-scale killing of Muslims. But now he felt empathy, inspired by the flower-throwing man's attempt to alleviate the moral injury that every vehicle driving over the sacred site inflicted all over again. Empathy was the emotional reaction I had previously accused him of lacking. I realized, not for the first time, that I did not entirely understand the relationship between Hindus and Muslims in Gujarat, how intimately they were connected to one another as internal and external objects by invisible strings. Somehow Bharat was able to see the Indian within the Muslim man, perhaps because of the fact that the practice of venerating dead saints by spreading flowers over their resting places is a ubiquitous local practice. It genuinely moved him and interrupted for a brief moment his strong identification with Hindutva (Hindu nationalism), which for him implied a rejection of Muslims (Ghassem-Fachandi 2012, 162–83).

In a sense, the barefoot Muslim man was the protagonist in an archetypal Hindu scene. This scene is the nightmare that Hindu nationalists frequently invoke as their own historical humiliation, a memory against which they rebel today: the destruction of their religious heritage, the humiliation of their gods (e.g., the placing of their idols underfoot at the entrances of mosques), and the alleged indifference by secularists and many others to these slights. During communal violence, residents and political actors alike invoked themes of past injury. Time and again, elaborate historical scenes were remembered to either justify or explain violence against Muslims in the present. Images of past injury provided the emotional stage for the rage that captured so many Gujaratis and made them complicit in violence. Bharat recognized what he could emotively relate to. As a Hindu, he had been told that he was at the receiving end of such wanton destruction, and yet seeing a Muslim suffer the same sort of injury led him to identify with his supposed oppressor. I was moved by the fact that Bharat was finally moved.

Doctor Shah's Silent Audience

A few weeks later, in August 2002, after I completed a blood test at a local laboratory, a friend recommended a particular Jain doctor at a middle-class locality in Ahmedabad to discuss the medical results. The doctor's practice was not far from Gujarat University, and a young Dutch man called Henk, an intern of a local NGO, asked to accompany me. Henk had just arrived in Ahmedabad, did not know the local language, and was puzzled when I told him about the facts of the pogrom. He had been entirely unaware of these events, as the locals with whom he worked had avoided mentioning anything to him—which, I admit, baffled me at the time.

The doctor's office was small, not larger than a living room, with a large, cream-colored, plastic divider separating desk and stretcher from the waiting area. After brief attention was given to our ailments, Doctor Shah wanted us to sit down for a talk. He nervously repositioned the divider, incompletely cordoning off his desk and our chairs from the rest of the waiting room. Irritated, he ordered his assistant to go downstairs and get some tea and asked his nurse to complete some paperwork.

"Let them wait," he said in English, in a dismissive tone, when I pointed toward the crowded waiting room. From then on, it was mostly he who spoke. Sipping our milky teas (he did not take any), Doctor Shah inquired as to our business in Ahmedabad. What were we doing here at this horrible time? Accustomed to such questions and not wanting to extend our stay, I opened tritely with a somewhat quadratic introduction, telling him that I was an ethnographer working on "Indian culture and values." He immediately stopped me short, scoffing at the whole idea before I could finish. I was able to add that I was specifically interested in the ancient concept of *ahimsa* (noninjury, nonviolence), a set of practices and notions that are strongly identified with Jain religious traditions. Usually, the mere mention of the Indian concept of nonviolence produced a particular kind of elaboration, from the value of dietary abstentions to assertions of the timeless beauty of "our Gujarati culture." Such lectures ended with admonitions of how much the West could learn from Gujarat and, more generally, from the Indian spiritual way of life. Few interlocutors at the time registered how bizarre and out of place such invocations might sound to an outsider like me, given the immediate context of their utterance after long weeks of violence and extremist rhetoric. It felt to me as if locals reassured themselves of something that suddenly had been put into question.

Not this time. Without any hesitation or circumlocution, in fact as if he had been waiting for this singular opportunity, Doctor Shah talked frankly about what had been happening locally for the past months (that is, since late February): after the Godhra incident, the city had been completely out of joint, bent on participating emotionally in violence with an enthusiasm rarely witnessed before. While he spoke of this, his voice became louder until, finally, it broke. He was short of breath, as if something had ruptured deeply within him. Ignoring my interjections, he asked dramatically: "What has happened to *this* Gujarat? Why is *this* happening?" Habituated to evasions and excuses whenever the subject was broached, I was stunned by his intensity, while Henk rocked uncomfortably back and forth on his chair.

Doctor Shah told us that the violent events in Ahmedabad had been suspicious. He had witnessed communal clashes before, but this time was different. "There was no fighting," he declared. He narrated scenes he had seen in residential areas: the looting and burning of Muslim homes, while the police were passive, complicit with the attackers—as were neighbors and onlookers. He also spoke of attempts to boycott the entire Muslim community, through flyers distributed at night and massive hate propaganda during the day in vernacular news media. Pointing toward the window, he drew attention to the charred remains of a former Muslim vegetarian restaurant just opposite his medical practice. With a hoarse voice, he asked whether we had seen it. Indeed, we had. Upon entering his office, I had pointed out the ruins to Henk. "It is a shame," the doctor exclaimed. He added that Jains, too, "took full part in it."

As Doctor Shah spoke, a silence settled over his waiting room. I looked past the makeshift divider that separated us from his patients, who were gazing God-knows-where in empty stares. The divider allowed for every word to be heard, loud and clear. Most of his patients were Gujarati women, usually no stranger to lively discussion. The doctor continued, ignoring Henk's attempt to ask a question, and exclaimed, incensed, "It is pure *himsa* [violence]," and "Pure shame, this city!" Finally, exhausted, Doctor Shah escorted us to the exit door of his practice, passing through the waiting area, where heads were lowered. No one met our gaze. We said our goodbyes, politely and somewhat subdued. Doctor Shah remarked loudly, as if he owed it to himself, that he felt ashamed to be a Jain. Surprised by this statement, which reminded me of a German locution, I almost forgot the prescription he had prepared for me. He brought it and repeated: "It is a shame what has happened here."

The doctor lost his composure when bringing up what he apparently had wanted to address for a while and could not. He was unable to speak of it in a calm manner. While he said little that I did not already know, the outburst took me by surprise. I suddenly became aware of how much I had missed such reactions in the preceding weeks, and how much I myself had begun to censor my behavior by evasions when interacting with locals. Paradoxically, Doctor Shah's indignation appeared to me like a moment of reason, even of *Menschlichkeit* (humanity). Today, I understand he was not really addressing us, his foreign visitors, but rather his silent patients seated behind the cream-colored divider in his waiting room—though he spoke in English, and I do not know how many of his patients actually comprehended the tirade. He knew that this violence had been committed in their name, and he wanted no part in it. I remember his body language when he handed me my prescription, the pained face and the tremble of his aged fingers. Here was a man whose own society had become strange to him. His memory has stayed with me. Henk was confused by the whole affair and mentioned it many times afterward.

Elision of Guilt and Reactionary Violence

When I initially began fieldwork in Gujarat in the mid-1990s, past episodes of Hindu-Muslim violence were rarely openly discussed. If they were, guilt was quickly apportioned to the British, who were held responsible for the dissection of the subcontinent and the animosities between Hindus and Muslims. What scholars had written of such matters found little resonance in the minds of my interlocutors. In inquiring with my subjects, I was rather quickly accused of summoning the tensions that I, in fact, simply wanted to learn about; a veil of silence lay over such issues, and foreigners were not supposed to be introduced to them.

Even in later years, I experienced much resistance to similar discussions in diverse contexts—among an urban, educated Brahmin class of professors at Gujarat University where I studied language, as much as in rural Gujarat, among farmers largely considered "illiterate." Communal violence was acknowledged but dismissed as an aberration, the machinations of corrupt politicians and criminal gangs. It did not reflect who Gujaratis thought they were, despite the fact that many were living witnesses of previous violence. Given the state's history, this avoidance behavior seems, in retrospect, to be a concerted form of denial.

What exactly was this denial avoiding?

The tale that Gujarat was a comparatively peaceful state is a cover story that Gujaratis, including many Muslims, tell themselves and others to this day. It is product of local pride and subnationalism, an impulse to omit whatever does not fit a smooth frame. The story is sustained in no small measure with references to Mohandas K. Gandhi, a son of Gujarat who is considered the symbolic father of the Indian nation. As a figure of identification, Gandhi in many ways constitutes the state's global moral face. For the longest time, this figure provided a convenient mask by which Gujaratis represented their state to the world and, more significantly, to themselves, despite the raging ambivalence they felt toward him. This also explains Daga's odd sudden reference to vegetarianism in the wake of the 2002 anti-Muslim pogrom in Ahmedabad, as in India the dietary abstention is more apodictically associated with nonviolence than in Europe. The experience of this pogrom made me aware of how retracted this routine avoidance had become, and how convoluted emotions rebelled against the inner moral command emanating from Gandhi. The days of violence in Ahmedabad were not only filled with fear, shame, and shock, but also with enthusiastic participation, enjoyment, and fascination. The psychological mobilization for violence resulted in miscellaneous forms of complicity. The tales that Gujaratis told themselves about themselves no longer made much sense.

The Gujarat pogrom seemed to follow a script that was anticipated yet curiously took everyone by surprise.[7] Violence after the Godhra incident was predictable, given the history of communal conflagration in the city of Ahmedabad and in the country at large. Construed as a provocation of Hindus by Muslims, the incident demanded a rebuttal. But then, as a local stereotype goes, Hindus are by nature passive, more cowards than bullies.[8] This understanding is the product of a sense that Gujarat as a state was characterized by a rational, economic ethos attributed to affluent merchant groups (*vepari*) associated with vegetarianism, nonviolence, and a particularly forward-looking peaceful character and culture.[9] Doctor Shah belonged to that affluent class; Bharat did not.

What occurred was nonetheless unusual because, as far as communal violence goes, the Gujarat pogrom exceeded in severity most of what urban residents had witnessed in their lifetimes. The alleged constitutional disadvantage of Hindus when confronted by aggression became an opportunity for Hindu nationalist organizations to fulfill their promise to defend the Hindu community against Muslims, in a delegation of violent labor. It is these organizations that committed the worst atrocities during the pogrom. When Uma Bharati and K. P. S. Gill referred oddly to "Hindu youth" and "ordinary Hindus" lacking remorse, they elided the fact that most atrocities were carried out by, or under supervision of, these organizations. The organizations had been recruiting, training, mobilizing, and planning for years, including among those segments of society that have suffered generations of discrimination and exclusion (such as Dalits and Adivasi). The omission is, therefore, not an innocent oversight, but a decisive displacement characteristic of the entire public discussion after the pogrom.

The most powerful man at the helm of the state at the time of the pogrom, Chief Minister Narendra Modi, had at his disposal the state machinery, the administrative apparatus, the intelligence services, and the police and security forces, along with close relations to the entire edifice of *Sangh Parivar* institutions—the family of Hindu organizations that espouse Hindutva ideology, such as the Rashtriya Swayamsevak Sangh (RSS), Vishwa Hindu Parishad (VHP), Bajrang Dal (BD).

It was from this mighty position that Modi, three days into the anti-Muslim violence, employed the term *svabhavik pratikriya* (natural reaction) to describe the anti-Muslim pogrom. This expression, uttered long before the incident in Godhra was understood or the post-Godhra violence had ended, suggests a particular relation of guilt to agency.

More than a mere description, this *pratikriya* (reaction) provided an interpretive frame for the unfolding violence; it combined the theme of revenge and retribution with the image of an automatic mechanism, suggesting the detachment of ritual procedure: a *pratikriya* is always preceded by a *kriya* (an act or a deed). If the Godhra incident was the deed, the initial *kriya*, then the pogrom was but its response, a counteraction to undo the bad deed. The widely circulated translation of the expression into English was, "Every action has an equal and opposite reaction," which references Newton's third law of physics. One can see here the invocation of the idea of an immediate karmic retribution in the prestigious garb of modern physics. Why Muslims in Ahmedabad and the central provinces should be on the receiving end of such retribution when the alleged acts were committed in the town of Godhra was never sufficiently explained. Such details did not matter at the time, nor do they today; Muslims were held collectively guilty for acts that the vast majority of them had never perpetrated, and, as a minority, they had to be taught a lesson. The Gujarat pogrom was interpreted as providing that lesson. Hindus had to overcome their constitutional cowardice in a supposed act of defense of their religion, their women, and their country.

Furthermore, by stressing detachment in the context of organized violence, Modi's words seriously obscured the question of agency. By using the term "natural," a sort of inevitability was suggested, as if organized collective violence were akin to a twitch following a pinch—an unavoidable and innocent impulse. While Muslims in Ahmedabad were held guilty for acts they had not perpetrated, violence perpetrated in the name of a generalized Hindu anger or fury (*krodh hindu*) was deemed an innocuous reflex (Eckert 2010, 159–61; Sundar 2004, 153–5). In other words, Hindus had reacted in anger but were not, so to speak, guilty; if *karma* strikes, there is no agency and hence no culpability. Muslims, however, were held collectively responsible, since they acted according to their violent nature; their suffering was the result of karmic retribution, invoking the automatic causality of *pratikriya*. While the expression "natural reaction" captured well the way in which many Gujaratis legitimized the violence, it entirely misrepresented the facts. In this way guilt became obfuscated for reasons of narrative coherence; namely, to uphold the myth that Hindus never strike first because they are, by nature, calm, passive, even fearful, and that they only strike out as a last resort, in moments of desperate self-defense. At the same time, the myth holds that Hindus, if provoked, could explode and vanquish their enemy in one fell swoop—akin to the image of the god Shiva, whose anger, once awakened, could make even the mighty Himalayas tremble (Makawana 2002, 11–7; see also discussion in Ghassem-Fachandi 2012, 185–212).

The term *pratikriya* was quickly picked up and came to be used widely to reference the pogrom. Other idiomatic expressions also circulated on the street, such as "*karvu j pade* (It must be done), which suggested a decisive imperative to violence. In this way, the Gujarat pogrom was accompanied by a distinctive collective atmosphere or mood for which the German noun *Stimmung* seems most adequate (Borneman and Ghassem-Fachandi 2017, 105–35). During the violence, on the basis of false information from rumors and vernacular newspapers, men on the street believed that Muslims had abducted Hindu girls reasons of *enjoi*.[10] Terminologies were used for killing that exceeded usual vocabularies and suggested a sacrificial script, summoning an imagery suffused with neo-Vedic terminology and a strange tendency to imagine and describe violent acts in gratuitous detail (Ghassem-Fachandi 2012, 66–78; 2017, 156–9).

The general psychological mobilization for violence was clearly palpable at sites where I ventured. Similar observation can be gleaned from various fact-finding reports. Normal folk were emboldened to engage in forbidden acts, such as looting and stone throwing, and, at other sites, there was rape and murder by organized gangs (Ghassem-Fachandi 2012, 31–57 and 93–122). During the days of violence, one ubiquitous saying was written on walls across the city: "It is an open secret, the Modi government is with us." The sentence was usually written in Hindi, India's national language, but occasionally also in Gujarati. The pronoun *us* here referred to Hindus, while the sentence implicitly addressed Muslims, the segment of society that was being collectively harmed. Most blatantly, this graffiti affirmed Gujarati approval of the criminal passivity of the police during the pogrom, and of the strong pro-Hindu bias of state administrators, services, government officials, and ministers alike. The expression *e to andar ki bat hai*, which I have translated here as "It is an open secret," also means "It is insider knowledge," suggesting a more conspiratorial twist; it expresses a general assumption that the government is complicit in the unfolding of events.

The consequence was a permissive atmosphere in the street that, aside from feelings of impudence or effrontery, can only be described as one suffused with sizzling excitement. At an improvised tea stall on Ashram Road, not far from what used to be Shiv Cinema (due to curfew, all venues were closed), a group of middle-aged men stood around a woman sitting on the ground selling tobacco, *paan* (betel leaf with areca nut), and plastic water pouches. I approached the group. The woman had set up the makeshift shop on the pavement to earn a few bucks by flouting the city's general lockdown. I joined the men, feeling safer in their company than alone, and bought a single cigarette. I wanted to overhear discussion and engage in a dialogue. The men spoke to one another in a joking manner, fooling around the way that groups of men often do when they have time on their hands. Silently standing to the side, I smoked and nodded, but they ignored me.

Disappointed at their disinterest, I planned to move along and find another corner from which to observe a city unhinged—but then a man on a bicycle passed with two overlarge, filled cloth bags that hung on both sides of his handlebars. It was a washerman (*dhobi*), apparently transporting his customer's freshly laundered clothes. He wore a large beard and a skullcap that identified him as Muslim. He approached relatively slowly and did not speed up when riding past us. Busy with one another, the men initially ignored him. After they noticed him, however, I observed a quick transformation of the atmosphere, into a sort of frenzy. It took a moment to build, but the dangerous climax felt abrupt: one man raised both arms, lamenting loudly into the air about how a Muslim dared to show his face under the open sky on this day. The cyclist turned his head and offered a faint smile while calmly pedaling on with his heavy load. Next, another of the group asked for a phone and called an acquaintance. After a few minutes, a small van arrived. I was dumbfounded by what happened quickly thereafter: the men boarded the van, gesticulating wildly while they drove off in the direction of the Muslim cyclist. They clearly wanted to harm or discipline the man. I do not know what happened afterward.

This, then, was the sort of "anger" invoked to explain and legitimize urban violence, a veritable contagion that had infused many residents. Such groups of men were not the organized goons of extremist Hindu organizations such as RSS, VHP, Shiv Sena, or Bajrang Dal that were roaming the streets killing and destroying at will, nor government officials or underclass criminal groups. These were regular citizens channeling feelings of indignation in a context of moral ambiguity, and authorizing themselves to engage in hooligan behavior that was characterized by a seemingly joyous antagonism toward Muslims.

What responsibility do such actors carry for surfing the permissive mood of the moment?

In 2002, most city residents with whom I spoke saw a clear involvement of the Gujarat government in the pogrom. However, this acknowledgment was made with a sense of pride and moral rectitude, not embarrassment or sorrow. Many individuals tended later to straight-out deny these forms of emotional participation, and I sometimes had the impression that they were sincere; that is, they were indeed no longer fully aware of their earlier statements and behaviors. This, together with the participation of subaltern communities in the attacks, constituted the reason for the brazen expression of triumph by *Sangh Parivar* leaders after the pogrom, the family of Hindu organizations. It also explains the sheer magnitude of humiliation for Muslims. The social engineering was successful: the obfuscation of the Godhra incident and its instrumentalization as a rationale for the violence of the pogrom, the repeated invocations of victimhood by way of a

Muslim minority in historical time, the constant erosion of the line that separates victim from attacker—these and other strategies eventually collapsed any notion of accountability that might have been invested in identifying a right versus a wrong (cf. Klaus Günter's chapter in this volume).

I fear that this sort of interaction is a rather common affair in Ahmedabad today. I know many Muslims who, for years now, circumvent all contact with Hindus and will hesitate to apply for jobs in Hindu areas or businesses in order to avoid humiliating and volatile experiences. These evasions disadvantage them even further in an already discriminatory context. A new hierarchy has established itself in the city. Locals refer to this hierarchy by employing an idiom derived from English: the city was characterized by the distinction between an "H-class" and an "M-class," the letters being colloquial abbreviations for Hindu and Muslim, respectively, in reference to an aggregate binary. The word for "distinction" that is usually employed here to qualify the relation between these two aggregates in the Gujarati vernacular is *bedbhaav*, a word that is also used in the context of caste distinction (*te bedbhaav raakhe che*, "he keeps distinction," which means to say that, "he will not drink tea with me").

It is in the context of such alienation between Hindus and Muslims that a surprising message reached me one day a few years ago while living in Fatehvadi, Juhapura. In a village not far from Sanand town, the news went, there lived a Bharvad, a Hindu, who was building a Muslim shrine, a *dargah*, on his own ancestral land.[11] This was unusual news, and it spread fast among the people with whom I was working. Many originally hailed from the rural areas surrounding Sanand and thus were immediately curious about this man. It is not uncommon for Hindus to visit or sometimes even to officiate at Muslim shrines; it is more rare for Muslims to officiate at Hindu shrines—although I have run into such cases over the years.[12] It is, however, rather curious for a Hindu to build a Muslim shrine from scratch and on his own initiative. The days of royal patronage of religious sites is but a faint memory, if such facts are remembered at all; if anything, Muslims today associate Hindus with the demolition of their religious heritage, be it the Babri Mosque in Ayodhya in 1992 or the shrine of Vali Gujarati and others in Ahmedabad in 2002.

At the time, it was said that this Bharvad was building his shrine on what had formerly been Muslim land that had been sold for a cheap price. The detail about the price immediately rang true for most residents present; it made sense because many Muslims living in the slum of Fatehvadi had hailed from villages in which their fathers were forced to sell their small plots of land for a pittance a generation ago. This was why they migrated to the city, where they gathered in ghettos after recurring bouts of urban violence had displaced them time and again. Now, these lost lands were suddenly worth manifold the price of prior years. *What would have happened if my father had not sold the land?* was a recurring question.

But the story of the Muslim shrine built by a Hindu turned out to be a little different. I met the man in Ahmedabad, where he engaged regularly in salaried work, while his residence remained in the village. I had no phone number but got information that he was working as a night guard (*chokidar*) on the grounds of the Ahmedabad Municipal Fruits and Vegetable Market. One night, I ventured there with two friends for a surprise visit. While climbing up the metal stairs to an empty office on the deserted grounds, we heard agitated voices engaged in an intense discussion in the dark. His name was Kalubhai Bharvad. When I entered he was talking to another night guard, a much younger man, who was listening intently and once in a while bombarded him with questions. He told of strange occurrences and appearances at the vegetable market that he had recently witnessed. After our first encounter, Kalubhai invited me to visit him in his home village and promised to show me the Muslim shrine he was building.

Kalubhai's Dreams

Kalubhai Bharvad is a man in his mid-40s, or so he thinks. It is common for rural folk not to know their exact age. Muscular and compact, he has a pleasant air about him, infused with much frenetic energy. Everyone greets him when he walks past in quick, determined steps. Villagers seem to appreciate him: his neighbors, work colleagues, caste brethren, and other acquaintances. Although vibrant, he displays no particularly unique external characteristics except one: when Kalubhai's legs rest, his mouth picks up the pace and speaks with the same rapidity as his legs do when walking. Each time I visited him he appeared somewhat annoyed that he had to interrupt whatever he was doing at the time in order to receive me properly as local protocols demanded.

Accustomed to speaking quickly, he rolls one word into another and strings together long chains of sentences without interruption—for long periods of time—often overwhelming his audience with allusions, and often also repeating what he said just shortly before. When I first tried to speak to him, I found myself mostly defeated. The impenetrable and continuous humming, overlaid by a strong locally inflected idiolect, allowed no clear identification of verbs or nouns, ends or beginnings of sentences, leaving me struggling to comprehend. I decided he was rapidly mumbling. It was as if he had to hurry in order to crowd as much as possible into one sentence, into one single breath, into one long monotonous sound. What did he just say?

I admit, this was hard to take, because I have been working in Gujarat for years, including in rural areas such as this village, and I was more or less proud of my linguistic abilities. Now I had to allow myself the expense of two research assistants to catch the intricacies of what he told me without irritating or fatiguing him through frequent, repetitious requests for clarification in the middle of my questioning—Kalubhai had little patience for a long sequence of questions. The works he engaged in were too important. I did not regret the decision to hire these assistants; both of them, although native speakers and from opposite ends of the Ahmedabad's social universe, acknowledged difficulties in understanding Kalubhai's phonology, which made me feel a bit better. "He speaks funny," Munavar commented dryly.

To be clear, Kalubhai's oddity is not the symptom of a handicap, but more a reflection of the interaction with his usual audience, consisting almost exclusively of local farmers, who apparently enjoyed and understood his references and allusions. Often quite monosyllabic themselves, they had gotten used to his way of singing a story in his humming mumble. Kalubhai's speech filled out the void that their distinctive silences left. He provided an energy with which his audience collaborated actively by listening attentively.

When we initially met, after what seemed like a brief moment of real surprise that someone like me had appeared at his nightly job, Kalubhai immediately took my presence as preordained and wove it into his general narrative (disappointing me, as I wanted to hear about the apparitions). For my own part, I was initially surprised as well, because I expected a man in traditional attire; Bharvads are often dramatic dressers, wearing large colorful turbans and traditional chemises, while Kalubhai, by contrast, looked more humble, more like a grocer or accountant than a shepherd. In busy Ahmedabad, he could easily disappear into any crowd.

Falling into a sequence of entangled stories over the course of several visits, Kalubhai eventually narrated to me the events and dreams that prompted him to begin constructing a Muslim shrine on his ancestral land in his residential village, being himself a Hindu. Some of these discussions were held in private, with only my two assistants and Kalubhai's wife Sheetal present; others were in larger groups of villagers who showed curiosity about watching "Kalu" sing to a foreigner, as if the conversation itself were a source of entertainment for them. They never asked questions of Kalubhai, but they did often contribute details to a story they already knew. "Tell him about *this* or *that*," they would often prompt Kalubhai. In this written text, the story will appear more systematic than how it was relayed to me, which is an unavoidable side effect of integrating a spoken, meandering, and associative oral rendering into a narrative textual form with a page limit: descriptive exposition on paper.

It began with unfortunate events and a series of accompanying dreams and visions, which Kalubhai calls alternately *sapnaa* (dream) and *drashti* (vision). The dreams began approximately seven years ago—about the same time Sanand irreversibly became a boomtown. At this time, things in Kalubhai's life were not going too well. Family members got sick and one almost died. Another had a serious accident. Kalubhai started having dreams that woke him and shook him up violently during the night. Those who habitually slept in close proximity to him became alarmed, and some refused to be at his side at night.[13] This was followed by periods of sleeplessness and general uneasy tension. A saintly figure began appearing in his dreams, calling himself *Peer*, demanding sternly the return of his land. Says Kalubhai:

> *Pir shu kidhu? Mara jagia api do, keh.*
> What did the Pir say? He said "Give me my plot," he said.

Note the imperative tone of voice, unembellished and direct. Kalubhai is addressed directly and repeats this command often when he speaks, more often than it is possible to reproduce in a written text. It is as if there should be no doubt, no ambiguity in the listener as to what the spectral entity was asking for. In front of his fellow villagers, Kalubhai speaks in a similarly blunt way. Sometimes he will use the word "plot" or "space" (both *jagia*), sometimes "land" (*jamin*), always indicating that which must be returned. In the logic of the story that Kalubhai relates, he is at this point not yet aware of the fact that the saintly or divine figure appearing in his dreams is a Muslim.

Kalubhai knew that his paternal grandfather, his *dada*, had illicitly enlarged his traditional agricultural plot, the one Kalubhai had inherited from his own father, by encroaching on adjacent land. On that adjacent land, which he refers to as "neutral" (specifically, he means "without an owner"), once stood what he calls a "cenotaph" (*paliyo*). Stone cenotaphs are strewn all over the rural landscape of central and northern Gujarat, stretching all the way into Rajasthan (Lehmann 2003). They are usually associated with the deeds of heroic local figures detailed in local legends. They include equestrian Rajput knights defending cows and kingdom from marauders, or, alternatively, chaste women who chose to immolate themselves on their husband's funeral pyre in chastity and sacrifice. Kalubhai's grandfather had razed the stone and enlarged his land by appropriating the unused part. Until recently, Kalubhai had not thought much of it, as illegal encroachment on adjacent or neighboring land is a common rural practice. But then unfortunate events began to happen, and Kalubhai became focused on these past transgressions.

Later, villagers affirmed this to me: it was business as usual to encroach on neighboring and adjacent agricultural land. Everyone would be doing it, as long as they could get away with it, and land and real estate registries in India are legendary for being knotty institutions. Information about land ownership was hazily recorded and can easily be manipulated through corrupt machinations. Encroachment constitutes a form of unscripted rural practice that is as common as it is kept secret and silent. Occasionally, land disputes can culminate in serious violent conflict, often between close relatives such as cousins or brothers.

In any event, a while later, the figure appeared again in Kalubhai's dreams and gave him an ultimatum, saying that in *sava mahina*—that is, in exactly five weeks—he *must* give back the plot, the place onto which his grandfather had encroached. In a third dream, finally, a face appeared and spoke. Again, he was addressed directly:

> *Hu Pir chhu, keh! Mara jagia mane api do!*
> "I am a Pir!" he said. "Give to me my plot!"

This time, it was the face of a small boy with a green turban. In the dream, the child began dancing and jumping wildly all over the place. For the first time, Kalubhai responded to the command: "I will give your land back, trust me." But the boy, instead of an answer, spread the fingers on one of his hands very wide. The fingers elongated unnaturally until the veins popped up from under his flesh and finally seemed to separate from one another in a horrific vision. Kalubhai was terrified. The image of limbs that stretch beyond any natural length is a typical characteristic for nightmarish apparitions associated with malevolent ghosts (*bhut*) in Gujarat. I have heard similar descriptions of ghostly apparitions, which are usually hard to bear and described as a disturbing, nightmarish ordeal. Kalubhai took the warning to heart.

Not understanding exactly who this unsettling figure might be, but sufficiently troubled by the warning, Kalubhai began planning for the building of a temple to Ramdev Peer Bhagwan, a locally known deity, on his land.[14] He naturally assumed that the nightly apparitions were of a Hindu divinity; the word *Peer* is often used in written Gujarati for what is considered a Hindu deity—however, even more commonly the homonym *Pir* denotes the presence of a dead Muslim saint in his shrine (mausoleum).[15] At the specified place, where the ancient stone had once stood, on the land he had inherited from his father and grandfather, Kalubhai made preparations to install, ritually, the picture of Ramdev Peer. Together with his cousin, he went to an appropriate shop in order to buy a small marble temple for housing the icon.

Then something strange happened. The shop owner, instead of welcoming the business, reacted with an inexplicable hesitation. Encouraging Kalubhai to sleep over his decision, he told him to go home and come back the next day. This was no ordinary behavior for a sales merchant, Kalubhai explained. Warned by the shop owner's strange behavior, for which Kalubhai can give no satisfactory explanation, the next morning the entity manifested itself again in Kalubhai's head and addressed him most forcefully:

> *Beta, hu mandir ma na hoi, hamara dargah hoi, keh. Me dargah na Pir chhe, keh*
> "Son, I am not in temple, we rest in *dargah*," he said. "I am a Pir of the *dargah*," he said.[16]

Now Kalubhai understood that these visitations were from a *Muslim* Pir, a dead Muslim saint, whom he began addressing as *Pir Dada* (Grandfather Pir). There was no other name at this time that he could have used. Kalubhai explained to me that he did not know enough about Islam or Muslim shrines, but in the villages of surrounding areas, Muslims shrines belong as much to the local, sacred landscape as do the ubiquitous mother goddess shrines or Ram and Hanuman temples. However, this village never had a Muslim shrine in living memory. Indeed, in the times of Kalubhai's grandfather, when the small stone was visible on the borders of his agricultural fields, villagers referred to it merely as *paliyo* (cenotaph) or simply *pattar* (stone). The stone was situated next to a tree, which somehow was associated with it. But no shrine existed on the spot at that time, and to date it remains unclear whether the entity believed to be dwelling there had ever been considered Muslim in the past.

This was now to change: news of Kalubhai's dreams sparked various memories among his rural neighbors and fellow villagers. They suddenly remembered that in the olden days, the spot around the tree was called the place of Chinthariya Pir—*chinthra* from "rags." Visitors brought pieces of used, colorful cloth and attach them to the tree as a form of *mannat* (vow).[17] Villagers began telling Kalubhai that many years ago, a white horse had been seen galloping in this area. Some claimed that it moved about alone through the brush; others saw a rider on the horse, a large male figure with a long beard and handsome features, wearing a turban (*pagdi*).[18] Some again confirmed seeing strange things moving about in the dark. Soon enough, Kalubhai saw the figure in his dreams in these very details, as it addressed him:

> *Hu Chinthariya Pir chhu ane hu Makkahna panch pir paikino ek chhu.*
> "I am Chinthariya Pir, and I am one of the Five Pirs of Makkah."[19]

The Pir rode on a white horse and carried a flame in his hand. Residents of the village told me that the flame had, in fact, often been seen at night: a light moving in in the dark through the fields in different directions, but always stopping at the approximate location of the shrine. This, then, was Chinthariya Pir, the star in Kalubhai's visions and nightly dreams.[20]

The shrine, whose infancy I witnessed four years ago when I began this ethnographic work, is now a large structure with corrugated ceilings and balustrades. It looks humble but sophisticated enough to withstand many years to come. It forms a high cement platform in the middle of nowhere, surrounded by brush and adorned by a large tree, located between agricultural fields of castor oil fruits (*eranda*). The tree has grown large, littered with hundreds of colorful pieces of ripped cloth from old garments like *saris*, *kurtas*, *shalwar kameez*—the strips of cloth acting as material reminders of issues, ailments, and desperate wishes. There is no paved road, nor is there any possibility to reach this shrine in any comfortable manner; it can only be reached by walking along fields and through thick underbrush pathways. During the monsoon season, the shrine is barely accessible except by wading through long stretches of muddy water.

The surprising speed with which Kalubhai Bharvad was able to furnish a proper Muslim shrine with limited funds speaks to the effectiveness of the local register in which he is playing to his village audience. Soon Kalubhai invited Alkhubhai, an elderly Muslim man from the Qureishi community of a neighboring village, to officiate as *mujavar* (caretaker) at the shrine, which is now officially named Chinthariya Pir Dada Dargah. In a dream, Kalubhai beheld that Alkhubhai's grandfather used to officiate for the forgotten Pir, and that Alkhubhai was actually born not in a neighboring village, as he himself had thought, but in Kalubhai's village, and only *afterward* was taken to the former to be raised. Kalubhai contacted Alkhubhai and told him this information from his dreams. They both relate to one another as old familiars and friends.

Alkhubhai now visits the shrine every Thursday. It is the day when Kalubhai falls possessed (*hajri*) and draws a small local audience, limited to surrounding villages and rural areas. Alkhubhai told me that he knew nothing about being a shrine caretaker before Kalubhai's invitation. He had to learn everything from scratch and impressed himself by managing to do this at such an advanced age. It seems that he has become positively infected by Kalubhai's energy and enthusiasm.

Conclusion

Why did the question of culpability never properly emerge in public discourse after the pogrom? One reason was Narendra Modi's political will to shape the electorate into a permanent Hindutva majority.[21] While the death of victims during the pogrom was deemed tragic, Hindu anger was ultimately rationalized by identifying a prior crime committed in Godhra that ostensibly offered moral legitimization for all that had followed. Construing the pogrom as a mere reaction to the Godhra incident confirmed the truth of Hindutva moral narratives that have defined Hindus as hapless victims of Muslim aggression since the arrival of Islam on the subcontinent. The aftermath of the pogrom was an obdurate morass of political scheming, public lying, and obfuscation of facts. Aided by the powerful influence of Hindu nationalist organizations in the state, after the violence Gujarat settled for weeks into a zombie-like disposition, simultaneously shocked and fascinated by the "naturalized" unfolding of the surreal events of the pogrom. The general silence of many Gujaratis crept up on me, too, like a paralyzing fog, lodging itself into my own speech, blurring the memories of many with whom I spoke, twisting tongues into so many make-believes, and finally obscuring everything that had been so obvious during the dark days of violence. What could have been an honest reckoning became a farce. It is that farce which eventually propelled the chief minister onto the national stage.

Modi pre-empted the time for Gujaratis to express sorrow or regret or any other form of compensatory action. The state itself orchestrated no mourning, which might have led to reflection on forms of culpability and complicity. In fact, instead of introspection and a reckoning, while Muslim relief camps were still operating, the chief minister began campaigning for elections to be held within months of the violence, in December 2002, a timing that the Election Commission of India criticized severely at the time. He denounced anyone questioning Gujaratis or their nonviolent credentials. In this way the Gujarat pogrom and its murky aftermath became the midwife to Narendra Modi's political success, which is why he can never properly address or denounce the events. The pogrom permanently stained his political career; whatever his political success may be in the future, he will never be able to whiten this bloody spot.

During and especially after the violence, Modi did not try to conceal his Hindu nationalist bias. This changed the rules of the game for India at large after which he later became prime minister. In Gujarat, he was the chief minister who need not show any apologetic contrition or hesitation because the pogrom defined what he considered a mere matter of fact: minority-Muslims are at the mercy of majority-Hindus. By declining to defend Muslims against the violence of the majority, he symbolically confirmed—openly—what had been implicitly understood for a long time. This was also one of the reasons why many considered him an authentic arbiter of truth. He did not hide his Hindu nationalist leanings and prejudices. This was a significant departure from the rhetorical contortion of prior generations of state politicians, who might have held similar beliefs, but tried to mold their communal strategies to fit democratic form and pay lip service to Mahatma Gandhi.

According to Ashis Nandy, the murder of Gandhi was a collective communiqué in which many participated, not least Gandhi himself (Nandy 1991, 70–98). Modi confirmed the sense that Gandhi must be "killed" again, so to speak, in the present; the Gandhi that Modi needed Gujaratis to dispose of was the troubling hesitation and doubt in their minds that had blocked them from blatantly affirming a majoritarian Hindu dominance in the past. Hindus were to emerge from the pogrom free to choose violence, free of the superego's torturous grip. Modi thus transformed the ambivalence that many Gujaratis felt for this figure into a form of Hindu righteousness and indignation; where Gandhi had invoked ancient Hindu ideals such as nonviolence (*ahimsa*) in order to renounce violent action, the new dispensation defined Hindus as vulnerable exactly because of their intimate association with the very same ideal. This vulnerability needed an organized, violent Hindu response, which was achieved by a delegation of violent labor to Hindu nationalist organizations, who, after the pogrom, went mostly unpunished—the leaders not held responsible, their institutional structures unchanged. The propaganda flyers, the statements by local politicians, the rumors and misrepresentations—nothing was ever walked back; nothing was addressed, reflected upon, or set right. Modi simply shifted gears, and most Gujaratis followed suit. The humiliation and stain of the victims remained.

This new strategy, wherein he had already shown himself in sync with the mood of the majority of voters, marked the beginning of what would become Modi's meteoric rise onto the national stage more than a decade after the Gujarat pogrom. He consequently has won every election in which he campaigned. In Ahmedabad, the absence of a discourse on guilt has petrified the relationship between the two aggregate categories, Hindus and Muslims, even further. It has now created the "H-class" and the "M-class" of citizens, a hierarchy as noxious as the one between caste-Hindus (*savarna*) and those traditionally excluded from the Hindu moral order (*avarna*). The logic of the former is significantly connected to the logic of the latter, in that this new hierarchy is to replace and substitute for the older division between caste and outcaste Hindus. In this new moral order, individuals who show remorse have not been able to transform such retrospective feelings into effective political opposition (Jasani 2020, 676–7).

Guilt, integrated into a landscape of memory, might have avoided this development, as it can, under the right circumstances, foster prosocial effects conducive to the expression of regret and responsibility, and can articulate a general concern for justice (Baumeister, Stillwell, and Heatherton 1994). (For more on guilt's prosocial effects, see Lisa Spanierman's chapter in this volume.)

In this chapter, I have described three ethnographic encounters wherein the identification of guilt becomes possible. In the first case, Bharat is stricken with sympathy when he witnesses a harm done to Muslims, because he can empathize—as a Hindu—with the experience and the cultural memory of seeing his own people's religious heritage destroyed. It is the only time I saw Bharat show something like hesitation or empathy for Muslims as a generalized, aggregate category. In the second case, Doctor Shah is indignant, shows feelings of anger and, most importantly, shame while opening up to two foreigners and simultaneously denigrating his native patients. For Doctor Shah, there is no question of culpability; there is only clearly defined guilt. The shame that he feels stems from the complicity of members of his own community (the Jains) in the Gujarat pogrom. (For more on the relationship of [collective] guilt to [collective] shame, see Matthias Buschmeier's chapter in this volume.)

The third case is the most complicated and the most theoretically challenging, as it suggests various levels of unconscious communication. (For more on unconscious communication, see John Borneman's contribution in this volume.) Kalubhai Bharvad denies any meaning that invokes contemporary politics (*rajkaran*) or sociological facts, such as the agrarian order of land tenure, acquisition, and loss; the marginalization of the minority-Muslim community; or the events of the pogrom. Yet Chinthariya Pir appears in his dreams and gives stern commands, which are almost exact replicas of what local Muslims say when they realize the gravity of the loss of their agricultural land in Sanand district and their corresponding economic decline. Kalubhai's shrine, in a way, takes us back to a prior loss. It manifests the need to make Muslims present again—indeed, quite literally, to *give them back their place*. The shrine of Chinthariya Pir represents an attempt to reconstitute the chain that links aggregate categories allegorically. It reconstitutes the complicated local concatenation of worship, belief, and co-presence to one another that characterizes the unique character of Indian cultural traditions. After waves of violence, Hindutva agitation, a takeover at the political center in New Delhi, and a hyped-up development agenda that has run aground, these links have been severed.

Kalubhai reinstates this chain by implicating his own father and grandfather in a transgression that could have easily gone unnoticed—but he makes their transgressions count productively in the present. His actions express a specific, generational guilt in somatic form, through disease and possession states in which land that belonged to a supernatural entity is reclaimed, and in which an ambiguous figure ultimately becomes defined in the form of a proper Muslim Pir. What is most astonishing in this gradual development is the support that Kalubhai receives from members of the village community, who actively participate by associating, remembering, and imagining a bearded equestrian figure trotting through the fields, holding a light in the dark. While Kalubhai might be considered an odd fellow, to my knowledge, no one really opposed him or tried to sabotage his ambitious project.

The Pir does not speak to Kalubhai as friend or ally, however, but as a grim figure of authority making demands, threatening and even taunting him. The orders are difficult for a Hindu to follow. And yet Kalubhai submits. This impresses his village audience, as well as visiting Muslims from neighboring villages or urban slum areas in Ahmedabad, adept at discerning fake from authentic moments of religious sublimity. What is the agency of this commanding voice? Who is speaking? It is hard to avoid the conclusion that Kalubhai speaks against all odds in the voice of a collective superego, a voice demanding the recognition of a transgenerational guilt vis-à-vis Muslims. And although this ethnographic exposition ends as an unfinished story whose further unfolding must be awaited, one cannot avoid the impression that, in the contemporary political climate, there lies an element of atonement in Kalubhai's eager willingness to submit to a Muslim Pir.

In 2019, while on a trip from Ahmedabad to Kalubhai's village, I ask my friend Habibabhai about Hindutva, a question I had not asked him for a long while. I wanted to know what it represents to him after 25 years of Bharatiya Janata Party (BJP) rule in Gujarat, nearly 17 years after the pogrom and five years of Modi rule at the Center in Delhi. As a Muslim living in the state at the margins of a large city, he had personally witnessed the violence of 2002. His survival story comes with many gruesome tales and experiences from which he never properly recovered. It seemed to me that he was the perfect person to ask. After pondering briefly, accustomed to my many questions, he gave me one of his shortest answers ever:

> What is Hindutva? All are all united, but Muslims remain separate . . .
> (hindutva etle shu? bada ek, musalman alag . . .)

The religion of Hindus and Muslims was *sanklaayelu*, intertwined or interconnected (also *jodaayelu*). In Gujarati, a *saankal* is a chain, the individual members of which are locked to one another. They cannot be separated without causing harm, without the chain losing its function of binding local communities together into a unity. Land stands metaphorically for the ancestral place of Muslims in Gujarat, for their link in the chain—their intimate belonging to the state. If they migrated from rural areas into the city, they did so out of dire need in years past because their plots of land yielded too little to survive. In cities like Ahmedabad, waves of violence have again displaced them into ghettos and slums. Muslims have lost their origins, their place in the chain. Kalubhai's quickly moving legs and busy voice seems to want to retrieve this missing link, to repair the chain, to compensate for their loss.

Notes

1. There are many large and small violent events in Gujarat that are routinely called "riots," although empirical events differ widely in form and motivation (cf. Ghassem-Fachandi 2012).
2. "Terror Within." *The Times of India* (Ahmedabad edition), May 29, 2002, 10. The interview was conducted by journalist Smita Gupta. Uma Bharati is a Hindu nationalist and currently national vice-president of the Bharatiya Janata Party (BJP), the ruling party of India.

3. "Psyche of the Aggressor: No Kalinga Effect in Gujarat." *The Times of India* (Ahmedabad edition), July 13, 2002, 10. The "Kalinga Effect" invoked in the title is a reference to the remorse and contrition attributed to Mauryan Emperor Askoka after the battle of Kalinga in 265 BCE.
4. Compare also Kingshuk Nag. "The Guilty of Gujarat: No One to Police the Police." *The Times of India* (Ahmedabad edition), June 1, 2002, 10.
5. One of the couplets attributed to Vali Gujarati, written in Urdu verse, goes: "My heart is full of thorns longing for Gujarat. Restless, frantic, wrapped in flames in the spring. In this world, there is no balm for its wound. My heart is split apart by the dagger of separation."
6. Compare Narendra Modi (2002, 33–40). The Godhra incident on February 27, 2002, remains highly obscure. For a detailed discussion of the unfolding of legal proceedings over the years in the wake of the 2002 violence until approximately 2011, see Jaffrelot (2012, 77–89) and Eckert (2010, 155–68).
7. The formulation is taken from Sudhir Kakar (1995, 51).
8. This stereotypical expression is very commonly used in Gujarat and elsewhere in India and can be traced to a text by Mohandas K. Gandhi, who wrote "(T)he Mussalman as a rule is a bully, and the Hindu as a rule is a coward." Compare "Hindu-Muslim Conflict, its Causes and Cure." *Young India*, May 29, 1924 (compare also Ghassem-Fachandi 2019, 83–98, and Kakar 1995, 438).
9. For an exploration of idealization in communal mobilization in India, compare Kakar (1995).
10. *Enjoi* is idiomatic Gujarati taken from the English word "enjoy" and employed in a Gujarati sentence with a local logic (Ghassem-Fachandi 2012, 74).
11. A *bharvad* is a member of a community whose traditional occupation was the caretaking of animals (shepherd caste: herding livestock).
12. These cases should be distinguished from the attempts at the reconversion of Muslims and Christians in the *ghar vapasi* campaigns by Hindu nationalist organizations, which peaked in 2014, twelve years after the pogrom.
13. In Gujarati villages, it is not uncommon for larger groups of men to sleep together during the night in a barn or even outside on cement platforms near a temple or shrine at a village square.
14. The choice of this Hindu folk deity is very significant but cannot be elaborated here for reasons of brevity. Note that it is somewhat astonishing that Kalubhai did not immediately understand the apparition of the boy to represent a Muslim. It makes sense, however, in light of the tradition of Ramdev Pir. For a discussion of Ramdev Pir (or Peer), the saint's relation to the Meghval community, to Islam and Ismailism, see the work of Dominique-Sila Khan (2003, 60–94).
15. Compare Khan (2003, 97–124). Note that in spoken and written Gujarati, the words *Pir* and *Peer* are indistinguishable.
16. A *dargah* is a Muslim shrine, the mausoleum of a dead Muslim saint, i.e., a Pir.
17. The tree onto which colorful pieces of cloth are attached when making a vow is a common sight in rural Gujarat. Not exclusive to one particular religious community, this is a traditional practice not mentioned in religious texts. The "rag uncle" is a traditional figure of folk tradition (Daya [1848] 1990, 19–21).

18. This description fits depictions of Ramdev Peer as much as various Muslim saints. It also resembles other apparitions that Muslims regularly mention to me of elongated, bearded figures that often appear in mosques. These latter apparitions are referred to as "white djinns." They visit mosques to pray during the night. People respectfully avoid them because their anger can have serious consequences. While a form of spirit being, they are not malevolent by nature and are considered pious.
19. Dominique-Sila Khan (2003, 63) mentions the current hagiography in which a Hindu Ramdev defeated five Muslim Pirs from Mecca, "who were forced to admit that he was more powerful than they." In Kalubhai's version, we see a transformation typical for the flexibility of folk traditions, where Ramdev is constructed as one of the five Pirs and not their rival.
20. So far, no one has been able to explain to me who or what the five Pirs of Mecca are. They appear in the mythology of Ramdev Peer, the Hindu deity. Ramdev Peer is depicted with a beard and sitting on a horse indistinguishable from the other five Pirs (or sometimes possibly even part of the five?). Compare Khan (2003, 65).
21. For a discussion of the Hindu majoritarian ascendency in recent years, see Chatterji, Hansen, and Jaffrelot (2019, 1–15).

References

Baumeister, Roy F., Arlene M. Stillwell, and Todd F. Heatherton. 1994. "Guilt: An Interpersonal Approach." *Psychological Bulletin* 115, no. 2: 243–67.

Borneman, John, and Parvis Ghassem-Fachandi. 2017. "The Concept of *Stimmung*: From Indifference to Xenophobia in Germany's Refugee Crisis." *Hau: Journal of Ethnographic Theory* 7, no. 3: 105–35.

Chatterji, Angana P., Thomas Blom Hansen, and Christophe Jaffrelot, eds. 2019. *Majoritarian State: How Hindu Nationalism is Changing India*. London: Hurst & Company.

Daya, Dalpatram. 1990. *Demonology and Popular Superstitions of Gujarat*. Translated by Alexander Kinloch Forbes. New Delhi: Vintage Books. First published 1848.

Eckert, Julia. 2010. "Kultur und Schuld: Narrative der Verantwortung." In Zurechnung und Verantwortung: Tagung der Deutschen Sektion der Internationalen Vereinigung für Rechts- und Sozialphilosophie vom 22.–24 September, edited by Matthias Kaufmann and Joachim Renzikowsky, 155–68. Stuttgart: Franz Steiner.

Frembgen, Jürgen Wasim. 1993. *Derwische: Gelebter Sufismus, Wandernde Mystiker und Asketen im islamischen Orient*. Cologne: DuMont.

Gandhi, Mohandas Karamchand. 1924. "Hindu-Muslim Conflict, Its Causes and Cure." *Young India*, May 29.

Ghassem-Fachandi, Parvis. 2012. *Pogrom in Gujarat: Hindu Nationalism and Anti-Muslim Violence in India*. Princeton, NJ: Princeton University Press.

Ghassem-Fachandi, Parvis. 2017. "In Conversation with Sindre Bangstad and Karin Kapadia." In *Anthropology of Our Times: An Edited Anthology in Public Anthropology*, edited by Sindre Bangstad, 149–82. New York: Palgrave Macmillan.

Ghassem-Fachandi, Parvis. 2019. "Reflections in the Crowd: Delegation, Verisimilitude, and the Modi Mask." In *Majoritarian State: How Hindu Nationalism is Changing India*, edited by Angana P. Chatterji, Thomas Blom Hansen, and Christophe Jaffrelot, 83–98. London: Hurst & Company.

Hacker, Paul. 1957. "Religiöse Toleranz und Intoleranz im Hinduismus." *Saeculum* 8: 167–79.

Hacker, Paul. 1983. "Inklusivismus." In *Inklusivismus: Eine Indische Denkform*, edited by Gerhard Oberhammer, 11–28. Vienna: Institut für Indologie der Universität Wien.

Jaffrelot, Christophe. 2012. "Gujarat 2002: What Justice for the Victims?" *Economic & Political Weekly* 47, no. 8: 77–80.

Jasani, Rubina. 2020. "Violence, Urban Anxieties and Masculinities: The Foot Soldiers of 2002, Ahmedabad." *Journal of South Asia Studies* 43, no. 4: 675–90.

Khan, Dominique-Sila. 2003. *Conversions and Shifting Identities: Ramdev Pir and the Ismailis in Rajasthan*. New Delhi: Manohar (Centre de Sciences Humaines). First published 1997.

Kakar, Sudhir. 1995. *The Colours of Violence*. New Delhi: Viking Penguin Books India.

Lehmann, Nicole Manon. 2003. "Über den Tod hinaus: *Sati*, das Ideal and Kshatriya Ehefrau." In *Selbstopfer und Entsagung im Westen Indiens*, edited by Nicole Manon Lehmann and Andrea Luithle, 9–277. Hamburg: Dr. Kovač.

Makawana, Kishor, ed. 2002. *Godhra Hatyakand, Kalam no Dharma ane Adharma*. Kishor Rajkot: Pravin Prakashan.

Modi, Narendra. 2002. "Lohi nahi, paani . . . " In *Godhra Hatyakand, Kalam no Dharma ane Adharma*, edited by Kishor Makawana, 33–40. Kishor Rajkot: Pravin Prakashan.

Nandy, Ashis. 1991. *At the Edge of Psychology*. Calcutta: Oxford University Press.

Sundar, Nandini. 2004. "Toward an Anthropology of Culpability." *American Ethnologist* 31, no. 2: 145–63.

Varadarajan, Siddharth. 2002. *Gujarat: The Making of a Tragedy*. New Delhi: Penguin Books.

12
The Guilt of Warriors

Susan Derwin

War Is a Moral Quagmire

In 2011, Karl Marlantes published *What It Is Like to Go to War*, a meditation on his combat experience as a Marine during the Vietnam War. It took Marlantes more than forty years to produce the book, and the timing of its publication was not coincidental: it appeared while the United States was heavily engaged in wars in Iraq and Afghanistan. For young people thinking about enlisting or on the verge of entering combat, Marlantes hoped that his reflections on war would serve as "a psychological and spiritual combat prophylactic" (Marlantes 2011, xi).

What It Is Like to Go to War focuses on the ethical, emotional, and existential consequences that result from the active infliction of death upon the enemy and the accompanying "compromises with, or outright violations of, the moral norms of society and religion" (xii). Its orientation is particularly noteworthy, in that it stands in sharp contrast to the major body of psychological writing that emerged after Vietnam on combat-related post-traumatic stress disorder (PTSD). Studies of PTSD understand war trauma as the outcome of a situation of incapacitating passivity, in which the warrior's mind is overwhelmed by a quantity of stimuli that it cannot process (see Freud 1963, 275).[1] In this regard, the diagnosis of PTSD advances an understanding of veterans as exceptional and damaged: they are survivors of situations in which they were victimized by forces they could not control. As a diagnostic category, PTSD reinforces the differences between (unwell) veterans and (healthy) civilians. Marlantes's book, by contrast, considers the psychological effects of killing not as a malady in need of a cure but as a normal reaction to a powerful experience. In this regard, it enjoins readers to place themselves in the situation of warriors, and, in so doing, to see veterans not as "Others" but as persons like themselves, whose differences are a question of experience only: they are different because they have been called upon to cultivate and exercise a capacity to kill that all people nonetheless share.

Most civilians, however, are resistant to recognizing their affinities with people who have killed. As a result, the moral "compromises and violations [that accompany wartime killing] are not generally discussed, and their impact on a warrior's mental well-being and soul is minimized or even ignored" (Marlantes 2011, xii).

Susan Derwin, *The Guilt of Warriors* In: *Guilt*. Edited by: Katharina von Kellenbach and Matthias Buschmeier, Oxford University Press. © Oxford University Press 2022. DOI: 10.1093/oso/9780197557433.003.0013

In Marlantes's view—and it should be noted that his frame of reference is primarily that of democratic cultures—this silence is unwarranted and harmful, because it enables civilians to evade taking their share of responsibility for the consequences of the wars carried out in their names. Hence, Marlantes's mission in writing a book that demystifies combat-related violence is twofold: to help prepare present and future warriors for battle, and to cultivate a greater feeling of responsibility among civilian citizens. He writes: "If the ideas discussed here help citizens and policy makers attain a clearer understanding of what they are asking of their warriors and of their own role in sending them into the moral quagmire and sacrificial fire called war, then the book will have succeeded, if not beyond my hopes, perhaps beyond my expectations" (xii).

War is a "moral quagmire" because it involves the complicated issue of guilt in relation to killing. Most recently, warriors' guilt has been a central focus in the treatment of veterans of the wars in Iraq and Afghanistan. Increasingly, mental health professionals are adopting the concept of "moral injury" to characterize the psychic wounds of war. The psychiatrist Jonathan Shay first used the term in his book *Achilles in Vietnam* (1994) to refer to the effect upon soldiers of the betrayal of their trust by figures of military authority. Shay argues that commanders fail their charges when they engage in poor, often self-interested decision-making and abusive leadership practices. According to Shay, warriors who suffer moral injury are victims of betrayal by military leaders whose actions destroy soldiers' trust, not only in their leaders but in society more generally. Today, though, mental health professionals have adapted the concept of moral injury to describe the guilt warriors carry in the aftermath of having actively perpetrated or witnessed acts that "transgress deeply held moral beliefs and expectations" (Litz et al. 2009, 697). If PTSD is an experience of suspended agency, moral injury is one of too much agency. The conflict between what warriors did or witnessed and their moral evaluation of those events leads them to experience themselves as agents of their own injury.

The point of overlap between Marlantes's exploration of the impact of killing upon warriors and the revised definition of moral injury concerns their common focus on the agency of the warrior. But whereas the concept of moral injury departs from the assumption that the warrior's feelings of guilt are justified—warriors have been involved in moral transgressions for which they must atone—Marlantes insists that the killing that warriors do must be understood within an amoral framework: in combat, killing is a necessary action, not a freely made choice; hence it is amoral. By the same token, Marlantes is aware that, once the warrior is removed from the battlefield, his moral sensibility reawakens, and with it, the possibility of belatedly feeling guilt for the past deeds, which, when they were committed, had no moral valence. In what follows I will trace how even warriors who believe the act of killing to be amoral, because it is an unavoidable part of combat, must manage the central contradiction of guilt, which derives from their dual membership in mutually exclusive worlds. Understanding the evolving significance of killing for warriors once they have left the battlefield and the contradiction of guilt they face affords insight into "the moral quagmire" (Marlantes 2011, xii) that confronts them in the aftermath of war.

Guilt Is a Misnomer

The question of guilt was central to journalist Bill Moyers's 2012 TV interview with Marlantes about his experiences in Vietnam. Their conversation was significant for what it revealed about Marlantes's combat history, and also because it laid bare the distinct, competing frames of reference that veterans and civilians bring to the table when they speak about war. Moyers began the interview with a question about returning to civilian culture:

> BILL MOYERS: When you have killed, for the rest of us, are you ever to feel that you are one of us again?
> KARL MARLANTES: That's a great question. There is a sense of alienation that I think most veterans deal with. And it is because you have done something that most people think is horrific. And the fact of the matter is, if you were a proud soldier or Marine, you felt good about it. And then you say, "Hm, maybe I should feel bad that I feel good about this." And so you get these moral reverberations going around in your head that you're sure that no one else is going to understand [...]. But you do have a sense of alienation. I think it's one of the issues of trying to reintegrate.
> BILL MOYERS: Talk a little bit about that further, about that estrangement. Is it guilt? Is it. . . What is it?
> KARL MARLANTES: Well, it's hard to say that it's guilt. Because I've killed people, more than one. And I feel sadness about it, a great deal of sadness. That's different than guilt. And—. But I think that if someone's going to talk about it, you have the fear that they won't understand what you're saying. And that they'll judge you badly. (Marlantes 2012)

Moyers first inquires about the impact of killing upon veterans' social reintegration. Implicit in his question about whether Marlantes feels like he belongs to the civilian world now that he has killed people is the assumption that the killing that Marlantes has done, even though it is "for the rest of us," nevertheless separates him from civilians. Marlantes's response suggests that the source of his alienation is not the act of killing, but rather the judgments civilians make about it, the fact that "most people think [killing] is horrific," while warriors may feel good about having killed. In his follow-up question, Moyers speculates about the cause of Marlantes's sense of alienation: "Is it guilt? Is it. . . What is it?" Invoking guilt as a possibility reveals more about civilian assumptions regarding what veterans feel, and should feel, about having killed, than it does about Marlantes's actual experience. For this reason, it is understandable that Marlantes does not accept Moyers's interpretation of his "estrangement" as evidence of feelings of guilt.

Even with his clear rejection of guilt as the explanation for his alienation, Marlantes is not satisfied with the outcome of the exchange; in the interview, he circles back to the topic, reiterating his initial response: "Again, I don't think the word is 'guilt.' Those people were trying to kill us. But [. . .] sadness that those people were there just like we were. They were 18. We were 19, 18. And there we are killing each other" (Marlantes 2012). The fact that he returns to the issue, unprompted, attests to its significance for Marlantes. In this regard, the language in which he discusses guilt is illuminating. Each of his rejections of guilt contains a qualification, expressed as a linguistic obstacle: the first time he states, "It is hard to say" he feels guilt, and the second time that "the word" for what he feels is not "guilt." He also makes an allusion each time to killing that alerts us as to why the issue of language is at the fore of Marlantes's thought process: in his initial response, he says, "I've killed people, more than one"; when he returns to the issue of guilt, he comments, "Those people were trying to kill us." I believe that Marlantes mentions that he killed men each time he states that he does not feel guilt, because his mind is engaged in the challenge of finding the right language to describe the impact of killing. That is, in rejecting Moyers's conjecture about "guilt," Marlantes is rejecting not the possibility that killing produces guilt, but rather he is rejecting the system of values that Moyers invokes in speaking about the impact of killing. "Guilt" is a term that references notions of moral probity, and using it in the context of battle glosses over the crucial difference between the world of the civilian and that of the warrior. Whereas civilian morality is predicated on the prohibition against killing, that prohibition is suspended in combat. Marlantes writes:

> War is the antithesis of the most fundamental rule of moral conduct we've been taught—do unto others as you would have others do unto you. When called upon to fight, we violate many codes of civilized behavior. To survive psychically in the proximity of Mars, one has to come to terms with stepping outside conventional moral conduct. (Marlantes 2011, 48)

Insofar as Marlantes's rejection of Moyers's invocation of "guilt" simultaneously calls attention to the challenge of finding the language in which to express the impact of killing upon the warrior, it is less a commentary on the presence or absence of guilt in the aftermath of killing than a rejection of the terms in which the alternatives are posed to him.[2] To have answered the question in the affirmative would thus have been to sidestep the problem inherent in using morally laden language to characterize the impact of actions undertaken in a world that is oriented around the absolutes of killing or being killed. In *not* taking that reality into account, Moyers's question tacitly assumes that Marlantes the warrior (whom Moyers presumes felt no guilt on the battlefield) must be at odds with Marlantes the veteran (whom Moyers presumes *does* feel guilt now). Marlantes's feelings of alienation may well derive from the fact that whenever he speaks about his experience with "the rest," he is confronted with the value-laden language of civilian morality. This enables us to understand why, when he enters into conversations with civilians, he is concerned that he will be judged poorly: in the eyes of civilians, who apply the values of civil society to combat, a warrior who does not feel guilty about killing must be without a conscience.

When Moyers asks Marlantes to talk further about his estrangement, Marlantes instead continues speaking about what killing makes him feel: not guilt, but sadness. Moyers never asks Marlantes to elaborate on his response, and his subsequent question indicates why: he is more interested in discussing the violence of war than Marlantes's sadness. Moyers states: "You write in *What It's* [*sic*] *Like to Go to War* that 'the violence of combat assaults psyches, confuses ethics, and tests souls.' This is the result, not only of the violence you suffer, but the violence you inflict?" (Marlantes 2012) Whereas, in the passage that Moyers reads aloud, Marlantes is discussing the warrior as the object of violence, "assaulted" by the violence of combat, Moyers's focus is on the warrior as an agent of violence.

At this point in the conversation, Moyers and Marlantes appear to be operating with different understandings of the warrior's subjectivity. Moyers's interest in guilt and in the warrior's infliction of violence bespeaks a conceptualization of the warrior as exercising the free will of autonomous agency. Marlantes's subsequent comments make it clear, however, that individual autonomy is not a part of war. He states: "Very important, is the ability to do ultimate teamwork. You lose your sense of being an individual. You are part of—I can only go back to—the hunting group—that feeling of 'It's "we," not "me."' And that you are willing to sacrifice your 'me,' both as a physical body and as an individual, to make sure that the 'we' of the unit gets through this okay" (Marlantes 2012). Marlantes's comment suggests that a warrior makes a double sacrifice: in becoming part of the group, he gives up his ego, his individual "me," and, in so doing, he becomes willing to make another sacrifice: of his physical body—his life—for the survival of the group.

From the cohesiveness of the military group, there develops a mutual reliance among warriors that both neutralizes the terror associated with combat and also enables the group to act with an invincibility that no individual can feel on his own, as the following exchange between Marlantes and Moyers reveals:

> BILL MOYERS: You describe a scene in which one of our helicopters has crashed on a hill, while trying to come to the relief of some of the Marines. And they're going to be taken out, the crew's going to be taken out, the enemy's at—. The North Vietnamese are after them. And all of you guys are below the summit of that hill. And suddenly, without even a command, you begin to act not as individuals, but as one. And you're all moving up the hill like an organism.
> KARL MARLANTES: Exactly. Yeah.
> BILL MOYERS: And you say it's not "me," it's "we."
> KARL MARLANTES: Exactly. We are—. We can't be killed. I can be killed, but that is an important part of the infantry. (Marlantes 2012)

The Ego at War

While surrendering one's ego to the identity of the group may serve to control fears about dying, for some warriors the struggle to give up the individual self persists into war. Iraq veteran Roy Scranton describes the terror that dogged him until he found a way to do so:

> In Iraq, at the beginning, I was terrified by the idea [of dying]. Every day I drove out past the wire on a mission, I looked in my Humvee's mirror and saw a dark, empty hole. Once the experience of war makes visible the possibility of death that lies locked up in each moment, our thoughts cannot travel from one day to the next without meeting death's face. I recognized that face in the dark of my Humvee's mirror. Its gaze almost paralyzed me. (Scranton 2015, 21)

Like Perseus before the Medusa, Scranton in Iraq confronts the image of his own death and is nearly paralyzed by it. Also, like Perseus, who protects himself from Medusa's deadly gaze by looking indirectly at the Gorgon's reflection in his shield, Scranton becomes inured to death by following the advice of an 18th-century Samurai's reflections on war: "Meditation on inevitable death should be performed daily" (21). Scranton writes:

> I took that advice to heart, and instead of fearing my end, I practiced owning it. Every morning, after doing maintenance on my Humvee, I would imagine getting blown up, shot, lit on fire, run over by a tank, torn apart by dogs, captured and beheaded. Then, before we rolled out through the wire, I'd tell myself that I didn't need to worry anymore because I was already dead. The only thing that mattered was that I did my best to make sure everyone else came back alive. (22)

Scranton envisions himself as dead in order to become deadened to his fears. When he manages to do so, only one thing still matters: ensuring the safety of the group.

Relinquishing the autonomy of the ego delivers another important effect. It enables the warrior to limit his capacity for empathy while in combat. Whereas in civil society, empathy is indicative of a person's humanity, on the battlefield, it is a liability. Literary and nonfictional accounts of the emotional numbness necessary for warriors to withstand killing often represent it as an experience of being dead to feeling.

For example, in the early pages of Eric Remarque's World War I novel *The Road Back*, the German soldiers returning from the front are represented as emerging from this insentient state. As Ernst, the protagonist, awaits orders to retreat, he reflects on the world he is leaving behind, and he realizes that he has been existing in a state of living death:

> They are many indeed that lie here, though until now we have not thought of it so. Hitherto we have just all remained there together, they in the graves, we in the trenches, divided only by a few handfuls of earth. They were but a little before us; daily we became less and they more, and often we have not known whether we already belonged to them or not. And sometimes too the shells would bring them back among us again, crumbling bones tossed up; scraps of uniform; wet, decayed heads, already earthy, to the noise of the drumfire issuing once more from their buried dugouts and returning to the battle. It did not seem to us terrible; we were too near to them. But now we are going back into life and they must stay there. (Remarque 1930, 20)

More than an empty figure of speech, Ernst's words about proximity to the dead are true in a literal sense; whenever enemy fire would reach its target, the trenches were transformed into graves for the soldiers who were hit. Conversely, the corpses of the dead soldiers would be ejected from their "shallow graves" by the force of the shells, as if suddenly come to life and mobilizing for action. Ernst's description of the dead as crossing back into life expresses a psychological truth: that soldiers in the trenches were physiologically alive yet emotionally "too near" to the dead. Hence the aptness of the description of returning to the civilian world as "going back into life," and leaving the dead to "stay there" in the trenches.

Vietnam veteran Tim O'Brien writes: "War makes you a man; war makes you dead" (O'Brien 1990, 120). To kill as a warrior, then, means to be dead to the death you are inflicting. When a warrior is in this state, the enemy appears to him as a non-self to be destroyed. O'Brien renders this experience in "Ambush," a story in *The Things They Carried*, in which the narrator describes what happened while he was on platoon watch and spotted a young Vietnamese soldier emerging from the fog and slowly moving up the trail in his direction. He writes:

> In a way, it seemed, he was part of the morning fog, or my own imagination [. . .]. It was entirely automatic. I did not hate the young man; I did not see him as the enemy; I did not ponder issues of morality or politics or military duty [. . .]. I was terrified. There were no thoughts about killing. The grenade was to make him go away—just evaporate—and I leaned back and felt my mind go empty and then felt it fill up again. I had already thrown the grenade before telling myself to throw it. (126)

As with the narrator in Remarque's text, O'Brien's narrator does not register the killing of the soldier as the latter's death, because, under the duress of war, killing becomes an automatic action carried out against an enemy in a desensitized state. The narrator perceives his target as just another part of the landscape—the morning fog—or as an extension of himself—a figment of his imagination, that is, as an entity either too distant or too proximate to be apprehended as a person in his own right. When the narrator throws the grenade, it is not with the intention of killing another human but of making him "go away—evaporate," as if the soldier were a liquid substance.

In *What It Is Like to Go to War*, Marlantes writes about his struggle to explain to civilians how sacrificing one's self to the survival of the group precludes feelings of empathy toward the enemy:

> When people come up to me and say, "You must have felt horrible when you killed somebody," I have a very hard time giving the simplistic response they'd like to hear. When I was fighting—and by fighting I mean a situation where my life and the lives of those for whom I was responsible were at stake [. . .] either I felt nothing at all or I felt exhilaration akin to scoring the winning touchdown [. . .] it makes me angry when people lay on me what I ought to have felt. More important, it obscures the truth. (Marlantes 2011, 26–7)

Persistent Humanity

The foregoing discussion has been meant to shed light on the reasons why guilt has no place on the battlefield. But in spite of the extraordinary pressure that killing and exposure to death place on warriors to numb themselves to their empathic responses, it is also the case that their humanity always threatens to break through. Consider the message communicated in August 2011 to troops from the 10th Mountain Division in Afghanistan, who had gathered to mark the deaths of five soldiers killed in an IED explosion. During the gathering, a military chaplain advised:

> This is a time of war and in times of war there is killing and dying [. . .] one thing that cannot happen effectively during war is grieving because this war will not stop to give us that time [. . .]. So the numbness you experience, the callous attitude you may have, your inability to cry like normal people during times of grief is not a burden or a curse that you should be concerned about. It is actually a blessing from God that allows you to continue in this fight [. . .]. You are warriors with hearts of steel that have allowed you to fight a brutal war day in and day out. You are not normal people, you are soldiers, American soldiers who go outside the wire and accomplish your mission in spite of the reality that every step, every movement mounted or dismounted could possibly be your last. (Rauch 2011, n.p.)

Were the floodgates of grief not in danger of opening, the chaplain would not have deemed it necessary to remind the troops of the importance of sustaining emotional hardness for the success of the mission.

Numerous accounts of wartime situations depict how the humanity of the Other threatens to break through the warrior's shield of indifference. Combat infantryman Chuck Newton, who served in Iraq, describes how he felt "a fleeting personal connection" with a person he killed:

> There's a split second, it's just him and me and . . . two guys with parents and brothers and sisters and . . . you know, families that care about them, walking around on the earth under the same sun, nothing to do with the war [. . .]. I know why executioners wear a mask and why the condemned always faces away from the executioner. Because the image of someone dying as you look him in the eye is nightmarish. It's something no one should ever do. (Wood 2016, 161–2)

In *What It Is Like to Go to War*, Marlantes recounts a firefight in which he killed a North Vietnamese soldier who was preparing to throw a grenade at him. Marlantes writes: "He cocked his arm back to throw—and then he saw me looking at him across my rifle barrel. He stopped. He looked right at me. That's where the image of his eyes burned into my brain forever" (Marlantes 2011, 29). Marlantes (or another Marine—Marlantes is not sure) killed the soldier, but not before the soldier had thrown the grenade. Marlantes recounts: "When I looked up after the explosion, the Vietnamese kid was dead. My feeling? I felt relief. 'Phew, no more grenades.' I churned up the steep slope to take on the next position and quickly forgot even that feeling. I didn't even think about the incident until years later"

(31). Forty years later, the memory of the soldier's "desperate fearful eyes [. . .] standing out like black pools in an exploding landscape of mud and dying vegetation" (27) returns to Marlantes. What he feels then is sadness. He writes:

> Oh, the sadness. The sadness [. . .]. What is different between then and now is quite simply empathy. I can take the time, I had the motivation, to actually feel what I did to another human being who was in a great many ways just like my own son. Back then, I was operating under some sort of psychological mechanism that allowed me to think of that teenager as "the enemy." I killed him or Ohio did and we moved on. I doubt I could have killed him realizing he was like my own son. I'd have fallen apart. This very likely would have led to my own death or the deaths of those I was leading. But a split occurred then that now cries out to be healed. (27)

At the time of the killing, Marlantes could not afford to feel (he had developed the capacity not to do so), because in this situation, empathy would have distracted him from the mission and thereby put him, and the men for whom he was responsible, at risk.

Even in the absence of an empathic reaction, both Newton and Marlantes retained the memory of an encounter with the Other through the gaze, the meaning of which was not yet known to them. Jean Laplanche's account of the material in the unconscious helps us understand why these experiences of empathic unresponsiveness remain present in the warrior's mind and surface years later. Laplanche describes what the unconscious mind registers and retains as "untranslated residues [. . .] uncoordinated, unbound elementary signifiers [. . .] heterogeneous to consciousness in the sense that the term implies unity [. . .]. That un-con-scious, that unbound, or non-bound, is what continues to attack us in the element of its strangerness" (Laplanche 2015, 141). For Laplanche, "unconscious" means lacking a framework of meaning, unbounded, and thus invasive. While these encounters with the Other take place during a time when the warrior cannot afford to acknowledge the humanity of the Other, that humanity is still registered. Once the warrior is no longer at risk, those "unbound" experiences—what the psychoanalyst Winifred Bion calls "beta elements" and "sense data"—emerge into consciousness and have "to be dealt with" (Bion 1984, 116–7). In Marlantes's case, unconscious "sense data" emerged years after he returned from war.

Often these elements take the form of decomposed images of bodies-in-pieces that hearken back to deaths witnessed. In *The Road Back*, for example, Ernst returns to his hometown and struggles to withstand the return of such nightmarish images. Evenings, he walks the deserted streets, when the night is "full of dark cries and indistinct voices, full of faces and things gone by, full of questions and of answers" (Remarque 1930, 86). Ernst's unprocessed memories appear as fragmented images and sounds and disarticulated questions and answers that hang in the air. On other nights, Ernst retreats to his rented room, where "the hours are long and strange thoughts often creep out of the corners—pale, wasted hands, beckoning, threatening; ghostly shadows of things past, but strangely changed, memories that rise up again of grey, sightless faces, cries and accusations" (232).

Ernst is helpless to prevent the unsolicited visits by these "residues," which refer back to his killing of an English soldier. As Ernst sleeps, he dreams of the soldier. Because the unprocessed material in his mind is in need of a conceptual framework to render it coherent, its elements appear to Ernst in a fragmented form, only assuming a discernible shape piece by piece: "The rim of a helmet shows over the bushes. A forehead, clear eyes, a firm mouth" (249). Eventually the soldier appears in a nightmare in which the mnemonic fragments coalesce into a meaningful whole:

> The red face grows pale and sinks in, and eyes and mouth are at last no more than black caverns in a swiftly decaying countenance that slowly inclines to the earth, sags and sinks into the dandelions. Finished!
>
> I worm myself off and begin to work my way back to our trenches. But I look around once more. The dead man has suddenly come to life again. He straightens up as if he meant to run after me [. . .]. The dead man is standing upright; he is showing his teeth! [...] He is running on his stumps, grinning, his arms stretched out toward me [. . .]. Two hands close around my neck, they bear me backwards, to the ground [. . .]. The dead man is dragging me toward the precious edge of the chalk pit. He is rolling me down into it. I lose balance, struggle to catch hold—I am slipping, I fall, cry out, fall endlessly. (249–50)

This dream conveys Ernst's inability to rest peacefully in the aftermath of his killing of the soldier and his fear that the soldier's death will "endlessly" possess him.

In a daydream that he has immediately after this nightmare, Ernst, who is now a teacher, imagines he is speaking to his pupils about the war, and his words shed further light on the meaning of the dream:

> I stand here before you, a polluted, a guilty man and can only implore you ever to remain as you are, never to suffer the bright light of your childhood to be used as a blow flame of hate. About your brows still blows the breath of innocence. How then should I presume to teach you? Behind me, still pursuing, are the bloody years.—How then can I venture among you? Must I not first become a man again myself? (253–4)

When read in relation to the imagined speech, Ernst's nightmare can be understood as expressing the guilt Ernst feels from these "bloody years," an emotion that pursues him now that he is no longer in battle. In one sense then, his guilt is symptomatic of the vicissitudes of his reintegration into society, and its occurrence can be understood as a signal that Ernst is regaining his individuality and, with it, his conscience.

The philosopher Jesse Glenn Gray, who served in World War II, describes the return of conscience in the soldier as the mark of his separation from the military group:

> So long as the soldier thinks of himself as one among many and identifies himself with his unit, army, and nation, his conscience is unlikely to waken and feel the need to respond. All awareness of guilt presupposes the capacity to respond as an individual to the call of conscience [. . .]. We respond to conscience only when we can separate ourselves from others and become conscious, often painfully so, of our differences. (Gray 1998, 175)

Gray's discussion of the soldier's guilt suggests that its painful emergence coincides with the returning warrior's emerging awareness of his individuality. Eventually, this process issues in the soldier's recognition that, far from distinguishing him from the other members of his society, his guilt connects him to them, because he shares it with all other citizens: "The reflective soldier on both sides of the conflict will see no escape from political guilt as long as he remains a member of the state [. . .]. [The soldier tells himself] 'I shall always be guilty as long as I belong to a nation at all. Yet there is no good life apart from some nation or other' " (198). Given that Ernst in *The Road Back* is in a state of transition, the fact that his feelings of guilt express themselves through hallucinations, dreams, and reveries suggests that his civilian self is not yet fully "awake." He has not yet ceased to be a warrior and "become a man again." This is not to say that his guilt is a figment of his imagination, but rather that the terror expressed in his imaginings indicates that his guilt is unmanageable.

When a fellow veteran is on trial for having shot a man in a bar, Ernst seizes the courtroom floor and delivers a speech that suggests he is beginning to find his footing with regard to his guilt. In solidarity with the accused, he inveighs against the civilians of his community for their negligence in not helping his comrade—or any of the veterans who returned to their community—reintegrate:

> You, every one of you, should stand before our tribunal! It is you, with your war, who have made us what we are. Lock us away too, with him; that's the safest thing to do. What did you ever do for us when we came back? Nothing, I tell

> you! Nothing! You wrangled about "Victory!" You unveiled war memorials! You spouted about heroism! And you denied your responsibility!
>
> You should have come to our help!—But no, you left us alone in that worst time of all, when we had to find a road back again. You should have proclaimed it from every pulpit; you should have told us so when we were demobilized; again and again you should have said to us: "We have all grievously erred! We have all to find the road back again!" (Remarque 1930, 327–8)

Ernst is pleading for collective engagement in the forging of a "we" that would bring together veterans and the community in the shared work of bearing the consequences of war. As he states about his friend on trial for murder: "Had this man not been trained to shoot men, he would not have shot one now" (323). Ernst's point is that society makes soldiers into killers in order to wage its wars, but it does not remake them into men when they survive them. He tells those in the courtroom: "You should have shown us again what life is! You should have taught us to live again! But no, you left us to stew in our juice. You left us to the dogs. You should have taught us to believe again in kindliness, in order, in culture, in love" (328).

Ernst expresses the idea that, without the support of the community, veterans cannot hold in the balance the antithetical experiences of, on the one hand, having been a warrior, charged with casting aside the injunction against killing, and, on the other hand, returning to society after war, only to be left alone to process their guilt-producing deeds and attempt to live peacefully in the civilian world. Ernst believes that his community would rather do "the safest thing," by which he means, protect itself against knowing about the costs of war, and against assuming the responsibility that such knowledge entails.

The Psychic Aftermath of War

Like Ernst in *The Road Back*, Marlantes recognizes that most warriors experience societal abandonment when they return from war. He writes:

> Killing someone without splitting oneself from the feelings that the act engenders requires an effort of supreme consciousness that, quite frankly, is beyond most humans. Killing is what warriors do for society. Yet when they return home, society doesn't generally acknowledge that the act it asked them to do created a deep split in their psyches, or a psychological and spiritual weight most of them will stumble beneath the rest of their lives. Warriors must learn how to integrate the experience of killing, to put the pieces of their psyche back

> together again. For the most part, they have been left to do this on their own. (Marlantes 2011, 26)

When soldiers return from war but are left alone with the challenge of healing the psychic split Marlantes describes, it becomes virtually impossible for them to discover a larger framework of meaning into which "combat's terror, exhilaration, horror, guilt and pain" (16) can be integrated. The veteran Phil Klay notes that when he returned from Iraq, civilians would ask him, "How are we doing over there?" (Klay 2014, 4). He explains:

> And I'd tell them. I'd explain in bold, confident terms about the surge and the Sunni Awakening. The Iraq I returned from was, in my mind, a fairly simple place. By which I mean it had little relationship to reality. It's only with time and the help of smart, empathetic friends willing to pull through many serious conversations that I've been able to learn more about what I witnessed. And many of those conversations were with friends who'd never served. (4)

Klay's experience attests to the importance of the community's participation in conversations about the meaning of war, for the warriors who fought it and for the nation that sent them to fight.

Ernst's return to his society reveals how guilt is the first sign of the reanimation of his humanity, and how recovery from war is heavily influenced by the social support, or lack of support, that veterans receive (Pietrzak et al. 2010). The war journalist Sebastian Junger, referring to the author and ethicist Austin Dacey, emphasizes the value of "shared public meaning" in helping soldiers integrate their wartime experiences into their larger life narratives: "Shared public meaning gives soldiers a context for their losses and their sacrifice that is acknowledged by most of society. That helps keep at bay the sense of futility and rage that can develop among soldiers during a war that doesn't seem to end" (Junger 2016, 97).

The wars in Iraq and Afghanistan, the longest in U.S. history, have been fought by an all-volunteer force, as a result of which, civilians have had the luxury of remaining detached from them and, in lieu of serious engagement with their social and individual impact, "prone to ricochet between lazy indifference to the warfighting and overwrought hero-worship of returning troops" (Wood 2016, 19). As Klay admonishes:

> You don't honor someone by telling them, "I can never imagine what you've been through." Instead, listen to their story and try to imagine being in it, no matter how hard or uncomfortable that feels. If the past 10 years have taught us anything, it's that in the age of an all-volunteer military, it is far too easy for

> Americans to send soldiers on deployment after deployment without making a serious effort to imagine what that means. (Klay 2014, 4)

More than forty years after Vietnam, Marlantes feels sadness when he thinks about wartime killing, because he has been able to develop the large framework of meaning necessary to integrate "combat's terror, exhilaration, horror, guilt and pain" (Marlantes 2011, 16) into his understanding of who he is and what he has experienced. He writes that "in war, we have to live with heavy contradictions. The degree to which we can be aware of and contain those contradictions is a measure of our individual maturity" (44). Marlantes refuses to accept "guilt" as the explanation for the alienation he felt when he came back from war, because it elides those contradictions and instead attributes to him the very moral agency warriors must suspend in order to kill. The explanation also fails to recognize the guilt that all members of society should rightfully bear, because they, along with the soldiers who do their bidding, have contributed to "the grief of evil in the world" (27) through the wars they wage.

Considered in light of the question informing this essay collection, Marlantes's experience of war and its psychological, emotional, and spiritual aftermath suggests that the guilt warriors feel can be a productive force on the condition that it is avowedly co-owned by both warriors and their society. If and when this happens, it becomes possible for warriors to regain access to, and assume, the full range of their humanity, not least of all their sense of belonging to the society in whose name they served.

Notes

1. Freud characterizes trauma as "an experience which within a short period of time presents the mind with an increase of stimulus too powerful to be dealt with or worked off in the normal way" (Freud 1963, 275).
2. Marlantes devotes a chapter to discussing the warrior's guilt in *What It Is Like to Go to War*, albeit in ways that exceed the conceptual bounds of Moyers's question.

References

Bion, Winifred R. 1984. "A Theory of Thinking." In *Second Thoughts: Selected Papers on Psychoanalysis*, 110–19. London: Karnac.

Freud, Sigmund. 1963. *Introductory Lectures on Psycho-Analysis. Part III: 1916–1917*. Standard Edition of the Complete Psychological Works of Sigmund Freud, vol. 16. Translated from the German under the General Editorship of James Strachey. London: Hogarth Press and the Institute of Psycho-Analysis.

Gray, Jesse Glenn. 1988. *The Warriors: Reflections on Men in Battle*. Introduction by Hannah Arendt. Lincoln: The University of Nebraska Press.
Junger, Sebastian. 2016. *Tribe: On Homecoming and Belonging*. New York: Hachette.
Klay, Phil. 2014. "After War, a Failure of the Imagination." *New York Times*, February 9. https://www.nytimes.com/2014/02/09/opinion/sunday/after-war-a-failure-of-the-imagination.html?.
Laplanche, Jean. 2015. "Response and Responsibility." In *Between Seduction and Inspiration: Man*. Translated by Jeffrey Mehlman, 123–44. New York: The Unconscious in Translation.
Litz, Brett, Nathan R. Stein, Eileen Delaney, Leslie Lebowitz, William P. Nash, Caroline Silva, and Shira Maguen. 2009. "Moral Injury and Moral Repair in War Veterans: A Preliminary Model and Intervention Strategy." *Clinical Psychology Review* 29, no. 8: 695–706.
Marlantes, Karl. 2011. *What It Is Like to Go to War*. New York: Grove Press.
Marlantes, Karl. 2012. "On the Mindset of a Modern Warrior." Interview by Bill Moyers, July 27, 2012. Transcript and video: https://billmoyers.com/segment/karl-marlantes-on-what-its-like-to-go-to-war.
O'Brien, Tim. 1990. *The Things They Carried*. New York: Houghton Mifflin.
Pietrzak, Robert H., Douglas C. Johnson, Marc B. Goldstein, James C. Malley, Alison J. Rivers, Charles A. Morgan, and Steven M. Southwick. 2010. "Psychosocial Buffers of Traumatic Stress, Depressive Symptoms, and Psychosocial Difficulties in Veterans of Operations Enduring Freedom and Iraqi Freedom: The Role of Resilience, Unit Support, and Postdeployment Social Support." *Journal of Affective Disorders* 120, no. 1–3: 188–92.
Rauch, Laura. 2011. "Memorial in Afghanistan Stirs Soldiers' Emotions as They Grapple with Combat Deaths." *Stars and Stripes*, August 20. http://www.stripes.com/news/middle-east/afghanistan/memorial-in-afghanistan-stirs-soldiers-emotions-as-they-grapple-with-combat-deaths-1.152717.
Remarque, Eric Maria. 1930. *The Road Back*. Translated by A. W. Wheen. New York: Ballantine.
Scranton, Roy. 2015. *Learning to Die in the Anthropocene: Reflections on the End of a Civilization*. San Francisco: City Lights Books.
Shay, Jonathan. 1995. *Achilles in Vietnam: Combat Trauma and the Undoing of Character*. New York: Simon & Schuster.
Wood, David. 2016. *What Have We Done: The Moral Injury of Our Longest Wars*. New York: Little, Brown.

PART IV

THE POLITICS OF GUILT NEGOTIATIONS

13

The Art of Apology

On the True and the Phony in Political Apology

Maria-Sibylla Lotter

> It is only in a metaphorical sense that we can say we feel guilty for the sins of our father or our people or mankind, in short, for deeds we have not done, although the course of events may well make us pay for them. And since sentiments of guilt, *mens rea* or bad conscience, the awareness of wrongdoing, play such an important role in our legal and moral judgement, it may be wise to refrain from such metaphorical statements which, when taken literally, can only lead into a phony sentimentality in which all real issues are obscured.
>
> —Hannah Arendt, "Collective Responsibility"

The Rise of Political Apology

Over the last 50 years, the political apology has evolved from an uncoded act of singular importance to a new global practice that spans the world. Former U.S. president Bill Clinton apologized for the U.S. violation of Hawaiian sovereignty in 1893, for the country's failure to stop the Rwandan Genocide, for slavery, and for the U.S. treatment of Africa in general. British prime minister Tony Blair apologized for British policy under Queen Victoria during the Irish Potato Famine. The Canadian and Australian governments apologized to Indigenous communities for policies aimed at destroying their cultures, and, in fact, since 1998, a National Sorry Day has been set up in Australia to remember and commemorate the mistreatment that the country's Indigenous peoples have suffered. Pope John Paul II asked forgiveness for no less than all sins and wrongdoings committed by the Catholic Church in the last 2,000 years. And, in 1996, a large group of Christians, including at times several thousand participants, set out on a pilgrimage of several years to the historical Crusade sites to repent and seek forgiveness from Jewish and Muslim authorities for the crimes of the Crusaders. These are just a few examples. According to the Stockholm Centre for the Ethics

Maria-Sibylla Lotter, *The Art of Apology* In: *Guilt.* Edited by: Katharina von Kellenbach and Matthias Buschmeier, Oxford University Press. © Oxford University Press 2022. DOI: 10.1093/oso/9780197557433.003.0014 of War and Peace, there have already been at least 250 political apologies since World War II (cf. Don 2019).

This new practice of public contrition for the deeds of former generations stands in striking contrast to centuries-old demonstrations of strength in power politics and of amnesty and amnesia after wars (as Ethel Matala de Mazza points out in her chapter in this volume). Why then did such a reversal of old political habits develop in the second half of the twentieth century?

At the turn of the millennium, Elazar Barkan answered this question with the thesis that a new and higher moral consciousness had replaced the old power politics. According to Barkan, apology (and other forms of restitution) "for historical injustices embodies the increasing importance of morality and the growing democratization of political life" (Barkan 2000, 308). It is in this spirit that the Stockholm Centre recently proposed—for many nations—"a regular day of apology on which the head of state publicly apologizes for a different past instance of serious misconduct by the state" (Don 2019).

However, political apologies, unlike interpersonal apologies, remain a controversial issue today. Perpetrators are expected to apologize to their victims. However, in the case of political apology, the apologizers most often are *not* the perpetrators of the historical wrongs in question (cf. Kodalle 2021). May or should others take blame by proxy for a wrongdoing in which they, as responsible individuals or as members of a living community, had no share? Who is authorized to do so, and in which situations? When is such a gesture morally desirable or necessary? Just as importantly, what good will it do?

While the literature on political apologies has been growing rapidly in recent years, these questions have not been satisfactorily answered. From a purely moral point of view, it would be appropriate to confess guilt if one had done something wrong, and to firmly refuse to make such an apology if one had not. However, regarding racism, genocide, mass murder, and other large-scale historical injustices, it is most often the case that the perpetrators do not confess guilt, not least because they do not even feel guilty; the prominent Nazi leader Hermann Göring, for instance, saw the Nuremberg Trials as a pure victor's justice. And those who *do* feel morally unsettled by the condemning gaze of others tend to hide their guilt behind pseudo-confessions of a more general, collective guilt, or behind less serious crimes. A very prominent example in Germany was Albert Speer, the Nazi minister of armaments, who spun an artfully presented pseudo-confession of guilt into a book in post-war Germany, in which he even denied his knowledge of the command to murder the Jews en masse in the concentration camps—and his book achieved great commercial success (cf. Serenyi 1996). If the perpetrators did not feel guilty, neither did many among those who had opposed National Socialism, such as Konrad Adenauer; nor was guilt communicated among passive bystanders, as Matthias Buschmeier points out in his chapter in this volume. As a matter of fact, very often, after wars (civil or otherwise), perpetrators are supported by the rest of their society in keeping silent about guilt, as the community is eager for peace and harmony to return after long-term violence. Only a later generation becomes willing, even eager, to admit to the injustice and to a kind of collective guilt, and this is not least because this later generation has a strong interest in being purified of a moral stain. One can perhaps attribute this new attitude to the fact that later generations assume that they cannot be personally accused of the crimes of their parents. This may explain why it is not an anomaly, but rather the rule, that only those who are not guilty (if anyone) will confess to guilt, and vice versa.

For these reasons, political apology differs from apology in the private sphere with respect to more than just the fact that the former is expressed on behalf of a collective. If we disregard apologies immediately after wars, such as the carefully coordinated apologies of heads of state after the Serbo-Croatian War, an apology is usually not expressed by a representative of the collective of perpetrators, but rather by a later generation that is not really under indictment. A collective that connects present and past peoples is mentally constructed by way of two types of identification: the identification of individuals with collectives, and the identification of present collectives with past collectives. However, performers and spectators of political apology not infrequently forget that it is a role-play, that the apologizers are usually not the "real" (historical and legal) perpetrators, and that those to whom the apology is addressed are usually not the "real" victims. Today, even in everyday conversations, collective admissions of guilt may be considered appropriate by individuals who feel doomed to identify with large cultural or political groups (e.g., wealthy countries, formerly imperialist nations, or "Western culture"), races, or even classes of people (e.g., white men).

Minds are divided regarding the rationality and the benefits of a contemporary culture in which a register of guilt and contrition sets the tone. On the one hand, a considerable number of people treat political apology as a moral duty and a general salve, applicable to injustices and atrocities of all kinds, and even to structural inequalities. They believe that the present generation has a moral duty to confess and compensate for the crimes and debts of the past, and that political apology, among other things, is a necessary condition not only for coming to terms with the past, but also for realizing one's one privileged position within society. On the other hand, there are more than a few skeptics who question the possibility of a meaningful political apology and worry about such apologies' influence on democratic culture. Even 20 years ago, Jean Bethke Elshtain expressed concern about the ever-increasing emphasis on guilt and victimization that was becoming observable in communication across various spheres, stating that "rectitude has given way to what one wag called 'contrition chic,' meaning a kind of bargain basement way to gain publicity and sympathy and even absolution by trafficking in one's status as a victim, a victimizer, a sinner, and so on" (Elshtain 2002, 14). Others, such as Michel-Rolph Trouillot, are particularly concerned about the new habit of transferring to collectives characteristics that are only applicable to individual guilt—for, by symbolically taking on the role of collective perpetrators and assigning the victim role to other collectives, political apologizers set the stage for the introduction of collective identities to which the ability to act and to feel repentance is attributed: "Steeped in a language of blood and soil collectivities are now defined by the wrongs they committed and for which they should apologize, or by the wrongs they suffered and for which they should receive an apology" (Trouillot 2000, 171). However, according to Trouillot, these practices cannot achieve the goal of transformation and reconciliation, which individual apologies are aimed at, because they do not fulfill the necessary conditions for the identity of apologizer and perpetrator; unlike individual apologies, they are doomed to be abortive rituals.

Although I share the critical view of an excessive contrition culture, in the following I will nevertheless pursue the question of whether political apologies, if they are not primarily understood in the moral register of self-accusation, do after all have an important political function. Of course, I do not mean to suggest that political apologies can have only one single and true function; political apologies can serve a variety of functions and aims, such as calming the anger of a group of people who considers themselves unfairly discriminated against, serving as barter-goods in diplomatic negotiations, initiating steps toward reconciliation, polishing one's image by removing a moral stain, distracting politically from other sensitive issues, putting other political powers on the moral sidelines, casting the government (or apologizer collective) in the most favorable light, saving money on more elaborate and expensive measures, and so on. As Nietzsche said, no historical phenomenon can be defined, and its functions are as varied as the occasions and interests involved.

However, whereas apologies are often discussed by philosophers as if they were purely moral interactions between perpetrators and victims—thus reinforcing the category confusion complained of by Trouillot—I will start from the observation that real political apologies are usually not voiced by the perpetrators *vis à vis* the victims, but rather by *third parties* after much time has passed. These third parties pursue, in the guise of a moral stance, their own political agenda, which does not coincide with the interests and concerns of either the perpetrators or of the victims (cf. Kodalle 2021). This raises the question of the role of moral guilt in political apologies.

Guilt in Moral and Legal Contexts

To what extent can political apologies by proxy be understood in terms of moral guilt? Morally speaking, expressions of remorse and contrition imply guilt. In our modern understanding—which, granted, has Jewish, Christian, and ancient Greek roots—guilt means authorship and blameworthiness pertaining to an individual person. A person can only be blamed for a *voluntary* act or omission by which one violates a moral norm. Consequently, one has to pay—in a direct or figurative sense. If we summarize the conditions and consequences of *guilt*, therefore, the term encompasses *authorship*, *blame*, and *liability*. At least the first two elements must be present if guilt is to be attributed in a more-than-metaphorical sense.

Blameworthiness implies control and voluntariness. What voluntariness entails, exactly, and whether this is a metaphysical or a practical question, has been the subject of intense debate in philosophy and law (cf. Watson 2003). Some believe that voluntariness presupposes free will in a metaphysical sense; that is, there is the assumption that a special psychological faculty exists called the "will," which exists beyond emotions and rational considerations, and from which our free decisions originate. However, the secular practice of blaming does not involve this model; it is based instead on the common understanding that it would not be fair to blame a person for an act or omission if it was not in one's power to have acted otherwise.[1] As early as the fourth century BCE, the principles implicit in this understanding were spelled out by Aristotle in his analysis of excuses: according to Aristotle, no one can reasonably be blamed for what he did not know, nor for what was not in his power (cf. Aristotle 1982). With regard to modern legal and moral excuses, the philosopher Thomas Nagel has summarized the condition of absence of excuses under the title of "condition of control" (cf. Nagel 1979, 25).

This does not mean that people could not be *liable without being to blame* for other people's wrongdoings, including those of former generations. However, in contrast to mere liability, the attribution of guilt does include blameworthiness, and it makes no sense to transfer *blame* to persons who were not even involved in the wrongdoing in question, nor to treat them as morally damaged. Thus, "it is only the liability that can be passed from one party to the other," according to Joel Feinberg, who warned against a confusion that is not uncommon when it comes to guilt: "In particular, *there can be no such thing as vicarious guilt*" (Feinberg 1991, 60). (This sharp distinction has not always been taken as self-evident, as Meinolf Schumacher shows in his chapter in this volume.)

How, then, do we arrive at the imaginary transfer of moral guilt from individuals to collectives, which Trouillot complained about? The motivation for this transference stems from the fact that the individualistic principle of guilt is insufficient as an instrument of justice and moral repair when it comes to collective crimes and crimes committed by states, as these are the result of the cooperation of a large number of individuals, whereby the causal share of each individual and their respective moral responsibility are unclear and difficult to decipher. This insight has recently led some philosophers to think that the concept of *collective guilt,* long used only polemically for confused and retrogressive thinking, is maybe not so flawed after all.

Naturally, the answer to the question of whether there is such a thing as collective guilt depends on how you understand and apply the term. Our individualistic understanding of guilt is not compatible with sweeping, collective accusations of guilt, which make no effort to differentiate the causal part in the event and the question of intention for each individual. At the same time, however, our individual understanding of guilt is not threatened if terms like "complicity" or "membership guilt" (cf. Gilbert 1997) are invoked to place the blame on individuals who have *contributed in some way* to a collective crime or can be blamed for a misconduct related to a historical injustice. For example, one could widen the circle of guilty parties in the Nazi era by arguing that the crimes of that era were only made possible by the silent consent of many people who were not active perpetrators but who nonetheless benefited from those crimes, and who, since they knew or could have known about the injustices (e.g., the expropriations of Jewish property), were thus somehow morally *complicit.* Concepts such as "passive guilt" can be introduced to refer to omissions, such as the failure to protest against a racist or anti-Semitic remark.

With regard to such passive behavior, Larry May distinguishes forms and degrees of moral taint by expanding the idea of moral responsibility to include not only behavior that warrants blameworthiness but also behavior "that warrants shame, remorse, regret and feelings of taint" (May 1992, 16). He distinguishes between moral guilt and moral taint: "Moral guilt is appropriate when there is something that a person brought about in the world. [. . . But] when the causal role one played did not make a difference in the world, then moral shame or taint may be the appropriate moral feeling" (May 1991, 248). And with regard to the guilt of later generations, Bernhard Schlink has argued that the way people react to the guilt of others can generate new guilt for themselves; thus, German guilt, according to Schlink, was generated anew in the second generation of Germans because the Nazis' descendants had not broken with their family members (cf. Schlink 2007, 13). Schlink did not mean to assert that the later generation actually inherited their parents' guilt; what he meant was that the subsequent generation was to blame for its own unwillingness to confront family members about their Nazi pasts.[2]

Now, while all these approaches are compatible with moral individualism, they provide no justification for an undifferentiated attribution of moral responsibility to collectives. However, the aftereffects of large-scale crimes can still be felt even when there are hardly any people left to whom—in the sense of May or Schlink—guilt or moral taint could legitimately be attributed. This is probably one reason why the willingness to plead guilty is, at least in the German example, inversely proportional to the legitimate attribution of *real* guilt, for it is rather those who are not personally accused who usually become receptive to acknowledging the moral catastrophe and to considering their own responsibility. Thus, it was precisely those members of the later generation who did not, as Schlink criticized, stick to their Nazi parents' worldviews, but rather sharply criticized them, who were willing to confess guilt. New problems, though, arise from this quite understandable attitude: as Hannah Arendt famously emphasized against concept creep with regard to matters of guilt, the function of the moral concept of guilt is to distinguish between those who are to *blame* for a misdeed or injustice and those who are not. "Where all are guilty, nobody is. Guilt, unlike responsibility, always singles out; it is strictly personal. It refers to an act, not to intentions or potentialities" (Arendt 1987, 43). For this reason Arendt saw no sense, and instead much harm, in the all-too-quick readiness to embrace collective guilt among the Germans of the Generation of '68: first of all, "with respect to what had been done by the Hitler regime to Jews, the cry 'we are all guilty' that at first sounded so very noble and tempting has actually only served to exculpate to a considerable degree those who actually were guilty" (43). Secondly, Arendt criticized the confusion of moral and political issues, as if the political issue were "not more than a special case of matters that are subject to normal legal proceedings or normal moral judgements" (43); specifically, she was concerned that the confusion of political responsibility for past injustices with guilt might prevent real historical understanding, for issues of collective responsibility with regard to the past "owe their relevance and general interest to political predicaments as distinguished from legal or moral ones" (44).

Following Arendt, I propose that we distinguish between the *political* and the *moral* meanings and functions of addressing guilt and responsibility. The question now becomes *why* the new phenomenon of political apology for the large-scale crimes and injustices of former generations *cannot wholly be dismissed as a category confusion or as purely sentimental*, even when we start from the assumption that the attribution of guilt (in the proper meaning of the word) only makes sense within a moral and legal framework that strictly distinguishes between the guilty and the innocent. Obviously, the rationality underpinning such declarations must be distinct from that of moral and legal blame.

The answer to this question is to be sought in the very political function of addressing guilt. It is to be sought, as Alice MacLachlan (2014, 13) has put it, in thinking of apologies as "a form of political practice, that is, a mode of doing politics." In what follows, I start from the assumption that in personal and political relations alike, it can sometimes be appropriate to behave *as if* one were guilty, in order to express empathy and respect, thereby turning guilt into a tool for repairing broken relationships and making new ones possible.

The Psychological and Social Function of Guilt in Intimate Relationships

I will approach the question of how political apologies differ from moral and legal attributions of guilt in the proper sense by starting from the psychological and social function that feelings of guilt play in interpersonal relationships.

Obviously, "admissions of guilt" have a function here that is different from attributing guilt before a court. For example, the phrase "I'm sorry," is used as a signal of politeness and is often meant to indicate neither "real" moral guilt nor personal feelings of guilt on the speaker's part. However, it is a well-known phenomenon that people may develop feelings of guilt and apologize, even if they have done nothing wrong. If the only rational reason for feeling guilty were that one *is* guilty (of having broken a moral norm), then we would have to classify such discordant feelings and behaviors as errors or even as pathological phenomena; in other words, if people feel guilty without having violated a moral norm, we would have to conclude that they either erroneously believe they have done so, or that they suffer from a guilt complex or other pathology. Nonetheless, despite the fact that these explanations might be helpful in individual cases, they are unsatisfactory as a sole explanatory pattern, because they mistakenly presuppose two things: first, that feelings of guilt and apologies causally follow the judgment that one has violated a moral norm, and secondly, that they should rationally follow such a judgment. These erroneous presuppositions are supported by the justifications we usually give. Of course, when asked, we tend to explain our feelings in a way that makes them appear understandable and arouses empathy; that is, we present reasons that fit into common narratives.

However, with regard to the motivation to feel guilty, the belief that we only feel guilty when we are aware of having committed some wrong is neither consistent with common experience nor with the findings of social psychology. As social psychologists have argued, feelings of guilt are not purely inner, mental reactions explainable by normative self-evaluation alone (cf. Baumeister 1994); instead, these feelings arise not within the isolation of the psyche, but rather within the realm of interpersonal communication, and therefore they have much to do with what we believe others think about us. Hence, for instance, when you get the impression that you are perceived with annoyance or irritation by people who are important to you, you will quite often feel somehow guilty; the guilt feeling gives rise to the idea that you might have done something wrong, even if you have no idea what that might be. In this respect, the so-called "Kafkaesque situation"—the feeling that you are judged, and rightly so, without knowing the accusation—seems to be a normal psychological reaction and not an anomaly.

Thus, we are better able to understand the role of guilt in personal relations if we do not proceed from the moral judgment as the *sine qua non* of the feeling, but vice versa, from the feeling itself. Of course, since communication about guilt usually takes place in terms of guilt's legal and moral criteria, these also have a retroactive effect on such feelings, but this impact is secondary and rather serves to clarify, absolve, weaken, or strengthen the feelings of guilt that arise from the disruption in the relationship. The self-imposed judgment that one has violated a moral norm is neither a necessary nor a sufficient condition for developing guilt feelings.

Another important aspect of guilt feelings is that they always involve empathy. The *intellectual* judgment that one is guilty of having broken a norm can be made even if one is unable to empathize with the victim of one's own behavior. By contrast, *feelings* of guilt arise from a certain empathy toward the person whom one thinks one might have hurt. This is why feelings of guilt often do not occur at all in the case of injuries and acts of violence against people who are strangers or are believed to be different.

If we assume that feeling guilt does not presuppose moral judgment, then the question arises as to whether feelings of guilt can be rational even when there is no "real" moral guilt. This question will sound strange if we assume that feeling guilt can only be rational to the extent that the feeling is an emotional reaction to the judgment that one has committed some wrong. However, we employ the term "rational" (among other meanings) for those attitudes or activities *that aim to promote our well-founded self-interest* (Arpaly 2003, 36–8)—and it is certainly in our own best interests to protect the social relationships that matter to us. In situations of irritation and conflict, you would likely not serve your friendship or marriage by insisting that you were not to blame for any inconsideration until your partner had enough evidence to prove the contrary—in other words, the "innocent until proven guilty" principle would not help to maintain trust, love, and respect. Instead, gestures of regret and contrition based on real feelings of guilt tend to be more beneficial, which raises the question of whether these guilt feelings fulfill a function distinct from the functions that moral and legal attributions of guilt fulfill.

The answer then is "yes"—at least that is what social psychologists claim to have empirically proven. As manifestations of empathy, feelings of guilt fulfill important social functions and thus can be "rational" even if there was no real wrongdoing or inconsideration (Baumeister 1994). This is especially true in close relationships where expectations are often very individualized. By feeling guilty, a person develops and shows empathy; she reveals that the partner is important to her and that she does not consider herself entitled to disregard her. Feeling guilty on the dominant side of a relationship can thus help to redress the other party's feelings of inferiority, humiliation, or relative powerlessness. In this function, guilt feelings may serve to strengthen or restore interpersonal trust, and they can even be called "political" in the very broad sense, which refers to the negotiation of one's own position of authority and power in a space shared with other parties who have competing claims to authority and power.

Truthfulness in Political Apologies

Likewise, in the political sphere, we must distinguish between the *legal* judgment of guilt by the international courts that have gradually developed since 1945, which make it possible to hold war criminals accountable, and the functions of *political* apologies, which, in certain respects, resemble the functions of guilt feelings in the private sphere. If political apologies were understood above all as self-attributions of guilt in a legal sense, then they would be based on categorical confusions, as has been argued (Trouillot 2000).

Now, apologies by representatives of states or big international organizations look quite different from what goes on in the realm of personal relationships, especially with regard to the following criteria: *first*, political apologies involve some form of symbolic gesture (e.g., kneeling, deep bowing) or speech (saying "sorry") that includes a carefully calculated display of remorse and contrition. Thus, such apologies are usually *staged events* and not involuntary or spontaneous expressions of emotion.

Second, they are representative and public: a representative of an organization or state apologizes on the collective's behalf, particularly on behalf of a group or institution that is considered the legal or moral successor of the guilty collective, or that is otherwise considered to have a relationship of some identity to the group or institution that perpetrated the wrongdoing.

Third, as *apologies by proxy*, political apologies differ from apologies in interpersonal relationships insofar as they are not expected to express the *sincere* feelings of the speaker. That is, the speaker may sincerely feel horror, regret, or empathy regarding the wrongdoing, but he or she is not expected personally to experience a lived sense of emotional guilt; observers know that such an apology is a symbolic political act. Therefore, since propriety calls for a certain exhibition of remorse, some "level of hypocrisy is no doubt intrinsic to the expression of political apology" (Griswold 2007, 131).

However, even when these three important differences are considered, one can see that felicitous apologies by proxy fulfill a *curative* function, which in some respect resembles the function of (discernible) guilt feelings in interpersonal contexts. Apologies by proxy, too, aim at repairing broken or disturbed relationships; one of their most important social and political functions lies in their suitability as *healing rituals* whose purpose is social reconciliation. They, too, do so by conveying regret, empathy, and respect. Also, as already stated, both political apologies and interpersonal guilt feelings are political in the broadest sense of serving to negotiate one's own position of authority and power in a space shared with other parties.

If these are the functions of political apologies, how can we distinguish between felicitous and abortive apologies?

Psychologists have observed that "apology effectiveness depends on trust in the apology itself and its perceived genuineness or sincerity" (Wenzel et al. 2018, 650). But how can political apologies be sincere and authentic at all? After all, they are not expressions of real remorse for one's own deeds but rather are apologies by proxy. They reflect the interests and the moral and political understanding of present-day groups, which are not to be confused either with those of the perpetrators or those of the victims. Moreover, as has been stated, they are not even expected to represent the subjective feelings of the representative—the party who is apologizing. For these reasons, Trouillot assumed that political apologies cannot be felicitous at all, because the representative has no choice but to pretend something that everyone knows is unrealistic: the representative must pretend to feel feelings of a collective.

So how *could* political apologies nevertheless achieve sufficient degrees of authenticity and sincerity? Can the belief in the sincerity of an apology perhaps be understood here in terms of a correspondence between the *displayed* attitude of the representative and the *real* mental attitudes of the many individuals belonging to the represented collective?

Let us start with an iconic but rather atypical apology, which has been widely recognized as felicitous: the Warsaw Genuflection. When German chancellor Willy Brandt took part in a ceremony held in Warsaw on December 7, 1970, to commemorate the Warsaw Ghetto Uprising, he knelt and remained silent in this position for a few minutes. Marek Edelmann, one of the few survivors and leaders of the uprising, later reported his response to this silent, apparently spontaneous gesture, saying, "The fact that he showed humility and respect in such a place proved to me that one can trust the Germans" (Zeslawski 2013, my translation).[3]

A *rebuilding of trust*[4] by showing *humility* and *respect*: Edelmann here summarized in just a few words what can be understood as maybe the most important function of political apology. The message conveyed here is not primarily about guilt, although an admission of guilt is naturally implied. Despite being a gesture of humble submission, a gesture like the Warsaw Genuflection could not be confused with the humble submission of a conflict's loser to its victor, whereby the loser acknowledges the supremacy of the victor. Rather, Brandt's gesture symbolically contradicted the claim to moral superiority, based on racial superiority, that "legitimized" the German attempt to annihilate the entire Jewish population of Europe. It should be understood against the historical backdrop that the crimes committed under Hitler, like many other mass murders of the 20th century, would hardly have been possible without an ideology that transformed such crimes into ostensibly heroic deeds (Lübbe 2003, 43–4); in this context, given Germany's erstwhile belief that it was possessed of a "higher" morality that overrode the rights of other humans, Brandt's gesture of humility conveyed Germany's symbolic submission now to a common morality shared with other nations (Celermajer 2009, 62–4).

As for the question of sincerity, nobody could have doubted that Brandt himself, a former exile during the Nazi era, was an opponent of the Nazi regime and its crimes. However, he was not acting as an individual during that historic trip to Warsaw, but as a representative. Regarding historical fact, the question of the extent to which Brandt was able, at the time, to represent a German collective that truly admitted its complicity in the Nazi crimes, and therefore also its obligations for apology, redress, and reparation (and whom thus, as Edelmann said, could be trusted again), is not so easy to answer. If a political apology could be considered authentic only insofar as it reflects the actual feelings and attitudes of the *majority* of the collective being represented, then the Warsaw Genuflection would constitute an "inauthentic" apology, since a close majority of Germans felt initially amazed and somehow alienated by Brandt's gesture. According to a survey by the Allensbacher Institut für Demographie commissioned by the magazine *Der Spiegel*, only one week afterwards, 48% of the German population at the time considered the gesture to be excessive, 41% considered it to be appropriate, and 11% had no opinion (*Der Spiegel* 1970). Brandt's political opponents even seized on what they saw as an opportunity to demand a vote of no confidence, which the chancellor narrowly won. However, Brandt won the next election in 1972 with a remarkable increase in votes, and this was because the German public now supported his politics of apology and reconciliation toward Poland and other eastern neighbors. This was hardly due to a moral reassessment of their own moral complicity. Rather, the majority, in the meantime, had become convinced that Brandt's policy had significantly improved Germany's international reputation. Thus, the Warsaw Genuflection was extremely productive as an avant-garde, symbolic gesture that, in the long run, prompted many Germans to revise their assessment of Willy Brandt's Ostpolitik and to welcome a policy of redress and reconciliation that involves a silent admission and later even a confrontation with the guilt—not only of the prominent Nazis, but also of the silent bystanders. In this way, too, political apologies can become a productive cultural force.

Let us return to the question of what is required to generate "trust in the apology itself and its [. . .] genuineness or sincerity." If the persuasive power of an apology lies neither in the presumed sincerity of the performer nor in the rather unknown attitude of those the performer represents, is it then to be understood analogously to a religious ritual as an " 'as if' reality, which shifts external relationships and affects internal and emotional conditions," as Katharina von Kellenbach points out in her chapter in this volume? This interpretation does not fit the bill either, because political apologies differ from regularly repeated rituals specifically in that they do not follow any rules of procedure and have no prescribed text; what is perceived as authentic and honest depends precisely on the fact that they are not coded. The gesture must be convincing in some other way, which presupposes a certain intellectual and artistic ability to bring what is required at a historic moment into a suitable form.

Drawing from the famous "impartial observer" concept of the philosopher Adam Smith, one could put it like this: the role of the representative is to express those feelings and attitudes that an *impartial observer* of the historical wrongdoing and its aftereffects would find appropriate to the situation (Smith 1976, 154–6). Thus, its persuasiveness ultimately depends on the convincing *artistic* way of expressing *the feelings one should expect* from the group in whose name the apology is being made, considering the type of atrocity in question.

However, while this is a necessary condition for the success of an apology, it is not a sufficient one; an apology's success depends on many factors that vary with the nature of the case and the cultural context. An apology may become somehow "falsified" in the eyes of the victims or international spectators if it is not accompanied by an offer of reparations. Even though such payments might be construed as offensive if not accompanied or preceded by an apology, they can be taken as evidence of the sincerity of the gesture (provided, of course, that the payments are substantial enough not to be dismissed as peanuts). Even in this case, though, an apology might *still* fail if it is perceived as inconsistent with other ways in which the nation or the successor group related to its past. Japan's formal apologies for its war crimes in World War II are often cited as an example of apologies that failed to fully regain trust, for the very reason that Japan's Asian neighbors considered the apologies "inconsistent with visits of Japanese public officials to the Yasukuni shrine honouring Japan's wartime dead (including fourteen Class A war criminals)" (Bovens 2008, 220) and with the Japanese coverage of the war in history textbooks and in the press.

A political apology may also fail to convince if it is too vague or too controversial in its implications to be considered meaningful, adequate, or truly representative of the nation (Lübbe 2003, 43–5). When President Clinton, on his 1998 tour of Africa, apologized for slavery, with the remark that "going back to the time before we were even a nation, European-Americans received the fruits of the slave trade and we were wrong in that" (Bennett 1998a), the apology was received as a sincere expression of goodwill. Nevertheless, it caused some guesswork and even ridicule worldwide because it was not too clear what exactly, and in whose name, he was meaning to say. Was he taking responsibility for the enslavement of Africans in the name of the U.S.A.? And who was included in "we"—all contemporary Americans, including the descendants of slaves? All Americans who were specifically of European descent? And why was he addressing this problem in Uganda and not somewhere in West Africa, a much more important place for the slave trade? Inevitably, Clinton's words invited criticism from different audiences—both from those who felt unjustly included into a collective of wrongdoers, and from those who considered his apology quite insufficient. While the *New York Times* complained that "no President in recent memory has spent so much time spreading contrition abroad [while having] precious little to offer in terms of new financial help" (Bennett 1998b), the political right wing attacked Clinton for "grovelling and pandering," invoking the facts "that on both sides of the continent, Africans themselves were also involved in the slave trade (President Museveni of Uganda himself acknowledged this), and that slavery still exists today in Sudan and Mauritania" (Ryle 1998). But the political left was also dissatisfied: it "was argued that if there was to be an apology it should be directed not at Africans but at black Americans, some of whom have long been demanding national reparations for slavery" (Ryle 1998).

How Even Sincere Apologies Can Flop

Philosophers have made repeated attempts to give general criteria on whose fulfillment the success or failure of apologies is supposed to depend. Even a critic of approaches that take for granted that there is an ideal type of apology, such as MacLachlan, assumes that an apology must fulfil certain criteria as a minimum:

> Central to the success of a given apology is the question of fit: do the elements of narrative (i.e., the story that is told by the apologizer), responsibility, and future commitment fit the seriousness and extent of the wrongful harms in question? Do they match up to what happened? To whom it happened? Do they accurately describe the relationships involved? An apology can misfire when its recounting of the wrong does not match the victim's own understanding, whether by downplaying the harm involved, offering excuses, casting the apologizer's intentions in a better light, or glossing over key aspects of the

> injury. Equally problematic are apologizers who address the wrong victim altogether, mischaracterizing what happened by rewriting to whom it happened. (MacLachlan 2013, 130)

There is much truth in this. However, in the end, whether an apology that calls a spade a spade and appears to conceal nothing turns out felicitous depends on other factors. Let us turn to the following example of a model apology in the sense of McLachlan.

As is known today, the Protestant Church in the German Democratic Republic, which had a reputation for offering refuge to many opponents and nonconformists, was infiltrated by the State Security Service in the German Democratic Republic's last decades, such that, in reality, many church authorities harassed and betrayed nonconformist church members.[5] Thirty years after the fall of the Berlin Wall, the Regional Church Council of the Protestant Church in Central Germany issued a declaration in which it apologized for not having opposed the German Democratic Republic state more clearly and uncompromisingly:

> (W)e confess: we have too often failed to withstand government pressure. [. . .] We have often not objected to injustice clearly enough. To this day, we have not given the required attention to our inadequate support for people who were expropriated [or] who were subjected to forced resettlement and expulsion, to those who were political prisoners in the German Democratic Republic and those driven to suicide. [. . .] We deplore the cases in which pastors and church employees conspired with government agencies, violated trust and harmed others. [. . .] To this day, we as a church do not take the necessary responsibility for people who have been imprisoned, humiliated, traumatized, or forced to leave church circles [. . .] after betrayal. (Protestant Church of Central Germany, EVM 2017)

The statement goes into more and more detail about the many different forms of individual misconduct and victimization. So one cannot fairly accuse the church of having concealed or downplayed any of its' authorities' misconduct in the German Democratic Republic times. However, the confession of guilt in the name of the church could be interpreted to establish an undifferentiated perpetratorship, shared as a kind of membership guilt by all church members—or so it was perceived by those who had been members of the church in the former German Democratic Republic. Thus, those who, according to their own understanding, had worked to the best of their knowledge and conscience at the time, felt that their personal integrity was unfairly called into question as a result of this blanket confession, which implicated them unfairly alongside those who had harassed and betrayed other church members. From their point of view, innocent parties were being assigned some form of undue collective guilt, while guilty parties had the advantage of seeing their own personal guilt presented as though it were a burden for a larger collective to share. For all these reasons, the apology was rather counterproductive for coming to terms with the past. Even though it was meant to be truthful, authentic, and sincere, it could not appear as other than a cover-up, even though this was clearly not intended. This example illustrates how a political apology may fail at the task of moral confrontation with the past by trying too diligently to confess to all wrongs, in contrast to Willy Brandt's ingenious gesture, which was so convincing precisely because he did not utter a single word.

These difficulties are somehow to be expected in the case of an apology for a wrongdoing whose perpetrators (all or some) are still alive. Political apologies concerning the very recent past always run the risk of being seen as a cover-up if they are not accompanied by an examination of *individual* guilt. This is the situation that Hannah Arendt had in mind when she criticized a certain German practice of guilt confessions.

The risks and failures of political apologies show that political apology is an art, a specifically *political* art, which cannot be practiced according to a standard model, but rather needs special skill and sensitivity. The performer requires a feel for timing, a certain empathy for the different circumstances and feelings of the members being represented, as well as those being spoken to; a special tactfulness of expression and the charisma of sincerity and authenticity are needed. The unthinking desire to signal one's own virtuousness by showing a willingness to repent without having sufficiently considered what this may mean for those still grappling with the past can ruin the political meaning of such a gesture. For this reason, political apologies cannot possibly be "standardized." What is proper varies from case to case, and from circumstance to circumstance. This sets apart the art of political apology from any legal procedure or religious ritual that is guided by fixed rules.

Notes

1. When I say that this insight is, in principle, accessible to every human being, I am not denying that in cultural history people have repeatedly been accused of actions they did not commit. Psychologically, there are strong incentives in most societies that trigger the so-called scapegoat mechanism.
2. May and Oshana have similarly applied the terms *moral taint* or *metaphysical guilt* to people who fail to criticize and to dissociate themselves from those who are associated with moral wrongs, such as racism (cf. Oshana 2006).

3. For the Polish hosts, the situation was more complicated because the gesture did not take place at a Polish national memorial, but rather at the memorial of the Jewish ghetto; that is, it was also specifically related to the Jews and could thus be understood as an indirect protest against the anti-Semitic riots in Poland at that time.
4. This element of trust hast been emphasized by Alice MacLachlan, too, as the primary object of political apology (cf. MacLachlan 2015, 441).
5. In what follows I draw from by Klaus-Michael Kodalle (2019).

References

Arendt, Hannah. 1987. "Collective Responsibility." In *Amor Mundi: Explorations in the Faith and Thought of Hannah Arendt*, edited by James Bernauer, 43–50. Dordrecht: Springer.

Aristotle. 1982. "The Nicomachian Ethics." In *Aristotle Volume XIX*, edited by G. P. Goold, translated by Harris Rackham, 17–187. Loeb Classical Library 73. Cambridge, MA: Harvard University Press.

Arpaly, Nomy. 2003. *Unprincipled Virtue: An Enquiry into Moral Agency*. Oxford: Oxford University Press.

Barkan, Elazar. 2000. *The Guilt of Nations: Restitution and Negotiating Historical Injustice*. New York: W. W. Norton.

Baumeister, Roy, Arlene Stillwell, and Todd F. Heatherton. 1994. "Guilt: An Interpersonal Approach." *Psychological Bulletin* 115, no. 2: 243–67.

Bennett, James. 1998a. "Clinton in Africa: The Overview; in Uganda, Clinton Expresses Regret on Slavery in U.S." *New York Times*, March 25. www.nytimes.com/1998/03/25/world/clinton-africa-overview-uganda-clinton-expresses-regret-slavery-us.html.

Bennett, James. 1998b. "The World: Sorry about That; Africa Gets the Clinton Treatment." *New York Times*, March 29. https://www.nytimes.com/1998/03/29/weekinreview/the-world-sorry-about-that-africa-gets-the-clinton-treatment.html.

Bovens, Luc. 2008. "Apologies." *Proceedings of the Aristotelian Society* 108, no. 3: 219–39.

Celermajer, Danielle. 2009. *The Sins of the Nation and the Ritual of Apologies*. Cambridge: Cambridge University Press.

Der Spiegel. 1970. "Kniefall: angemessen oder übertrieben?" *Der Spiegel*, December 14, 1970. https://www.spiegel.de/spiegel/print/d-43822427.html.

Don, Allison, and Per-Erik Milam. 2019. "The Case for Regular Political Apology." Stockholm: Stockholm Centre for the Ethics of War and Peace (blog), February 8. http://stockholmcentre.org/the-case-for-regular-political-apology/.

Elshtain, Jean Bethke. 2002. "Politics and Forgiveness." *Consensus* 28, no 1: 13–31.

EVM. 2017. "Erklärung des Landeskirchenrates im Gottesdienst der 6. Tagung der II. Landessynode der Evangelischen Kirche in Mitteldeutschland am Bußtag 2017." Translated by Maria-Sibylla Lotter. https://www.ekmd.de/attachment/aa234c91bdabf36adbf227d333e5305b/2cfd2666fb184a8791a5a641f829337b/busswort_herbst-synode_2017.pdf.

Feinberg, Joel. 1991. "Collective Responsibility." In *Collective Responsibility: Five Decades of Debate in Theoretical and Applied Ethics*, edited by Larry May and Stacey Hoffmann, 53–76. New York: Rowman & Littlefield.

Gilbert, Margaret. 1997. "Group Wrongs and Guilt Feelings." *Journal of Ethics* 1, no. 1: 65–84.

Griswold, Charles L. 2007. *Forgiveness: A Philosophical Exploration*. Cambridge: Cambridge University Press.

Kodalle, Klaus-Michael. 2019. "Bußfertigkeit: Die evangelische Kirche Mitteldeutschlands stellt sich ihrem Versagen während der DDR-Zeit." In *Universitas: Ideen, Individuen und Institutionen in Politik und Wissenschaft; Festschrift für Klaus Dicke*, edited by Manuel Fröhlich, Oliver W. Lembcke, and Florian Weber-Stein, 77–94. Jenaer Beiträge zur Politikwissenschaft, vol. 18. Baden-Baden: Nomos.

Kodalle, Klaus-Michael. Forthcoming. "Amnesty—Amnesia—Anamnesis: Temporal Relations and Structural Antagonisms in the Moral Economy of Forgiveness and Reconciliation." In *Guilt and Forgiveness*, edited by Saskia Fischer and Maria-Sibylla Lotter. London: Palgrave Macmillan.

Lübbe, Herrmann. 2003. *"Ich entschuldige mich": Das neue politische Bußritual*. Berlin: Siedler.

McLachlan, Alice. 2013. "Gender and the Public Apology." *Transitional Justice Review* 1, no. 2: 126–46.

McLachlan, Alice. 2014. "Beyond the Ideal Political Apology." In *On the Uses and Abuses of Political Apologies*, edited by Mihaela Mihai and Mathias Thaler, 13–31. London: Palgrave Macmillan.

McLachlan, Alice. 2015. "'Trust Me, I'm Sorry': The Paradox of Public Apology." *The Monist* 98, no. 4 (October): 441–56.

May, Larry. 1991. "Metaphysical Guilt and Moral Taint." In *Collective Responsibility: Five Decades of Debate in Theoretical and Applied Ethics*, edited by Larry May and Stacey Hoffman, 239–54. Lanham, MD: Rowman & Littlefield.

May, Larry. 1992. *Sharing Responsibility*. Chicago: University of Chicago Press.

Nagel, Thomas. 1979. "Moral Luck." In *Mortal Questions*, 24–38. Cambridge: Cambridge University Press.

Oshana, Marina A. L. 2006. "Moral Taint." *Metaphilosophy* 37, no. 3/4: 353–75.

Ryle, John. 1998. "A Sorry Apology from Clinton." *The Guardian*, April 13. https://www.theguardian.com/Columnists/Column/0,5673,234216,00.html.

Schlink, Bernhard. 2007. *Vergangenheitsschuld: Beiträge zu einem deutschen Thema*. Zürich: Diogenes.

Serenyi, Gitta. 1996. *Albert Speer: His Battle with Truth*. New York: Vintage Books.

Smith, Adam. 1976. *The Theory of Moral Sentiments*. Oxford: Clarendon Press.

Trouillot, Michel-Rolph. 2000. "Abortive Rituals: Historical Apologies in the Global Era." *Interventions* 2, no. 2 (July): 171–86.

Watson, Gary. 2003. *Free Will*. Oxford Readings in Philosophy. Oxford: Oxford University Press.

Wenzel, Michael, Ellie Lawrence-Wood, and Tyler G. Okimoto. 2018. "A Long Time Coming: Delays in Collective Apologies and Their Effects on Sincerity and Forgiveness." *Political Psychology* 39, no. 3: 649–66.

Zeslawski, Margot. 2013. "Willy Brandt: Kniefall am Abgrund der deutschen Geschichte." *FOCUS*, November 15. https://www.focus.de/politik/ausland/tid-20661/willy-brandt-kniefall-am-abgrund-der-deutschen-geschichte_aid_579245.html.

14

Relationships in Transition

Negotiating Accountability and Productive Guilt in Timor-Leste

Victor Igreja

Introduction

Sometime in August 2009, a group of residents in Suai, a district in Timor-Leste, spotted someone they believed to be Maternus Bere, the former second commander of the Laksaur militia, a ferocious militia group supported by the Indonesian military. Bere's group had allegedly massacred more than 200 people in a church in Suai 10 years earlier, following the United Nations' (UN) sponsored referendum for independence. Although the now defunct Serious Crimes Unit, established in 2000 by UN Security Council Resolution 1272, indicted Bere for the 1999 massacres, he had not been brought to justice because he was hiding in Indonesia under the protection of the Indonesian authorities. Hence, when Bere re-entered Timor-Leste in August 2009, for presumably personal reasons, those who saw and recognized him rushed to undertake a people's arrest.

Instead of being consumed by the desire for revenge and exacting justice in situ, the people handed Bere over to local law enforcement agents, expecting that criminal accountability would eventually be attained for the victims of the Suai massacre. However, when word of Bere's arrest swirled around in Indonesia, officials at the Indonesian Ministry of Foreign Affairs used backstage diplomatic channels to pressure the Timorese authorities to release and return Bere to Indonesia. The Timorese government fell into turmoil as a result of the pressure to extradite him, which culminated in political interference in the nascent independent judiciary, violated the Timorese Constitution, and compromised Timor-Leste's obligations to the UN to bring to justice those responsible for crimes against humanity—and other serious crimes—that had been committed in Timor-Leste in 1999.

While the pressure exerted by the Indonesian authorities cannot be discounted in the Bere affair, equally significant is the perspective of realism that has shaped the position of some prominent Timorese politicians toward Indonesia and the legacies of the violent colonial past. Following, as a point of departure,

Victor Igreja, *Relationships in Transition* In: *Guilt*. Edited by: Katharina von Kellenbach and Matthias Buschmeier, Oxford University Press. © Oxford University Press 2022. DOI: 10.1093/oso/9780197557433.003.0015

Friedrich Nietzsche's idea as expressed in *On the Genealogy of Morality* [1887] (2017, 37) that in order to grasp reality, one must first learn to distinguish between what happens by accident and what by design, there are still no certitudes about the best approaches for addressing a number of ontological challenges at a given historical juncture. For example, how to tame an enemy that was never totally defeated? How to celebrate victories that were never complete? And how to hammer home the truths of innocence and guilt when the powers for such pronouncements are so frail? The Timorese government, particularly spearheaded by Xanana Gusmão, a survivor and hero of the resistance struggle and former president (2002–2007) and prime minister (2007–2015), appears to have understood that the multiplicity of experiences acquired through accumulated histories of violence makes it difficult to classify events as having been based on either accident or design. Equally important is the conscious realization of the gap, for such a divided people and fragmented nation, between available justice instruments and the possibility that justice can be achieved at all. Thus, Gusmão and other Timorese political leaders have rejected calls for what I have termed "reproductive guilt" as the sole approach for handling the legacies of the Indonesian military occupation and the violence, and of those Timorese who were deemed culprits in Timor-Leste's fall in 1975, given their vows in 1975 to integrate Timor-Leste into Indonesia.

Reproductive and Productive Guilt Approaches

The notions of reproductive and productive guilt are adapted from Paul Ricoeur's phenomenology of human action and creativity. While Ricoeur wrote extensively on the topic of guilt (Ricoeur 2006), however, he did not approach guilt in terms of reproductive dimensions and productive possibilities; therefore, instead, I have adapted and extended his argument by analogy to his analytical model of ideology and utopia (Ricoeur 1986). In his extensive model—and counter to the dominant approaches of the Marxist school—Ricoeur explored the positive and perverse dimensions of ideology and utopia. A dialectic relationship between the two, whereby ideology consists of a form of reproductive social imagination, was Ricoeur's approach. This is the case because ideology, or reproductive imagination, constitutes a replica of existing mechanisms for the legitimation of political authority (Taylor 2006). Accordingly, the history of Western thought is one of attention to reproductive imagination only, and to address this gap, Ricoeur proposed his new model, inspired by Jean-Paul Sartre's idea that "the imagination is the necessary condition for [human] freedom," as well as by Sartre's focus on the unreal as a basis for fecund thoughts (Taylor 2006, 95).

Ricoeur (1986, 16) considered that the credibility gap is embedded in all systems of authority, which is why ideology or reproductive imagination compensates by working as an integrative mechanism for the legitimacy of authority. In contrast, utopia implies a nowhere and an "unreal," in that it is not a copy of a pre-existing image or of reality. From a functional perspective then, utopia signals a field of what is possible, *beyond* what is actual; it is a field, therefore, for alternative ways of thinking and living (16). When compared to ideology, utopia obliterates existing mechanisms of legitimacy while expanding people's sense of reality and reality's possibilities (Taylor 2006, 96). Thus, the development of new, alternative perspectives defines utopia's or productive imagination's most basic function (Ricoeur 1986, 16): it is utopia that shapes how we radically rethink concepts such as—and the meanings of—family, consumption, authority, religion (16), the body, pain, and accountability (Igreja 2019b). In this chapter, then, I propose a rethinking of guilt and accountability in ways that allow for a better understanding of the legacies of violent collective processes in diverse non-Western settings (Igreja, Colaizzi, and Brekelmans 2021).

From this perspective, reproductive guilt consists of a set of legal and quasi-legal accountability mechanisms, which, since the Nuremberg Trials 70 years ago, have been variously adapted and have circulated globally as mainstream mechanisms for laying blame on those who are suspected of having perpetrated mass political violations and crimes. Particularly in post–Cold War times, reproductive guilt approaches have been applied in numerous countries in transition, with the expectation that criminal accountability resolves the legacies of grisly pasts and contributes to the legitimacy of the new regime. Unfortunately, such approaches—also known as transitional justice—when hastily applied, often decouple justice needs from the historical and lived realities, and from the complexities of social transformation, in nations that are emerging from decades of mass political violence. One of the golden rules of reproductive guilt is due process, along with a clear separation of the categories of "perpetrators" and "victims" (and sometimes also "bystanders"). Yet these types of neat separations, along with the goal of due process, are extremely difficult to attain in communities in conflict, which have been marred by serious violations across multiple eras (e.g., colonial, anti-colonial, and post-colonial violence aggravated by the Cold War schisms) (Igreja 2012a); intersecting cultures of guilt formation (e.g., domination and resistance, complicity and passivity, and betrayal and sacrifice); cultural and intersubjective notions of the body, including serious violations and embodied accountability; and the involvement of a multiplicity of disparate state and supranational actors (in the current case, Timor-Leste, Portugal, Indonesia, the United States, Australia, and the UN Security Council).

The UN intervention in Timor-Leste was predicated upon a simplification of these complex realities and historical processes, by creating hybrid Special Panels, which held few trials (Reiger and Wierda 2006). The UN intervention also created a new state and new constitution for Timor-Leste and set up a quasi-legal institution that had been previously implemented in various countries around the world under the rubric of "truth commissions" (Boraine 2000). In Timor-Leste, this commission was named the *Comissão de Acolhimento, Verdade e Reconciliação* (CAVR—Commission for Reception, Truth, and Reconciliation) (Lambourne 2010).

Despite the meaningfulness of these legal and quasi-legal dispositions as indicators of the birth of the new state, some prominent Timorese political leaders have, in contested ways, de-emphasized reproductive guilt as part of the formation of the nascent state. Instead, they have underscored what I have termed "productive guilt approaches." These approaches entail a prolific process that separates the alleged criminal actions from their actors. Such a separation differs from the notion of "agency without actors" whereby "non-human actors contribute to the shaping, maintenance, disruption, change as well as the breakdown of social order" (Passoth, Peuker, and Schillmeier 2012, 4). Nor is this separation consistent with the idea that actors were not behaving rationally for having lacked control of their own actions—nor with the idea of collective agency, whereby such actors could not be held accountable as individuals. Instead, such a separation is a rehearsal in radical thinking; it attempts to expand people's sense of reality's possibilities, particularly relationality with one's former bitter and unwavering enemy. This approach is equally constitutive of an anthropology of guilt, which focuses on the ways that deeply estranged individuals within and across nations reconfigure their relationships and interact with one another without effacing past resentments or making the guilt central, yet still negotiating for the resolution of meaningful disputes, both present and ongoing (Igreja 2012b).

In the Timorese context, it is not that the past actions of the alleged perpetrators are not condoned; rather, their past serious misdeeds are kept in "reserves of memory" (Ricoeur 2006), while the same individuals are taken seriously as interlocutors in everyday exchanges to address other significant contemporary issues for the country. In this way, the separation between actions and actors has facilitated the ability for the Timorese to maintain amicable relationships with former enemies, to engage in theatrical displays of affection, and even to enact ambiguous performances of forgiveness (for an analysis of theatrical performances and guilt, see Saskia Fischer in this volume) (Murdoch 2009). Nonetheless, these friendship acts do not constitute exoneration for the serious violations and crimes committed by members of the Indonesian military regime and their Timorese militias; the public pronouncements of forgiveness are ambivalent, as the same Timorese political leaders approved the Decree-Law no. 19/2009, which constituted the country's new criminal code, in the national parliament. This new code has reinforced the notion that impunity is legally impermissible, given that war crimes, crimes against peace, and crimes against humanity and freedom have no statute of limitations for criminal prosecution and related penalties (Timor-Leste Government 2009).

In this way, these relationships are characterized by a sense of ongoing transition, in which new interactions and negotiations can take place even though the legacies of past serious violations are not necessarily legally resolved (Igreja 2019c). Sucing negotiations of order have focused on trying to address several significant post-independence conundrums for Timor-Leste, such as the fact that thousands of Timorese live in diaspora across the border in Indonesia, of whom Maternus Bere is just one. In particular, bilateral mechanisms with the Indonesian authorities have attempted to ensure that these Timorese diasporas do not mobilize themselves and return to instigate political instability in Timor-Leste. It would be a paramount challenge to deal proactively with such issues without engaging the former Indonesian military generals, who once fought in Timor-Leste's war and now occupy prominent political positions in Indonesia, even though some of them were indicted by the UN-sponsored hybrid tribunal in Dili. Another significant factor that has shaped the strategies of some influential Timorese political and military leaders lies in the fact that the powerful advocates of reproductive guilt in the international community have avoided pressing the Indonesian authorities to take their share of the responsibility for dealing judicially with the legacies of the political violence.

Overall, the productive guilt approach has in an ambiguous way proved instrumental for regional political stability and for the post-colonial socio-economic and physical reconstruction of Timor-Leste. The role of Indonesia for the economic recovery and socio-political stability of Timor-Leste cannot be underestimated. This fact has contributed to an informal competition among the Timorese political elites to develop better public rapports with their Indonesian counterparts, and "adds a layer of legitimacy for the Timorese politicians."[1] These outcomes complicate the global accountability discourses, which suggest that a lack of criminal prosecution amounts to impunity (for further discussion on impunity, see Dominik Hofmann in this volume) (Dexler 2011). Instead, the current justice outlook is that there is no prosecution — but there is no impunity either. Moreover, the separation of the alleged criminal actions from their actors has given rise to complex subject positions whereby former perpetrators have not faced criminal justice, yet they do not live in freedom either. Instead, they live under the constant shadow of their own unresolved past of serious violations and can become bargaining chips in untold political deals. The inconvenience of this type of subjectivity was visible when the U.S. government in 2009 refused an entry visa for the former Indonesian army general Sjafrie Sjamsoeddin, who at the time was a close advisor of the former Indonesian president Susilo

Yudhoyono (2004–2014); the American government had blackballed the general because of his alleged involvement in war crimes in Timor-Leste (Dorling 2011); in this case, the visa rejection and the public justification given for that action constituted a form of public reminder of the existing unsettled debt. Thus, the separation of the alleged criminal actions (legally defined) from their actors (who are political operatives) creates an ongoing tension of subjectivity, which can be properly grasped through the dynamics of reproductive and productive guilt. This dynamism and tension makes sense when located within an analysis of local histories of violence; of the regional power asymmetries between Timor-Leste and Indonesia; and of the global geopolitics of avoidance, fear, and control.

The current chapter is based on fieldwork conducted in Dili in 2009, 2010, 2013, and 2019. I interviewed 55 politicians, military personnel, judges, former officials of the CAVR and of the *Comissão de Verdade e Amizade* (CVA, Commission of Truth and Friendship, 2005–2008), and members of civil society organizations. I also conducted research in the Archives of the National Parliament and the Centro Nacional Chega (Igreja 2021).

Formation of Guilt Relations in Timor-Leste and Betrayal as Hope

Timor-Leste has seen various historical periods of serious violations perpetrated by numerous actors, along with multiple forms of complicity and resistance. The country went through almost five decades of violent colonial occupations, wars, and persecution, on top of a lethal famine; in total, all of these events are estimated to have killed around 100,000–200,000 people, according to different sources (CAVR 2003; Kiernan 2003; Mattoso 2012). However, it was the political violence among the Timorese factions, along with the protracted Indonesian military occupation, that gave rise to the most tragic events and experiences, significantly affecting people's lives on a broad scale and destroying the country's physical infrastructure (Lambourne 2010). In the aftermath, the formation of guilt relations took on a different dynamic, focused on attempts by various actors to shape and even to control the memories of the country's previously divided relations.

The coup in Portugal in April 1974 abruptly ended the colonial power's dictatorship and precipitated major political events in most of Portugal's former colonies. As such, a hasty process of decolonization commenced—which, in turn, amplified serious internal disagreements and violent persecution among various national groups in the former colonies (Igreja 2013). In Timor-Leste, the unanticipated decolonization gave rise to a set of political parties, each composed of various factions, which were involved in both intergroup and intragroup violent confrontations. At the center of this conflict (at least until the Indonesian military invasion in 1974–1975) was the island's future political status (Igreja 2021).

From a regional perspective, the transition at this juncture was equally very complex. On the one hand, the world was at the height of the Cold War, which implied that the politico-ideological choices made by several post-colonial governments had tremendous consequences for the country's stability. Timor-Leste was surrounded by two Cold War allies in the regional eradication of communism, namely Australia and Indonesia. On the other hand, internally, there were serious rifts involving mostly fragmented Timorese groups, each making fierce claims of being the only true representative of the people. Ultimately, however, these divisions became subsumed under the clashes between communists and anti-communists.

The main political movements were the União Democrática de Timor Leste (UDT), and the Associação Social-Democrata Timorense (ASDT), which evolved into the Frente Revolucionária de Timor Leste Independente (FRETILIN) (Jolliffe 1978; Pires 1991). In the midst of the political disputes, in August 1975, the UDT surreptitiously and violently seized control of state powers through an armed coup against the colonial Portuguese administration. The coup heightened the milieu of uncertainty and chaos, with arrests of FRETILIN members in Dili, while other leaders of the FRETILIN central committee sought refuge in the country's interior, in fear of their lives (Igreja 2021).

On the issue of historical agency and the formation of guilt relations, Gusmão (2006) has sometimes been uncompromising in rejecting self-attributions of victimhood by Timorese political actors, and he has rhetorically asked, "Can we just ascribe guilt to others, without assuming our own too?" (Gusmão 2004, 110, my translation). Gusmão assumes this position because, while acknowledging that the Indonesian military regime committed abominable acts, they did not create Timor-Leste's divisions or violent conflicts during the transition from Portuguese colonialism to the country's declaration of independence in 1975. Additionally, over time, Mário Carrascalão, one of the founders of the UDT, became Indonesian governor to Timor-Leste. For some segments of the Timorese population, Carrascalão's acceptance to serve the Indonesian military oppressive regime as governor for Timor transformed him into a symbol of the external aggressors and perpetrators. Yet, while governor, Carrascalão betrayed Indonesia too, by secretly supporting the Timorese Resistance under the leadership of Gusmão (Carrascalão 2006). This betrayal was crucial in helping the Timorese resistance at a critical juncture in 1983 and ultimately contributed to the long-term survival of the resistance under Gusmão's leadership (Gusmão 2004, 107).

In this context, acts of betrayal were constitutive of hope and heroism, in ways that complicate present-day attempts to draw stark distinctions between heroes and villains, traitors and patriots, and victims and perpetrators.

Unilateral Declaration of Independence: Mushrooms of Political Discord

In response to the UDT's armed coup, FRETILIN gained the balance of power through a counter-coup. This permutation of power relations allowed FRETILIN to declare independence unilaterally. In the aftermath, key figures from FRETILIN adopted revolutionary violence, hoping that they could hastily halt the reproduction of political discord across the country and take national control. A comparative analysis of various post-colonial transitions of the 1960s–1970s demonstrates that the revolutionary leaders and their followers often relied less on facts and more on rumors in their violent suppression of insubordinate factions, or even of ordinary individuals that had rejected the dominant factions' ideology. In such post-colonial contexts, the so-called reactionaries or anti-revolutionaries were persecuted, tortured, and killed, thus spreading the violence and guilt from the center to the peripheries of their countries (see Igreja 2010). As Xanana Gusmão expressed in a letter to his sister Felismina regarding FRETILIN's revolutionary justice in the wake of the disputed independence in 1975, "There was the case of Uncle Fernando Sousa (in my region), there was the case of Aquiles, for example, and other patriots (of neighboring regions) that did not accept the communist ideology and they were murdered" (cited in Carrascalão 2006, 300–1, my translation).

All told, revolutionary justice led to waves of bloodshed, and many Timorese fled to Indonesian West Timor (Kiernan 2003), thereby expanding the ranks of those who called for Timor-Leste's integration into Indonesia.[2] Thus, the UDT-Movimento Anti-Comunista (MAC), in alliance with the Associação Popular Democrática Timorense (APODETI), announced through the Declaration of Balibo the integration of Timor-Leste into Indonesia (Pires 1991, 317). Ever since, the Indonesian authorities have used the Balibo Declaration to claim (controversially) innocence by arguing that the alliance of UDT, MAC, and other small parties invited them to militarily invade Timor-Leste and contain the breakdown of social order and the spread of violence in the country.

The Indonesian Military Invasion and the Mountains of Resistance

In December 1975, the Indonesian military regime illegally invaded Timor-Leste. FRETILIN and its military wing, the *Forças Armadas da Libertação*

Nacional de Timor-Leste (FALANTIL), mounted some initial resistance, yet the power of the Indonesian army, which was heavily supplied with U.S. armaments, gradually contributed to the weakening of local resistance (Arnove 2008, 188). Timor-Leste's other military giant and neighbor, Australia, turned a blind eye to this illegal Indonesian military occupation.

The invasion expanded the formation of guilt relations in Timor-Leste as FRETILIN/FALANTIL created *Bases de Apoio* (support bases), by taking entire populations from rural villages to the mountains to bolster the resistance alongside the soldiers. However, this strategy was frail, as both the Timorese soldiers and the Timorese populations became easy targets for Indonesian military violence. Furthermore, this strategy also reinforced a generalized perception of guilt by association, in that the Timorese people who had stayed in "the cities were considered as pro-Indonesia," even though "this notion was false."[3]

In 1977, two years following the invasion, the Suharto regime announced amnesty for all FRETILIN members who would surrender (see Igreja 2015a). Francisco Xavier do Amaral, a FRETILIN founder and the first president of independent Timor-Leste in 1975, advocated ending the *Bases de Apoio* strategy and mobilizing the Timorese populations to accept Suharto's call to surrender in exchange for amnesty. Do Amaral's position created serious fissures among the FRETILIN/FALANTIL leaders and culminated in a spiral of mutual, intragroup accusations of treason. Xavier do Amaral was arrested by members of his own group and fired as the country's president, amid accusations of "creating divisions between the leadership and the people" (Acácio 2005, 72, my translation). Other FRETILIN members were arrested, tortured, and executed, under suspicion and accusation of sabotaging the resistance (Mattoso 2012, 70). Nicolau Lobato assumed the leadership of FRETILIN/FALANTIL, until the Indonesian army killed him in combat in 1978, an event that marked the end of the first generation of FRETILIN leaders (Gusmão 2000, 63).

Critical Junctures: Expanding Formation of Guilt Relations

Following the death of Lobato, the *Bases de Apoio* strategy was abandoned. The surviving populations returned to their villages, and the Indonesian army killed many FALANTIL soldiers. Some soldiers surrendered while others continued with FALANTIL under the leadership of Xanana Gusmão to re-establish the resistance. Similar to the violent conflicts over the future political status of the country in 1974 and 1975, the process of re-establishing the resistance from 1980–1984 was marred by serious intragroup disputes, mutual suspicion and violent persecution, and disappearances and killings from among the FRETILIN/FALANTIL ranks in the mountains. Hitherto some of these alleged disappearances and killings have not been clarified, which continues to fuel new cycles of discord among the survivors and their offspring (Igreja 2021).

Over time, the Indonesian security forces attempted to consolidate their control of the social life in the cities through the use of excessive violence against many civilian populations, a strategy with the unintended effect of increasing dissatisfaction among the Timorese who had initially supported the Indonesian regime and occupation (Carrascalão 2006). As a result, many members of the UDT, APODETI, and KOTA (East Timor Popular Movement) on the side of the Indonesians flipped and joined the FRETILIN-FALANTIL resistance on clandestine bases. Such flips created vast networks of informants, spies, and collaborators, which Xanana Gusmão, the resistance's leader, referred to as "the art of coexisting with the enemy" (Gusmão 2000, 117, my translation). While this "art" was instrumental in sabotaging the overwhelming power of the Indonesian military forces and secret services, it "was paid with too many imprisonments, tortured bodies and death(s) of Timorese people."[4] This complicated predicament is a testament to the fact that the Timorese resistance was continuously made and remade on foundations of guilt. Significantly, acts of infiltration of Timorese individuals into the Indonesian military and security networks and of betrayal, euphemistically called "coexistence," were deemed necessary to save the resistance's collective cause.

Furthermore, the art of coexisting with the enemy was ambiguous, and over time it complicated attempts in the post-Indonesian occupation era to set marked distinctions between victims, perpetrators, and bystanders (as required by the golden rules of the reproductive guilt approach). The multiplicity of group affiliations and secret identities from 1986 implied that the thousands and thousands of clandestine members, in many instances, were spread throughout the country and did not know each other. This created great uncertainty regarding how to handle some of the legacies of past collusions, as was clearly illustrated when a member of parliament (MP) representing the *Conselho Nacional da Resistência Timorense* (CNRT) was accused by other MPs of having a tainted past for her alleged collaboration with the Indonesian regime; she defended herself, arguing, "You don't know what I gave to the resistance as a clandestine collaborator, and the resistance soldiers who received it didn't know where it came from either."[5] The multiple subject positions that involved relationships of complicity with the Indonesian authorities and former integrationists, combined with the perspectives that find the UDT guilty of Timor-Leste's collapse in the 1970s, have, over the years, fueled intense political disagreements and conflicts among the Timorese political and military elites. These conflicts have revolved around the best strategies to handle the legacies of the Indonesia military violence as well as legacies of Timorese intragroup violence (Igreja 2021).

Economies of Guilt: Local and Global Moralities

In November 1991, the Santa Cruz massacre (also known as the Dili massacre), perpetrated by the Indonesian military against pro-independence demonstrators, sent shock waves around the world. (For another type of communal violence, particularly involving mass participation of civilians in acts of extreme violence and its legacies, see Parvis Ghassem-Fachandi in this volume.) This renewed the hope of the Timorese people that the world was finally paying attention to the crimes against humanity perpetrated by the Suharto regime. The Indonesian president, B. J. Habibie, who had just replaced President Suharto, agreed through negotiations with Portuguese and United Nations diplomats to hold a referendum in Timor-Leste in 1999.

The fact that the majority of the Timorese voted for independence ignited a spiral of extreme violence, perpetrated by militia groups who had been supportive of Timor-Leste's integration into Indonesia and who were backed by Indonesian military forces. To restore order following the post-referendum violence, the UN established the Transitional Authority in East Timor (UNTAET) and the Serious Crimes Unit (SCU). At this juncture, the formation of guilt relations took on different dynamics, focused on attempts to shape and control the memories of past bitter relations. Thus, some Timorese participants have made accusations of guilt and claims of innocence since the UN initiated the local footprints of the reproductive guilt approach.

Yet other Timorese political leaders wanted to move beyond the use of "memories as weapons" (see Igreja 2008), on the grounds that Timor-Leste could only lose by getting stuck in memory wars and attempting to pursue reproductive guilt approaches openly with respect to the Indonesian authorities. Thus, despite the lure of globalization and the global flows of reproductive guilt approaches, human beings are both makers and products of specific places, conflicts, and forms of imagining resolutions for deep-seated divisions (see Igreja 2019a). This can be seen through the ways that some influential Timorese political leaders have insisted on finding local solutions to deal with the legacies of bitter divisions and enmities with the Indonesian counterparts.

On this issue, Xanana Gusmão admitted that he disagreed with the creation of a Special Tribunal for Timor-Leste—"But because the rule was based on consensus, we had to concede" its creation (Gusmão 2004, 120). The nature of Gusmão's disagreement partly revolved around the temporal scope of the reproductive guilt approach that the UNTAET officials set through the hybrid Special Panels (Dexler 2011; Jeffery 2016; Kent 2011; Reiger 2006; Reiger and Wierda 2006). The temporal focus was strictly on the 1999 criminal events, and thus sidestepped the crisscrossing histories and specificities of place that were involved in the formation of guilt relations in Timor-Leste (see Igreja 2012b, Igreja 2018). On this issue, Gusmão retorted:

> Is it fair that we set a Special Tribunal to try the acts committed in 1999, which according to the Rome Statute only constitute 5% of the crimes committed in Timor-Leste, whereas the remaining 95% were committed 24 years prior to 1999? Is this fair? Is it fair that we ascribe guilt to the Timorese here in the

> Special Tribunal just because we have the capacity to try them and let free those that committed crimes from 1998 back? (Gusmão 2004, 121)

While Gusmão and his political allies could not stop the creation of the Special Tribunal, by 2005 the SCU and the Special Tribunal in Timor-Leste had managed to hold 35 trials and indict around 400 people; 48 people were convicted and 2 acquitted (Reiger and Wierda 2006, 3). Yet these cases involved just a fraction of the suspected criminals, because the majority of the 1999 perpetrators from the militias fled to Indonesia, including Maternus Bere, and the Indonesian authorities did not share responsibility by extraditing them back to Timor-Leste to face trial. Such outcomes reinforced the perception of an economy of guilt whereby the less powerful (Timorese) perpetrators would pay a price for their actions, whereas the suspected individuals from the powerful (Indonesian) military and their allied militias would not face, for the time being at least, the consequences of reproductive guilt.

Alongside the instruments of reproductive guilt, the UN sponsored the creation of a modernist constitution with a global legalist promise for Timor-Leste. The constitutional Article 160 (Serious Crimes) extended the need for criminal accountability all the way back to 1974 (Timor-Leste Government 2002, 59). The Timorese constitution was equally in tune with the necessity to give voice to victims through legalizing the creation of the *Comissão de Acolhimento, Verdade e Reconciliação* (CAVR—Commission for Reception, Truth and Reconciliation) (Timor-Leste Government 2002, 60).

Thus, principles of reproductive guilt approaches ensured a process of transition in Timor-Leste that encompassed both judicial procedures and popular testimonies as supposed sources of truth and moral repair (Hayner 2001). The CAVR (2002–2005) was able to read off the names of individuals, institutions, and countries responsible "for both serious and less serious crimes" committed from 1974 to 1999 by the Indonesian military forces and their backed militias, as well as Timorese factions (Burgess 2006; CAVR 2003, xxiii; Lambourne 2010).

Overall, however, as stated earlier, the national and international transitional justice actors were unable to bring to justice the Indonesian military leaders and their backed militias, who were mostly responsible for the serious human rights violations and crimes. As a result, the constitutional article came to appear ambivalent in time, and the former Timorese attorney general told me that the adoption of reproductive guilt approaches in the country's constitution was the result of "perhaps a lot of our own ingenuity."[6] Along a similar line of reasoning, the former president of the CAVR told me that the constitution's inclusion of Article 160 was due to the fact that "everyone was euphoric about justice,"[7] but that this justice, as time went on, was not thoroughly possible for a variety of reasons at national, regional, and international levels (Dexler 2011; Jeffery 2016; Kent 2012; Lambourne 2010; Reiger 2006).

Furthermore, the CAVR report is yet to be discussed in the Timorese national parliament. A member of parliament representing the Conselho Nacional da Resistência Timorense, who had been personally accused of collaborating with the Indonesian authorities, explained why the CAVR report had not been discussed in parliament:

> We have some colleagues in parliament who are veterans. They fear that if we discuss this report in the parliament they will not be considered as just victims but also actors of crimes. But this is something that we have to accept, it was a long process of struggle.[8]

Ten years later, in 2019, I was told again that the CAVR report had not yet been discussed by the national parliament:

> Because in all legislatures (parliamentary sessions) we have important laws to deliberate. . . This does not mean that the CAVR report is not important. The issue is that we have the life of the country, financial investments so we cannot focus on everything in five years.[9]

This economy of guilt plunged the reproductive guilt approach into serious doubt, to the extent that when calls were made for the creation of an international tribunal to replace the special tribunal, various high-profile Timorese political and military leaders were again adamantly against such an idea. "I will be the first to reject it," stated the former president and prime minster Xanana Gusmão (Gusmão 2004, 121). Gusmão's position gradually gained broad support among other local leaders. José Ramos-Horta, a Nobel Peace Prize laureate (1996) and former president (2007–2012), repeated demands to "the United Nations to stop gathering evidence against the killers of hundreds of Timorese" and argued that the "Timorese must forgive Indonesians who committed heinous crimes against us" (Murdoch 2009; Dexler 2011, 55).

Regional Moralities: Productive Guilt and Friendships

The Indonesian authorities had promised a reproductive guilt approach in dealing with the legacies of violence in Timor-Leste. They held ad hoc trials of their own, but the overall result was unsatisfactory (Hirst 2009, 6). Despite the failure of the reproductive guilt approach in Timor-Leste and Indonesia, the Timorese government adopted the strategy of regional friendships by creating and implementing a bilateral institution involving Indonesia and Timor-Leste: the aforementioned *Comissão de Verdade e Amizade* (CVA, Commission of Truth and Friendship, 2005–2008). This initiative, as its name suggests, aimed at achieving truth in order to create friendship.[10] The issuing friendship was fecund, as the Timorese authorities separated the alleged criminal actions from their actors. In everyday life, they embraced the Indonesians as contemporaneous political actors, but not their past gruesome actions when they were members of the Suharto military regime. This division was predicated upon the uncompromising position of the Timorese laws (constitution and criminal code), which rejected the statute of limitations for war crimes, crimes against peace, or crimes against humanity and freedom.

In this context, the Timorese political leaders reached out to their Indonesian counterparts without changing the status of their past actions. With these bold initiatives, some of the key Timorese leaders paved the way to "radically rethink" (Ricoeur 1986, 16) their relations with the Indonesian authorities, which in turn favored the Timorese agenda of "social justice as the best remedy for the national trauma" (Gusmão 2004, 120). Ultimately, friendly relations with the Indonesian authorities facilitated Timor-Leste's post-independence state reconstruction, trade relations, and a gradual socio-economic development. The unresolved land border issue between the two countries is also an important factor in the careful management of their relationship.

However, this productive guilt approach was heavily criticized by the UN, the main sponsor of the transition process, on the grounds that the Timorese leaders were prioritizing relationships with alleged perpetrators to the detriment of achieving reproductive guilt (Strating 2014). Nevertheless, while the UN openly criticized Timor-Leste's leaders for their apparent soft stance on justice, it seemed hesitant—even unwilling—to put any concerted international pressure on Indonesian authorities to cooperate with criminal justice processes (Reiger and Wierda 2006, 1). The UN's unwillingness to pressure Jakarta's political and military elites with respect to Timor-Leste could be linked to geostrategic considerations and fears; since the terror attacks of September 11, 2001, the U.S.-led global war on Islamist terrorism has rehabilitated the defunct Cold War alliance and included Indonesia, the world's biggest Muslim country, as an important regional partner (Ramakrishna 2005).

At the same time, Timor-Leste's political and military leaders have been realistic about their own geostrategic limitations and fearful of the potential consequences of miscalculations. They told me that "Indonesia is a giant neighbor,"[11] and Timor-Leste imports most of its basic goods from Indonesia: "rice, *supermi* (noodles), cigarettes, cooking oil, butter, and we also have a large community of Timorese youngsters studying in Indonesian universities."[12] They were assertive in their refusal to pursue criminal justice against some members of the Indonesian military, because "[t]he Indonesians consider these war generals their heroes,"[13] and some of them have remained political operatives who have launched serious candidacies for the Indonesian presidency (Marks 2004). For state survival reasons, then, "it was essential to maintain a sincere relationship of friendship with our neighbors. Not only for us but also for the stability of the region."[14] The sincere relations in this case imply open and constant venues to dialogue with their counterparts.

At the time of the presentation of the CVA report, Susilo Yudhoyono, a former Indonesian president and former Indonesian army general who fought the war in Timor-Leste in the 1980s, expressed regret for Indonesia's history of violence toward Timor-Leste. (For discussion on apology, see Maria-Sibylla Lotter's chapter in this volume.). In response, former Timorese prime minister Xanana Gusmão declared, "I am satisfied," and added, "Now, in our country instead of crying every day, we have to make policies, instead of crying, instead of saying we are victims" (*Sidney Morning Herald* 2008). It was partly this conviction—that it was necessary to engage constructively with the current Indonesian political operatives without condoning their violent past actions—that drove Gusmão to violate the country's own constitution by illegally releasing Maternus Bere from prison and returning him to Indonesia. Regardless, the Timorese political leaders did not change their constitution; it still contains the article that permanently reminds of an existing serious debt (see Igreja 2008). Furthermore, and as stated above, following Yudhoyono's expression of regret for Indonesia's previous conduct, the Timorese parliament approved in 2009 the Decree-Law no. 19/2009, the country's new criminal code, which officially rejects impunity. In this train of thought, Gusmão's statements—expressing a readiness to reset relations with former enemies—mirror a mindset of separating the injurious actions from their actors.

Concluding Remarks

This chapter analyzed the dynamics of reproductive and productive guilt in Timor-Leste, following almost five decades of political violence, which encompassed colonial, anti-colonial, and post-colonial violence and resistance, aggravated by Cold War schisms. In order to grasp the dynamics of reproductive and productive guilt in Timor-Leste, as well as the ongoing tensions over issues of truth and blame—tensions that led to the arrest and subsequent release-without-trial of Maternus Bere (suspected of committing international crimes)—this chapter considered two interrelated factors.

First, it considered how permutations in local histories of violence internally thwarted clear-cut definitions of the concepts of "innocent," "traitor," and "culprit." Building upon Ricoeur's phenomenology of human action and creativity, the chapter developed the approaches of reproductive and productive guilt. These two categories allow an understanding of how the Timorese political leaders—some of whom were euphoric for international criminal justice while others were more cautious and pragmatic—adopted a modernist constitution that is deeply embedded in global accountability discourses and the era of the witness. The reproductive guilt approach through the country's new constitution included legal provisions for criminal accountability, as well as popular mechanisms of bearing witness for past violations (of varying degrees of severity).

Second, alongside reproductive guilt, some influential members of the Timorese political and military elites adopted the approach of productive guilt. This approach does not derive from any previously recognized legal or political playbook. In the Timorese context, the productive guilt approach, which is based on a separation of violent actions from their actors, has led to the establishment of amicable relations with political and military elites who have been linked to the Republic of Indonesia, without condoning their past violent actions. Some of the individuals in question either served in Timor-Leste as Indonesian military commanders during the violent military occupation of the island (e.g., Susilo Yudhoyono, the former Indonesian president, 2004–2014) or were officially indicted for perpetrating serious human rights violations and crimes in Timor-Leste, such as General Wiranto. On a number of occasions, the former Timorese president and prime minister Xanana Gusmão was photographed in public with General Wiranto, and both were smiling like close and sincere friends. Yet these seemingly rituals of reconciliation between Gusmão and Wiranto were not acted out as a demonstration of solidarity with the Indonesian military. This appears to be the case, as such public signals of friendliness have neither resolved nor impacted the legal status of the crimes of the past in the eyes of the Timorese authorities. Instead, Timorese authorities approved a new criminal code in 2009, which reiterated the existence of an unresolved debt and stipulated that, for such debts related to serious crimes perpetrated in the country's history, there is no statute of limitations.

In this regard, this ambiguous status—consisting of neither criminal accountability nor legal impunity—has opened the way for Timor-Leste to resolve maritime border disputes with Australia and Indonesia without antagonizing these powerful neighbors.[15] In turn, the stability of relations with Indonesia has created the political space for Timorese political and military elites to continuously negotiate exit strategies for the disturbing legacies of their own internal, violent conflicts and the disputes over legitimacy that date back to Timor-Leste's transition from Portuguese colonialism to independence in the mid-1970s.

Notes

1. Personal Interview, Mário Carrascalão. Dili, December 10, 2010. All personal interviews are translated by the author.
2. Personal Interview, José Soares, former executive assistant at the Council of Ministers and former CAVR Commissioner. Dili, July 30, 2009.
3. Personal Interview, Maria Lay, Parlamento Nacional de Timor Leste (PNTL), MP for CNRT. July 17, 2009.
4. Personal Interview, David Mandate, FRETELIN MP, Member of the Commission for Defence and Security and External Relations. Dili, July 4, 2009.
5. Personal Interview, Maria Lay, PNTL. Dili, July 24, 2009.
6. Personal Interview, Dr. Ana Pessoa, attorney general of Timor-Leste. Dili, July 9, 2009.
7. Personal Interview, Aniceto Guterres Lopes, PNTL. Dili, July 16, 2009.
8. Personal Interview, Maria Lay, PNTL. Dili, July 17, 2009.
9. Personal Interview, Maria Lay, PNTL. Dili, October 3, 2019.
10. Personal Interview, Dr. Mari Alkatiri, FRETILIN general secretary and former prime minister. Dili, December 14, 2010.
11. Personal Interview, José Soares, former CAVR official. Dili, July 30, 2009.
12. Personal Interview, Virgílio Smith, Secretário de Estado da Cultura. Dili, July 6, 2009.
13. Personal Interview, Jose Soares, former CAVR official. Dili July 30, 2009.
14. Personal Interview, Dr. Ana Pessoa. Dili, July 9, 2009.
15. Personal Interview, Maria Lay, PNTL. Dili, October 3, 2019.

References

Acácio, Manuel. 2005. *A última bala é a minha vitória*. Cruz Quebrada, Portugal: Oficina do Livro.

Arnove, Anthony. 2008. *The Essential Noam Chomsky*. London: The Bodley Head.

Boraine, Alex. 2000. *A Country Unmasked*. Oxford: Oxford University Press.

Branscombe, Nyla, and Bertjan Doosje. 2004. "International Perspectives on the Experience of Collective Guilt." In *Collective Guilt*, edited by Nyla Branscombe and Bertjan Doosje, 1–15. Cambridge: Cambridge University Press.

Burgess, Patrick. 2006. "A New Approach to Restorative Justice—East Timor's Community Reconciliation Processes." In *Transitional Justice in the Twenty-First Century*, edited by Naomi Roht-Arriaza and Javier Mariezcurrena, 173–205. Cambridge: Cambridge University Press.

Carrascalão, Mário. 2006. *Timor antes do futuro*. Dili: Livraria Mau Huran.

CAVR (Comissão de Acolhimento, Verdade e Reconciliação). 2003. *CHEGA! The Final Report of the Timor-Leste CAVR.* Vol. 1. Jakarta: PT Gramedia.

Dexler, Elisabeth. 2011. "The Failure of International Justice in East Timor and Indonesia." In *Transitional Justice*, edited by Alexander Hinton, 49–66. New Brunswick, NJ: Rutgers University Press.

Dorling, Philip. 2011. "Yudhoyono's Top Adviser a Timor War Crimes Suspect." *Sydney Morning Herald*, March 11, 2011. https://www.smh.com.au/world/yudhoyonos-top-adviser-a-timor-war-crimes-suspect-20110311-1brc7.html.

Government of Timor-Leste. 2009. Decreto Lei 19/2009: Código Penal. *Journal da República.* Dili: Government of Timor-Leste.

Gusmão, Xanana. 1994. *Timor Leste: Um Povo, Uma Pátria.* Lisbon: Edições Colibri.

Gusmão, Xanana. 2000. *Resistir é Vencer.* Victoria, Australia: Aurora Books.

Gusmão, Xanana. 2006. *Discurso ao Povo Amado e Sofredor e aos Lideres e Membros da FRETILIN.* Dili: Archives Parlamento Nacional.

Gusmão, Xanana. 2004. *A construção da nação timorense.* Lisbon: LIDEL.

Hayner, Priscilla. 2001. *Unspeakable Truths.* New York: Routledge.

Hirst, Megan. 2009. *An Unfinished Truth.* New York: International Center for Transitional Justice.

Igreja, Victor. 2008. "Memories as Weapons." *Journal of Southern African Studies* 34, no. 3: 539–56. https://doi.org/10.1080/03057070802259720.

Igreja, Victor. 2010. "Frelimo's Political Ruling through Violence and Memory in Postcolonial Mozambique." *Journal of Southern African Studies* 36, no. 4: 781–99. https://doi.org/10.1080/03057070.2010.527636.

Igreja, Victor. 2012a. "Multiple Temporalities in Indigenous Justice and Healing Practices in Mozambique." *International Journal of Transitional Justice* 6, no. 3: 404–22. https://doi.org//10.1093/ijtj/ijs017.

Igreja, Victor. 2012b. "Negotiating Order in Postwar Mozambique." In *The Dynamics of Legal Pluralism in Mozambique*, edited by Helene Kyed, João Coelho, Amelia Souto, and Sara Araújo, 148–66. Maputo: Centro de Estudos Aquino de Bragança.

Igreja, Victor. 2013. "Mozambique." In *Encyclopedia of Transitional Justice.* Vol. 2, edited by Lavinia Stan and Nadya Nedelsky, 305–11. Cambridge: Cambridge University Press.

Igreja, Victor. 2014. "Memories of Violence, Cultural Transformations of Cannibals, and Indigenous State-Building in Post-Conflict Mozambique." *Comparative Studies in Society and History* 56, no. 3: 774–802. https://doi.org/10.1017/S0010417514000322.

Igreja, Victor. 2015a. "Amnesty Law, Political Struggles for Legitimacy and Violence in Mozambique." *International Journal of Transitional Justice* 9, no. 2: 239–58. https://doi.org/10.1093/ijtj/ijv004.

Igreja, Victor. 2015b. "Media and Legacies of War." *Current Anthropology* 56, no. 5: 678–700. https://doi.org/10.1086/683107.

Igreja, Victor. 2018. "Negotiating Temporalities of Accountability in Communities in Conflict in Africa." In *Time and Temporality in Transitional and Post-Conflict Societies*, edited by Natascha Mueller-Hirth and Sandra Oyola, 84–101. London: Routledge.

Igreja, Victor. 2019a. "Frames and Intersections of Studies of Place, Conflict and Communication." In *The Nexus among Place, Conflict and Communication in a Globalizing World*, edited by Pauline Collins, Victor Igreja, and Patrick Danaher, 1–16. Singapore: Palgrave Macmillan.

Igreja, Victor. 2019b. "Negotiating Relationships in Transition: War, Famine, and Embodied Accountability in Mozambique." *Comparative Studies in Society and History* 61, no. 4: 774–804. https://doi.org/10.1017/S0010417519000264.

Igreja, Victor. 2019c. "'What Made the Elephant Rise Up from the Shade?'" In *Truth, Silence and Violence in Emerging States*, edited by Aidan Russel, 88–110. Abingdon, UK: Routledge.

Igreja, Victor. 2021. "Negotiating the Legacies of Intragroup Violence in Timor Leste." *International Journal of Transitional Justice*, ijab007. https://doi.org/10.1093/ijtj/ijab007.

Igreja, Victor, Janna Colaizzi, and Alana Brekelmans. 2021. "Legacies of Civil Wars: A 14-Year Study of Social Conflicts and Well-Being Outcomes in Farming Economies." *British Journal of Sociology* 72, no. 2: 426–47.

Jeffery, Renée. 2016. "Trading Amnesty for Impunity in Timor-Leste." *Conflict, Security & Development* 16, no. 1: 33–51. https://doi.org/10.1080/14678802.2016.1136139.

Jolliffe, Jill. 1978. *East Timor*. Brisbane: University of Queensland Press.

Kent, Lia. 2011. "Local Memory Practices in East Timor: Disrupting Transitional Justice Narratives." *International Journal of Transitional Justice* 5, no. 3: 434–55. https://doi.org/10.1093/ijtj/ijr016.

Kent, Lia. 2012. "Interrogating the 'Gap' between Law and Justice." *Human Rights Quarterly* 34: 1021–44. https://doi.org/10.1353/hrq.2012.0059.

Kiernan, Ben. 2003. "The Demography of Genocide in Southeast Asia." *Critical Asian Studies* 35, no. 4: 585–97. https://doi.org/10.1080/1467271032000147041.

Lambourne, Wendy. 2010. "Unfinished Business: The Commission for Reception, Truth, and Reconciliation in East Timor." In *The Development of Institutions of Human Rights*, edited by Lilian Barria and Steven Roper, 195–207. New York: Palgrave Macmillan.

Marks, Kathy. 2004. "Indonesian General Accused of War Crimes Moves a Step Closer to the Presidency." *The Independent*, April 22, 2004.

Mattoso, José. 2012. *A Dignidade*. Lisbon: Temas e Debates.

Murdoch, Lindsay. 2009. "Forgive Indonesian Crimes, Ramos-Horta Urges East Timorese." *Sydney Morning Herald*, August 31, 2009.

Nietzsche, Friedrich. (1887) 2017. *On the Genealogy of Morality and Other Writings*. Edited by Keith Ansell-Pearson. Translated by Carol Diethe. Cambridge: Cambridge University Press.

Passoth, Jan-Hendrik, Birgit Peuker, and Michael Schillmeier, eds. 2012. *Agency without Actors*. London: Routledge.

Pires, Mário. 1991. *Descolonização de Timor*. Lisbon: Dom Quixote.

Ramakrishna, Kumar. 2005. "'The Southeast Asian Approach' to Counter-Terrorism." *Journal of Conflict Studies* 25, no 1: 27–47.

Reiger, Caitlin. 2006. "Hybrid Attempts at Accountability for Serious Crimes in Timor Leste." In *Transitional Justice in the Twenty-First Century*, edited by Naomi Roht-Arriaza and Javier Mariezcurrena, 143–70. Cambridge: Cambridge University Press.

Reiger, Caitlin, and Marieke Wierda. 2006. *The Serious Crimes Process in Timor-Leste*. New York: ICTJ.

Ricoeur, Paul. 1986. *Lectures on Ideology and Utopia*. Translated by George Taylor. New York: Columbia University Press.

Ricoeur, Paul. 2006. *Memory, History, Forgetting*. Translated by K. Blamey and D. Pellauer. Chicago: Chicago University Press.

Sidney Morning Herald. 2008. "Apology Accepted Now It's Time to Move On, Gusmão." *Sidney Morning Herald*, July 16, 2008.

Strating, Rebecca. 2014. "The Indonesia-Timor-Leste Commission of Truth and Friendship." *Contemporary Southeast Asia* 36, no. 2: 232–61.

https://www.smh.com.au/world/apology-accepted-now-its-time-to-move-on-gusmao-20080716-3ga9.html.

Timor-Leste Government. 2002. *Constituição da República Democrática de Timor Leste*. Last modified March 5, 2021. Constituicao_RDTL_PT.pdf (timor-leste.gov.tl).

Taylor, George. 2006. "Ricoeur's Philosophy of Imagination." *Journal of French Philosophy* 16, no. 1/2: 93–104.

15

Disputes over Germany's War Guilt

On the Emergence of a New International Law in World War I

Ethel Matala de Mazza

Civilization is the organization of human responsibilities.

—Georges Clemenceau (1919)

The Treaty of Versailles and Its "War Guilt Clause"—A Case of Victor's Justice?

World War I ended on November 11, 1918, in a copse near Compiègne, close to the French lines on the Western Front. Three days earlier, at the crack of dawn, the supreme commander of the Allied forces, Marshal Foch, had received the small German peace delegation in a railway carriage and confronted them with demands that they accepted only with extreme reluctance (cf. Krumeich 2018, 139–45). The German envoys hoped to negotiate more advantageous terms at the peace conference scheduled to begin two months later in Paris.

Among the advisory panel that traveled to the peace conference alongside the German diplomats was the liberal sociologist Max Weber, who was active in the newly founded "*Arbeitsgemeinschaft für eine Politik des Rechts*" (Committee for Legal Policy) (cf. Jäger 1984, 30; Heinemann 1983, 34). Weber had made repeated public statements arguing that a defeated Germany ought not to incur greater losses than necessary in order to preserve its prominent role among the great powers. At a highly publicized lecture given in Munich in January, Weber had admonished his student audience to proceed with discernment and circumspection: "Instead of searching, like an old woman, for the 'guilty party' after the war," the task now, he declared, was to preserve "a manly and unsentimental bearing," to face up to the enemy and say: " 'We lost the war—you won it. The matter is now settled. Now let us discuss what conclusions are to be drawn in light of the *substantive* (*sachlichen*) interests involved and—this is the main thing—in the light of the responsibility for the *future* which the victor in

Ethel Matala de Mazza, *Disputes over Germany's War Guilt* In: *Guilt.* Edited by: Katharina von Kellenbach and Matthias Buschmeier, Oxford University Press. © Oxford University Press 2022. DOI: 10.1093/oso/9780197557433.003.0016 particular must bear' " (Weber 1994, 356). Weber cautioned against an ethics of conviction that considered itself morally superior, but which in fact would only end up raising the cost of peace by looking for guilty parties to blame. With this appeal to the primacy of *raison d'état*, Weber anticipated positions that are once again—and increasingly—being heard in contemporary debates. In their introduction to this volume, Matthias Buschmeier and Katharina von Kellenbach give an overview regarding current adherents to similar arguments, and Maria-Sibylla Lotter echoes these in her chapter as well. Laying blame, it is claimed, serves only to discredit the opposite side in moral terms. Charges against individuals fall short of addressing the complexity of politics, where such charges are out of place and indeed unhelpful in reaching sensible solutions. Politics, the argument goes, has more urgent problems to address than raking over "ancient" history.

Following World War I, such arguments derived their urgency from the fact that the case for the prosecution (unlike in such cases discussed today as "white guilt," which Lisa Spanierman deals with in her chapter) was made by the victors. The defeated Germans accused the British, French, Americans, and Italians of not being satisfied with the punishment that defeat inevitably entailed, and, still further, of feeling the need to blame the vanquished side for having started the war in the first place. The approach that the Allies took was indeed unprecedented, a departure from established convention.

Why they chose this path and—for the first time in history—explicitly tied the peace settlement to the prosecution of offenses remained a controversial issue for decades. As late as the 1960s, German historians uniformly treated this as a case of victor's justice, an abuse of the treaty to give a moral justification for excessive demands. By contrast, even at the time of the Weimar Republic, English and French lawyers pointed out that the treaty's controversial "war guilt clause" had a technical function, testifying above all to the new significance accorded by the Allies to international law (cf. Binkley and Mahr 1926, 398–400; Binkley 1929, 294–300; Bloch and Renouvin 1932, 1–24). According to this argument, explicitly making international law the sole guarantor of rule-based relations between states set new standards of civility.

This chapter picks up this thread by considering the key role played by debates over guilt and responsibility within the process of modernization that culminated in the Paris Peace Conference of 1919–1920 and its five principal peace treaties. Guilt turned out to be productive in two ways. First, it became a category of legal importance in international law. Second, it opened up new avenues in foreign policy for representatives of the young German democracy, at least in the eyes of Kurt Eisner, then prime minister of the republic of Bavaria. His initiative to publicly admit the guilt of imperial Germany was heavily combatted at that time. Yet it deserves, as I argue, recognition as an early effort to accept political responsibility and thereby foster new bonds between former enemies.

Indicative of my first point, the question of German war guilt had been internationally discussed almost from the war's outbreak. In the polemics that erupted around the conduct of the German forces, the conflict was framed as being fought over the very foundations of civilization. Consequently, the narratives that gained currency not only generated considerable energy for mobilization, but also engendered normative expectations by which the negotiators at the peace conference felt bound. By appealing to inviolable principles of international law, the victorious Allies had committed themselves, before the gaze of a global public, to a particular normative foundation. Rather than compromise it by leniency toward the accused, the Allies felt bound to consolidate it by ensuring that German offenses were punished by legal means, and the debates over guilt and responsibility helped to bring into being a new legal culture that refused to tolerate impunity for punishable offenses in war or peace. (As Dominik Hofmann points out in his chapter, impunity is nowadays, in fact, widely considered a breach of normative expectations, though autocratic rulers still claim it as their privilege.) The new legal culture also politically universalized standards of civil law, thereby ensuring that the same rules obtained between states as between individuals, and that the great powers of the West, at least in principle, were no less bound by statutory law than was the rest of the world. Just how decisive a transformation this was can be seen by taking a brief look back at the amnesty provisions that had previously been the norm in peace treaties.

Yet this paradigm shift did not necessarily entail a diminution of Germany's scope for political activity, as Weber had suspected it might in view of Germany's probable discrediting at the Paris Conference. In the second half of my chapter, I turn to Kurt Eisner, a left Social Democrat, who regarded the guilt clause—if accepted by the Germans—as a starting point for future international relations. Eisner made a bold declaration of guilt in the public forum of the press. He was concerned with making a credible statement of self-renewal, and, this time, with doing so at a national level. Whereas the armistice on the Western Front had brought about the first German Republic (the Weimar Republic), "as it were, in the passive voice" (Schivelbusch 2001, 276, translation jpk), Eisner strove to relegitimize the abolition of the monarchy and to underscore that this young democracy was a state that, unlike that of the Kaiser, set great store in the tenets of international law. Since this freely made confession was widely understood in Germany as an all too freely granted concession to the Allies, it was met with fierce resistance in Germany, where Weber's opinion that peace talks ought to steer clear of questions of guilt was widely shared. This attitude had a long tradition upon which to draw.

Amnesty and Guilt in International Law

Until World War I, it had been the long-established custom among warring parties that a truce should be accompanied by mutual guarantees to dispense with any further argument over the recent events. To this end, European peace treaties of the 17th and 18th centuries would contain an "oblivion clause." This interdiction on recalling past wrongs was premised on the assumption that peace was only possible under an amnesty, and that only a declared intent to forget could keep conflicts from being rekindled.

Such rules were already widespread in antiquity,[1] but they acquired new significance in early modern Europe, in the form of reciprocal assurances between European sovereigns who recognized one another as foes equal in rights and standing. In the international system as it was consolidated under the Peace of Westphalia, the medieval distinction between just and unjust wars had already, de facto, ceased to be central in the 14th century, even though it remained in force de jure. Following the Thirty Years' War, the church lost its monopoly as the arbiter of international law (cf. Keen 2015, 63–82; Lesaffer 2004, 9–44). This development was paralleled, in the peace treaties, by a conceptual dissociation of forgiving and forgetting (cf. Fisch 1979, 92). Where absolution had once been granted by one side and, in accordance with the Christian idea of sin, referred to the opposite party's explicitly condemned sins, treaties now came to contain a sweeping mutual assurance not to bear grudges over past wrongs and to consign to oblivion all that might threaten the balance and harmony between the parties—as in the proverbial "Concert of Europe."

The most conspicuous indicator of this shift was the new, now entirely technical meaning taken on by the concept of amnesty (cf. Fisch 1979, 95). Etymologically, "amnesty" and "amnesia" derive from the same Greek root—*amnēstia* (ἀμνηστία)—which the *Etymological Dictionary of English* defines as "forgetfulness." It was in this sense that the concept had been used in the 5th century BCE, when the Athenian civil war ended with all citizens being prohibited from "recalling the misfortunes of the past" (περί του μη μνησικακείν) (Loraux 2002, 145–70).

The early modern concept of amnesty then translated this prohibition into a legal guarantee of indemnity. None of the warring parties were to be punished for starting the conflict nor held liable or obligated to pay reparations for the destruction wrought in its course. The relevant amnesty clause of the Peace of Westphalia reads as follows:

> [A]ll that has pass'd on the one side, and the other, as well before as during the War, in Words, Writings, and Outrageous Actions, in Violences, Hostilitys, Damages and Expences, without any respect to Persons or Things, shall be

> entirely abolish'd in such a manner that all that might be demanded of, or pretended to, by each other on that behalf, shall be bury'd in eternal Oblivion. (Israel and Chill 1967, 9–10)

By renouncing all claims to prosecution and restitution, the contractual parties assured each other at the same time of their mutual recognition as powers for whom warfare was a legitimate means of pursuing their political objectives. The sovereignty claimed by each of them in equal measure freed them of the obligation to justify their actions to their opponents (cf. Dickmann 1964, 4–5; Schmitt 1997, 91).

Yet in spite of this long precedent, the victorious powers of World War I, in dispensing with such clauses, were neither being ignorant nor vindictive, but rather were following the trend of the discussion among scholars of international law since the Enlightenment. Jurists had already begun to direct their efforts not only toward securing peace by means of treaties, but also to using law to prevent wars from breaking out in the first place. Conferences to promote such efforts were held at The Hague in 1899 and 1907 and were attended by delegates not only from the great European powers, but also from Canada, Mexico, China, Japan, Persia, and Siam. These conferences were clear political statements, and the failure of the most ambitious initiatives—for global disarmament or the creation of an international criminal court—did not detract from the determination with which the leaders of these movements appealed to a transnational consensus, which held that there were sanctionable as well as justifiable means of settling conflicts (cf. Payk 2018, 27–78). If states (the idea went) were to commit themselves to deploying their forces only in self-defense, international law might assume the role of the "gentle civilizer of nations" (Koskenniemi 2004) and bring about a lasting pacification of international relations. The sanctity of contracts and a commitment to abide by the law would become not only the general rule of international conduct, but would also serve to distinguish civilized from uncivilized countries.

What was new about these conceptions of international law was not the principle of *pacta sunt servanda* (agreements must be kept), which was already part of medieval canon law, so much as their claim to universal validity. It was no longer just the "backward" countries of southeastern Europe, Asia, and elsewhere that were measured by the civilizational norm of lawful behavior—against which they were likely to fall short, thus providing a justification for their conversion and/or colonization (cf. Osterhammel 2006; Barth and Hobson 2020, 1–20)—but now also the great powers of Western Europe themselves, regardless of whether they were signatories to the relevant treaties or not. In view of this exalted position accorded to international law as the definitive guarantor of orderly relations between states, the withholding of consent was tantamount to a state's breach of the global social contract, and thus that state risked marginalizing itself. A reluctance to commit itself to such principles—as Germany was reluctant to do at The Hague—signaled a lack of reliability on the state's part.

Such shifts in international-level discourse aside, it also became increasingly difficult at the national level to make the prospect of dying for one's country appealing to young men and, moreover, to win the support of parliaments (for the requisite funds) and the public at large (for the inevitable sacrifices and losses). The French armies during the revolutionary and Napoleonic eras had shown how successful troops might be on the battlefield if armies consisted no longer of mercenaries, but rather of citizens imbued with patriotism. If monarchs were to keep up with this development and ensure public consent to political decisions, even those monarchs not otherwise dependent on popular consent were nonetheless under pressure to find good reasons for embarking on a war and to justify it, even if only in official pronouncements, as a worthwhile and necessary enterprise.

With regard to World War I, this helps explain why the question of war guilt arose not only at its conclusion but even at its outset, and why the mobilizations in the east and west were accompanied by a *Krieg der Geister* (war of minds) (Kellermann 1915), in which scientists and academics could be found making no less inflammatory pronouncements than writers and journalists.

When, in August 1914, five German armies invaded neutral Belgium en route to northern France, they broke a treaty that had been drawn up to protect this small and only quite recently independent country from suffering a second battle of Waterloo on its own territory (cf. Clark 2013, 547–51; Horne and Kramer 2001, 9). Intellectuals in neighboring countries were outraged, bolstering their condemnation of Germany's violation of Belgian neutrality with grand narratives in which calls to national resistance took second place to appeals for a coalition of nations acting in defense of international law. By rhetorical sleight of hand, this coalition was made to include prospective Allies not yet bound to France by treaty. Not only were such supporters welcome for their shared ideals; they could also expect a degree of cultural acceptance that went beyond the immediate context in which they expressed their solidarity.

Around the notion of German war guilt thus grew up a new world, one in which the joint defense of international law against violations created relations—distinct, albeit still informal—across cultural and geographical boundaries, and wherein these relations could be used for political purposes even after the war had ended. The narratives deployed both in the Allied countries and in Germany played a key role in creating this world. In these narratives, the contested front lines were reconceptualized as cultural chasms. The task at hand was to redraw the political map in such a manner as to make "Europe" no longer co-extensive with "civilization."

The First World War as Cultural War

Among the first intellectuals to raise their voices immediately following the German invasion of Belgium was the French philosopher Henri Bergson. Speaking with the authority of the president of the Académie des Sciences morales et politiques, Bergson declared the war against Germany, which had just begun, to be nothing less than "the very struggle of civilization against barbarism." (translation jpk). Bergson stressed that Germany's invasion was an attack not just on Belgium, but also on the fundamental values that the latter shared with many other countries.[2] Similar statements could be heard from England, not least in view of daily press reports of new German atrocities, such as the mass shooting of civilians or the destruction of such cultural monuments as the university library of Leuven (cf. Horne and Kramer 2001; Münkler 2013, 248–60). The newspapers also made public that the German chancellor had, in the presence of the British prime minister, dismissed the guarantee protecting Belgium as "a scrap of paper" (Hull 2014). This revelation imbued the provisions of the 1839 Treaty of London with greater significance than even political insiders had previously accorded them, turning them into a symbol of a civilization now imperiled by Germany.

Under pressure from such widespread outrage, German intellectuals found that offense was the best defense, hitting back with a harsh critique of civilization that, in many cases, simply reappropriated or inverted the charges brought from the Allied side. A notable example was Thomas Mann, whose essay *Gedanken im Kriege* (Thoughts in Wartime) was published only a few weeks after Bergson's address to the academy and welcomed the war as an act of "cleansing" and "liberation" from a decadent "world of peace" of which he professed himself to have been "tired, so very tired": "Was it [the world] not in ferment, stinking with the decomposition of civilization?" (Mann 1914, 1474–5, translation jpk). To the philosopher Max Scheler, the war was a decisive battle in a global cultural struggle that would reveal whether the future belonged to the "capitalist bourgeois spirit" (Scheler 1915, 74, translation jpk) or rather to warlike and honorable soldiery. The economic historian Werner Sombart, too, held "heroes" to be made of altogether different stuff than "merchants," who valued peace only because war interfered with their business. Sombart thus found the true enemy to reside neither in France nor in Russia, but rather in England, where profit counted for more than the kind of bravery by which Germany's heroes were distinguished:

> To make advantageous treaties [. . .] and—which is directly related—to neutralize opposing forces by pitting them against each other, thereby reducing the danger to oneself: that alone has been England's purpose since time immemorial. [. . .] This idea of "balance" now in turn is clearly born of the mercantile

> spirit: its image is that of the scales in which the grocer weighs raisins and pepper. (Sombart 1915, 39, translation jpk)[3]

In such broadsides against humble grocers and their love of peace, readers of Nietzsche would have recognized the creed of a *Lebensphilosophie* (philosophy of life) that saw no contradiction in glorifying death as long as such sacrifices gave birth to communities that performed heroic deeds, thereby testifying to the vital force with which the martial spirit had imbued them (cf. Aschheim 1992). In his essay, Thomas Mann went so far as to proclaim the soldier a kinsman of the artistic genius and to exalt both as emblems of a German *Kultur* in which creation and destruction were inextricably connected. Mann measured the distance of this *Kultur* from the *Zivilisation* of Germany's neighbors, both historically and geographically, by comparing Germany with Mexico and crediting the ancient Aztecs with having attained a higher level than the various "Kyrgyz, Japanese, Gurkhas, and Hottentots" who had been enlisted to fight Germany on the Allied side—which, to Mann, was nothing short of an "affront" (Mann 1914, 1483):

> Nobody would deny that Mexico, at the time of its discovery, was possessed of culture, but nobody would claim that it had been civilized. Culture is manifestly not the opposite of barbarism; often enough it is merely wilderness tempered with style. [. . .] Culture may comprise oracles, magic, pederasty, [the Aztec god] Huitzilopochtli, human sacrifice, orgiastic cults, inquisition, *autos-da-fé*, St. Vitus' dance, witch trials, poisoners galore, and other colourful atrocities. Yet civilization is reason, enlightenment, becoming mild, mannerly and skeptical; [it is] dissolution. (1471)

While such pamphlets may read today as testaments to a baffling kind of "lapse into intellectual sin" (Münkler 2013, 248, translation jpk), foreign readers at the time would have been scandalized above all by the manner in which war was celebrated as a return to a state of nature where might equaled right and questions of guilt and innocence were rendered otiose by the ever-present possibility of struggles to come. These counternarratives to the international front—a front that stood in defense of civilization—made their authors complicit in the disregard of international law by legitimizing the *lack of any awareness of guilt* and gave a blanket dispensation to German troops without stopping to ask what exactly they were doing. No one doubted that *any* act of war was a heroic act, justified by service to the fatherland.

Following the armistice, the victorious side was all the keener to differentiate. The form of the peace treaty itself was to leave no doubt that the Allied victory had also been a victory for international law. For the German deputies, this meant that they not only faced staggering bills for the collateral damage caused by their own military action, but also that they could not even be sure of having a say in the negotiations. After all, they had discredited themselves as "barbarians," both by force of arms and by the statements made in the war of minds.

Not least among the factors militating against Germany's swift return to the ranks of the civilized nations was the poor reputation of Turkey, which had fought in the war as Germany's ally. The Ottoman Empire was notorious among the powers of Western Europe for its despotic regime, and already in the 19th century had been the target of a series of humanitarian interventions aimed particularly at protecting its minority Christian populations, but also at enforcing free trade or an independent judiciary (cf. Rodogno, 32–62). The massacres committed against the Armenian population in 1915—which set the precedent for the subsequent coining and definition of the term *genocide* (Barth 2006, 8; Dadrian 2004)[4]—only served further to horrify the Allies, who, in a joint declaration of intent, undertook to hold all members of the Turkish government criminally liable for these "new crimes of Turkey against humanity and civilization."[5] Nonetheless, and despite being enshrined in the provisions of the peace treaties, these good intentions remained without consequence; both the Kaiser and his Turkish allies escaped prosecution for the charges made against them.

Another matter involved the financial compensations that the victorious powers demanded for the breaches of international law. The Allies agreed that Germany must pay, and their disagreements concerned only the overall tally as well as the specific items to be added to the bill (cf. Kent 1989; Payk 2018, 496–543). As far as the German public was concerned, these were incendiary demands. The debates that erupted following the armistice picked up where those of 1914, over guilt and blame, had left off, and the heated war of minds continued as a cold war in which the parties were armed with facts and documents. This was important not least because resistance had come first from within Germany itself. Even in 1917, once the war in the east had come to an end, the point of continuing the fight in the west no longer seemed plausible to all. The *Burgfrieden* (party truce) concluded at the war's outset could no longer be upheld, and the Social Democrats themselves split in the process. Following the revolution of 1918–1919, the party's moderate wing held power in Berlin, while the radical minority formed a government in the state of Bavaria. It was from this left Social Democratic Party that the German public was surprised by a step that was "as spectacular as it was unique" (Heinemann 1983, 25, translation jpk) in the history of the Weimar Republic. Instead of continuing to attack the Allied side, one man took to the press to assert that the Allies were right.

Eisner's Politics of Confession

On November 24, 1918, the liberal daily *Berliner Tageblatt* published four incendiary excerpts from confidential dispatches of the Bavarian legation. They concerned "documents on the origins of the war" (translation jpk) that had hitherto been kept under wraps and were made public based on the "realization that only the whole truth might succeed in restoring that relation of trust between the peoples that is the condition for a peace founded on reconciliation." The "diplomatic documents" from July and August 1914 revealed how impatiently the German side had awaited Vienna's "strong and effective intervention" against Serbia and how irritated it was at Austrian "irresolution and confusion." Another telegram, sent around the same time from Berlin to Munich, had announced an "assault upon France" that could be made "only along the Belgian line;" German forces would, hence, be "unable to respect"[6] that country's neutrality.

Behind these revelations stood Bavaria's new prime minister, Kurt Eisner. Only three weeks earlier, Eisner had led a bloodless coup with the aid of war-weary soldiers (cf. Mitchell 2015, 92–109; Gurganus 2018, 367). The "Free State of Bavaria" (Eisner 1919, 5, translation jpk), as Eisner proclaimed it on the night of November 7, 1918, was the first republic of the new Germany. A native of Berlin, Eisner had only been living in Bavaria for eleven years, where he had made his name as a writer for socialist newspapers before taking to politics himself. Within the Social Democratic Party, his literary inclinations and years of trying to establish a worker's *feuilleton* earned Eisner a reputation as an aesthete. He was himself well aware that somebody who could "make a poem might be suspected of not knowing the first thing about politics" (Eisner 1975, 114, translation, jpk).[7]

After the call to the January strikes in Munich—for which Eisner served several months in prison—and the November revolution, the publication of these documents was Eisner's third big splash within a year. Berlin's political scene was in an uproar. While the foreign ministry aimed at damage prevention, Max Weber went for Eisner directly. In a newspaper article, Weber lambasted the "pathetic" way in which "*littérateurs*" might find comfort for their tender souls "by rummaging through feelings of a supposed 'war guilt.'" While such treasonous activities might "win the world's attention," they showed little sense of the necessities of far-sighted *realpolitik*. To interpret "defeat" as a consequence of prior "guilt" was an idea that, according to Weber, could occur only to "weak natures unequal to facing up to the aspect of reality" (Weber 1988, 488, translation jpk).

To Eisner's mind, however, indicting the imperial elites was not an end in itself, but rather a necessary part of the democratic change on which the country as a whole had just embarked through its proclamation of a brand new republic. "We have been made new, and I set no store by claiming that we have remained the same" (Eisner 1980, 232, translation jpk), Eisner declared to a congress of socialists from across Europe assembled in Bern. The documents aimed to enable everyone to see that the previous government knew what it was doing in riding roughshod over the tenets of international law. In any case, the moral verdict was more important to Eisner than one delivered in a court of law. Legal justice, in his eyes, was not a sufficient or adequate response to Germany's war guilt, because he worried that Germany might be tempted to believe it had no further obligations of any sort once it had agreed to bring certain individuals to trial; believing itself wholly absolved, the country might then backslide into moral laxity and eventually just create more guilt for itself again. Thus, Eisner felt that indicting the old regime must, on no account, mean letting the new Germany off the hook, but rather must represent an earnest signal of the country's willingness to meet any claims to compensation that might be imposed on it. The economic liabilities incurred by such a commitment would soon, Eisner hoped, forge a social bond; in this way, an admission of guilt should result in financial *and* moral ties with the former enemies. Germany's obligations, he thought, offered a chance to move beyond the differences that had been expressed in such fierce polemic and instead to recast the country's common destiny.

In the months following the armistice, however, Eisner was largely alone in making such proposals. The publication of the documents had the opposite effect to that intended, and public indignation was directed only at Eisner himself. Had he been more realistic, he might have anticipated this failure and not been so obstinate in overestimating the value of Kantian ethics as a guide to political praxis—that, at least, was the opinion of many later historians who concurred with Weber's estimation.[8]

Yet to describe Eisner's position as an ethic of conviction devoid of any sense of responsibility would barely do justice to his stance. In fact, Eisner did take responsibility—in the sense of liability—very seriously indeed. Though he explicitly excluded German civilians at large from his charges and accused only the (military) leadership of criminal conduct, he was fully aware that a verbal break with the *Kaiserreich* was not enough. What was called for was instead a change of attitude with regard to the demands of the Allies, a recognition that these demands were founded on principles of justice that were *in themselves* worthy of respect and universally applicable. To admit Germany's guilt freely was to admit the legitimacy of foreign demands for German self-criticism. This would, furthermore, entail admitting that merely to *recognize* the Allied victory was insufficient for reconciliation, so long as the losing side demanded indemnity from the political and economic costs of their own warfare.

By publishing these documents, Eisner conceded that the victorious Allies were *justified* in demanding that precedence be given to the standards of civil society, even in relations between states. In doing so, they insisted on a principle whose validity was in no way disqualified by the benefits they might derive from them in the present case. However, in the view of the German government at the time, this was going too far; its delegates at Versailles took a confrontational stance instead, looking to the battle over rights and wrongs as an opportunity to make up for the defeat suffered on the battlefield. Their goal was to prove that the Allies were unjustified in the demands they were making of Germany. The notorious war guilt clause, Article 231 of the peace treaty, provided a welcome opportunity for doing so, basing as it did the victorious powers' claims on a bald, formal, legal definition of Germany's liability. It stated:

> The Allied and Associated Governments affirm and Germany accepts the responsibility of Germany and her allies for causing all the loss and damage to which the Allied and Associated Governments and their nationals have been subjected as a consequence of the war imposed upon them by the aggression of Germany and her allies.[9] (translation jpk)

This idea of "responsibility" as the legal basis for the case against their country was, in the opinion of the German delegation, too thin. In memorandum upon memorandum, they demanded proof, pressing the opposite side to offer justification and complaining that "the demand for atonement of wrongs committed" was being "intertwined, for political ends, with the opponent being stigmatized and ostracized" (Deutsche Regierung 1919, 82, translation jpk). The Allies retaliated with the famous "jacket note," in which they accused Germany of having pursued a policy of aggressive imperial expansion by which, even before the war, it had aimed "to dictate and tyrannise to (sic) a subservient Europe." (U.S. Department of State 1946, 926).

One lesson from this dispute is that even "thin" concepts of responsibility are prone to being misunderstood. Recent philosophers have repeatedly preferred such "thin" concepts (e.g., responsibility) over "thick" ones (e.g., guilt), arguing that the former come with the advantage of carrying little cultural ballast. Since such "thin" concepts do away with the practice of public shaming, they could expect greater acceptance and would translate more easily to other cultural contexts (cf. Lotter 2012, 21–5, 317–28). What the war guilt debate after World War I shows, however, is that the schismatic power of such concepts does not so much reside in their cultural semantics as it accrues to them in the arenas where they are put to performative use. No "thin" concept is inherently above suspicion of being a "thick" one in disguise, and as such it is liable to be despised for strengthening the weak and challenging the predominance of the strong. Nor is it possible to compel people to accept sacrifices and restrictions against their own interest in holding on to comfortable positions. Such acceptance would require, as Eisner had realized, a freely made decision to set new priorities, an awareness of *Schuld* not so much in the sense of "guilt" as of what is owed to others if conflict-free relations are to be possible. Without such an awareness—and regardless of whether it is sustained by the "thick" concept of a moral obligation or the "thin" one of a convenient way of overcoming differences—there can be no solidarity.[10]

For his part, Eisner did not live to witness the escalation in the peace negotiations at Versailles. The publication of the documents gave a boost to the far right, which was emboldened in spreading the so-called stab-in-the-back legend. In the conservative press, the Jewish Eisner was smeared by the false rumor that he was in fact a dubious migrant from Galicia living in Germany under a false name (cf. Grau 2001, 373; Gurganus 2018, 418). In Bavaria's first post-war elections, the left Social Democrats received a mere 2.5 percent of the vote, and on February 21, 1919, Eisner, who was on his way to the Bavarian state diet to tender his resignation, was shot twice by Anton Graf von Arco auf Valley, a 22-year-old monarchist and anti-Semite (cf. Mitchell 2015, 271–2; Gurganus 2018, 423–4). "The idiocy of Arco's deed at this moment must be immediately obvious, but many people are delighted by it," a perturbed Thomas Mann noted in his diary, though he had previously been far from pleased that the "Jewish scribblers" (Mann 1982, 19, 34) had come to power. Thus was squandered the first chance to establish a "guilt culture" in Germany. The likelihood of the Weimar Republic being borne ahead by a cultural change had dwindled sooner than many observers realized.

Translated by Joe Paul Kroll

Notes

1. On corresponding practices in Syria, Egypt, China, Greece, and Rome, see Close 2019, 11–45; Fisch 1979, 57–72.
2. "La lutte engagée contre l'Allemagne est la lutte même de la civilisation contre la barbarie. Tout le monde le sent, mais notre Académie a peut-être une autorité particulière pour le dire. Vouée en grande partie à l'étude des questions psychologiques, morales et sociales, elle accomplit un simple devoir scientifique en signalant dans la brutalité et le cynisme de l'Allemagne, dans son mépris de toute justice et de toute vérité, une régression à l'état sauvage" (Bergson 1972, 1102).
3. Numerous studies have shown that such claims revived pre-war patterns of cultural criticism that were as widespread among the literary avant-garde as they were in conservative circles. Examples include (but are not limited to): Beßlich 2000; Horn 1999; Hepp 1987.
4. A detailed overview of the Armenian genocide can be found in Dadrian 2004.
5. See Armenia: Joint Allied Declaration [1915] 2019, 164; Tusan 2014; Garibian 2010. On the efforts to establish an international criminal justice system at the Paris Peace Conferences, see Lewis 2014, 27–63.

6. The documentary materials published at the time are published verbatim, with excisions from the original restored, in Dirr 1925, quoted here: pp. 3–4, 6, 15. Whether these excisions were meaningless or they distorted the documents' meaning was later the subject of heated debate among historians, a reconstruction of which can be found in Dreyer and Lembcke 1993, 63–77.
7. The phrase "make a poem" probably alludes to the song "Gesang zur Feier," which Eisner had written especially for the celebration of the revolution at the Munich National Theatre on November 17, 1918. See also Gurganus 2018, 390.
8. Eisner had studied philosophy in Berlin and later worked as a political editor for the *Hessische Landeszeitung* in Marburg, where he joined the circle around the neo-Kantian philosopher Hermann Cohen. More nuanced appraisals can be found in Gurganus 2018, 34–50; Grau 2001, 115–29; F. Eisner 1995; Mitchell 2015, 41.
9. Quoted according to Payk 2018, 528. The consensus view in international scholarship is "that no imputation of guilt was intended, since there is no mention of 'war guilt' in Article 231. Liability for civilian damage was a responsibility that the German government had freely and explicitly recognised during the Armistice negotiations. The Article was a well-intentioned attempt to set a ceiling for Germany's financial liability. That is to say, it was designed to protect Germany rather than to punish her" (Gomes 2010, 27). See also Payk 2018, 528–9; Dreyer and Lembcke 1993, 132; Heinemann 1983, 45.
10. This might be argued against Max Weber, who accused Eisner's love of the plain truth to be no more noble in motivation than the victorious powers' veiling their self-interest in the rhetoric of justice. Weber argued that "people lose sight of the inevitable falsification of the whole problem by very material interests—the interests of the victor in maximising the gain (whether moral or material), and the hopes of the defeated that they will negotiate advantages by confessing their guilt. If anything is '*common*' (*gemein*) it is this, and it is the consequence of using 'ethics' as a means of 'being in the right'" (Weber 1994, 356–7). The idea of solidarity as the practice of "bridging differences," by contrast, is emphasized by Lessenich (2019, 100, translation jpk).

References

"Armenia: Joint Allied Declaration, May 24, 1915." 2019. In *Modern Genocide: A Documentary and Reference Guide*, edited by Paul R. Bartrop, 164. Santa Barbara, CA: Greenwood.

Aschheim, Steven E. 1992. *The Nietzsche Legacy in Germany: 1890–1990*. Berkeley: University of California Press.

Barth, Boris. 2006. *Genozid: Völkermord im 20. Jahrhundert: Geschichte, Theorien, Kontroversen*. Munich: Beck.

Barth, Boris, and Rolf Hobson. 2020. "Civilizing Missions from the 19th to the 21st Centuries, or from Uplifting to Democratization." In *Civilizing Missions in the Twentieth Century*, edited by Boris Barth and Rolf Hobson, 1–20. Leiden: Brill.

Bergson, Henri. 1972. "8 Août 1914: Discours prononcé à l'Académie des sciences morales et politiques." In *Mélanges, textes publiés par André Robinet*, edited by Henri Bergson, 1102. Paris: Presses universitaires de France.

Beßlich, Barbara. 2000. *Wege in den Kulturkrieg: Zivilisationskritik in Deutschland 1890–1914*. Darmstadt: Wissenschaftliche Buchgesellschaft.

Bloch, Camille, and Pierre Renouvin. 1932. "L'art. 231 du Traité de Versailles: Sa genèse et sa signification." *Revue d'histoire de la guerre mondiale* 10: 1–24.

Binkley, Robert C. 1929. "The 'Guilt' Clause in the Versailles Treaty." *Current History* 30, no. 2: 294–300.

Binkley, Robert C., and A. C. Mahr. 1926. "A New Interpretation of the 'Responsibility' Clause in the Versailles Treaty." *Current History* 24, no. 3: 398–400.

Clark, Christopher. 2013. *The Sleepwalkers: How Europe Went to War in 1914*. London: Penguin.

Close, Josepha. 2019. *Amnesty, Serious Crimes and International Law: Global Perspectives in Theory and Practice*. London: Routledge.

Dadrian, Vahakn N. 2004. *The History of the Armenian Genocide: Ethnic Conflict from the Balkans to Anatolia to the Caucasus*. 6th revised ed. New York: Berghahn.

Deutsche Regierung. *Die Gegenvorschläge der Deutschen Regierung zu den Friedensbedingungen: Vollständiger amtlicher Text*. 1919. Berlin: Reimar Hobbing.

Dickmann, Fritz. 1964. *Die Kriegsschuldfrage auf der Friedenskonferenz von Paris 1919*. Munich: Oldenbourg.

Dirr, Pius, ed. 1925. *Bayerische Dokumente zum Kriegsausbruch und zum Versailler Schuldspruch*. 3rd ed., 3–16. Munich, Berlin: Oldenbourg.

Dreyer, Michael, and Oliver Lembcke. 1993. *Die deutsche Diskussion um die Kriegsschuldfrage 1918/19*. Berlin: Duncker & Humblot.

Eisner, Freya. 1995. "Kurt Eisners Ort in der sozialistischen Bewegung." *Vierteljahrshefte für Zeitgeschichte* 45, no. 3: 407–35.

Eisner, Kurt. 1919. "Aufruf aus der Nacht zum 8. Nov. 1918." In *Die neue Zeit*, edited by Kurt Eisner, 5–7. Munich: Georg Müller.

Eisner, Kurt. 1975. "Die Stellung der revolutionären Regierung zur Kunst und zu den Künstlern: Rede in der Sitzung des provisorischen Nationalrats vom 03.01.1919." In *Sozialismus in Aktion: Ausgewählte Aufsätze und Reden*, edited by Freya Eisner, 113–23. Frankfurt: Suhrkamp.

Eisner, Kurt. 1980. "Rede vom 04. Februar 1919." In *Die II. Internationale 1918/19: Protokolle, Memoranden, Berichte und Korrespondenzen*, vol. 1, edited, introduced, and annotated by Gerhard A. Ritter, 230–43. Annotation assisted by Konrad von Zwehl. Berlin: Dietz.

Eisner, Kurt. 2018. "Gesang zur Feier" [1918]. In *Dichtung ist Revolution: Kurt Eisner, Gustav Landauer, Erich Mühsam, Ernst Toller: Bilder, Dokumente, Kommentare*, edited by Laura Mokros, 35. Regensburg: Friedrich Pustet.

Fisch, Jörg. 1979. *Krieg und Frieden im Friedensvertrag: Eine universalgeschichtliche Studie über Grundlagen und Formelemente des Friedensschlusses*. Stuttgart: Klett-Cotta.

Garibian, Sévane. 2010. "From the 1915 Allied Joint Declaration to the 1920 Treaty of Sèvres: Back to an International Criminal Law in Progress." *Armenian Review* 52, no. 1–2: 87–102.

Gomes, Leonard. 2010. *German Reparations, 1919–1932: A Historical Survey*. New York: Palgrave Macmillan.

Grau, Bernhard. 2001. *Kurt Eisner 1867–1919: Eine Biographie*. Munich: Beck.

Gurganus, Albert Earle. 2018. *Kurt Eisner: A Modern Life*. Rochester, NY: Camden House.

Heinemann, Ulrich. 1983. *Die verdrängte Niederlage: Politische Öffentlichkeit und Kriegsschuldfrage in der Weimarer Republik*. Göttingen: Vandenhoeck & Ruprecht.

Hepp, Corona. 1987. *Avantgarde: Moderne Kunst, Kulturkritik und Reformbewegungen nach der Jahrhundertwende*. Munich: Deutscher Taschenbuch Verlag.

Horn, Eva. 1999. "Krieg und Krise: Zur anthropologischen Figur des Ersten Weltkriegs." In *Konzepte der Moderne: DFG-Symposium 1997*, edited by Gerhart von Graevenitz, 633–55. Stuttgart: Metzler.

Horne, John, and Alan Kramer. 2001. *German Atrocities, 1914: A History of Denial*. New Haven, CT: Yale University Press.

Hull, Isabel V. 2014. *A Scrap of Paper: Breaking and Making International Law during the Great War*. Ithaca, NY: Cornell University Press.

Israel, Fred L., and Emanuel Chill, eds. 1967. *Major Peace Treaties of Modern History 1648–1967*, vol. 1. New York: Chelsea House.

Jäger, Wolfgang. 1984. *Historische Forschung und politische Kultur in Deutschland: Die Debatte 1914–1980 über den Ausbruch des Ersten Weltkrieges*. Göttingen: Vandenhoeck & Ruprecht.

Keen, Maurice Hugh. 2015. *The Laws of War in the Late Middle Ages*. London: Routledge. First published 1965.

Kellermann, Hermann, ed. 1915. *Der Krieg der Geister: Eine Auslese deutscher und ausländischer Stimmen zum Weltkriege 1914*. Dresden: Duncker.

Kent, Bruce. 1989. *The Spoils of War: The Politics, Economics, and Diplomacy of Reparations 1918–1932*. Oxford: Clarendon Press.

Koskenniemi, Martti. 2004. *The Gentle Civilizer of Nations: The Rise and Fall of International Law 1870–1960*. Cambridge: Cambridge University Press.

Krumeich, Gerd. 2018. *Die unbewältigte Niederlage: Das Trauma des Ersten Weltkriegs und die Weimarer Republik*. Freiburg: Herder.

Lesaffer, Randall. 2004. "Peace Treaties from Lodi to Westphalia." In *Peace Treaties and International Law in European History*, edited by Randall Lesaffer, 9–44. Cambridge: Cambridge University Press.

Lessenich, Stefan. 2019. *Grenzen der Demokratie: Teilhabe als Verteilungsproblem*. Stuttgart: Reclam.

Lewis, Mark. 2014. *The Birth of the New Justice: The Internationalization of Crime and Punishment, 1919–1950*. Oxford: Oxford University Press.

Loraux, Nicole. 2002. *The Divided City: On Memory and Forgetting in Ancient Athens*. Translated by Corinne Pache and Jeff Fort. New York: Zone Books. First published 1997.

Lotter, Maria-Sibylla. 2012. *Scham, Schuld, Verantwortung: Über die kulturellen Grundlagen der Moral*. Frankfurt am Main: Suhrkamp.

Mann, Thomas. 1914. "Gedanken im Kriege." *Neue Rundschau* 25, no. 11: 1471–84.

Mann, Thomas. 1982. *Diaries 1918–1939*. Translated by Richard and Clara Winston. New York: Abrams.

Mitchell, Allan. 2015. *Revolution in Bavaria, 1918/19: The Eisner Regime and the Soviet Republic*. Princeton, NJ: Princeton University Press. First published 1965.

Münkler, Herfried. 2013. *Der große Krieg: Die Welt 1914–1918*. Berlin: Rowohlt.

Osterhammel, Jürgen. 2006. *Europe, the "West" and the Civilizing Mission*. London: German Historical Institute.

Payk, Marcus M. 2018. *Frieden durch Recht? Der Aufstieg des modernen Völkerrechts und der Friedensschluss nach dem Ersten Weltkrieg*. Berlin: De Gruyter.

U.S. Department of State. 1946. "Reply of the Allied and Associated Powers to the Observations of the German Delegation on the Conditions of Peace [Paris, June 16, 1919]." In *Papers relating to the Foreign Relations of the United States: Paris Peace Conference,* vol. 6, 926–35. Washington, DC: U.S. Government Printing Office.

Rodogno, Davide. 2011. *Against Massacre: Humanitarian Interventions in the Ottoman Empire, 1815–1914.* Princeton, NJ: Princeton University Press.

Scheler, Max. 1915. *Der Genius des Krieges und der deutsche Krieg.* Leipzig: Verlag der weißen Bücher.

Schivelbusch, Wolfgang. 2001. *Die Kultur der Niederlage: Der amerikanische Süden 1865—Frankreich 1871—Deutschland 1918.* Berlin: Fest.

Schmitt, Carl. 1997. *Der Nomos der Erde im Völkerrecht des Jus Publicum Europaeum.* 4th ed. Berlin: Duncker & Humblot. First published 1950.

Sombart, Werner. 1915. *Händler und Helden: Patriotische Besinnungen.* Munich: Duncker & Humblot.

Tusan, Michelle. 2014. "'Crimes against Humanity': Human Rights, the British Empire, and the Origins of the Response to the Armenian Genocide." *American Historical Review* 119, no. 1: 47–77.

Weber, Max. 1988. "Zum Thema der 'Kriegsschuld.'" In *Gesammelte Politische Schriften.* 5th. ed. Edited by Johannes Winckelmann, 488–97. Tübingen: Mohr. Originally published in *Frankfurter Zeitung,* January 17, 1919.

Weber, Max. 1994. "The Profession and Vocation of Politics." In *Political Writings,* edited by Peter Lassman and Ronald Speirs, 309–69. Cambridge: Cambridge University Press. First published 1919.

16

The Absence of Productive Guilt in Shame and Disgrace

Misconceptions in and of German Memory Culture from 1945 to 2020

Matthias Buschmeier

It has long been a common assumption that guilt is deeply ingrained in Germany's memory culture, politics, and public discourse[1] (cf. Berger 2012, 35–82). The German scholars Aleida Assmann and Ute Frevert argued that the post-1945 guilt debates had effects that have endured in the controversy about German history even until today (Assmann and Frevert 1999, 97).

Most recently, however, political science scholars have fiercely challenged the assumption that German society has, in one way or another, been entangled in guilt (feelings) since the end of World War II. More to the point, Samuel Salzborn's book *Kollektive Unschuld* (2020) (*Collective Innocence*) argues that, contrary to much of the world's assumption, Germans have aggressively tried everything to hide from their historical guilt, to minimize it, and to twist the narrative of the Shoah such that Germany appears not as a collective of perpetrators but as a collective of victims.

The resistance to fully acknowledging historical guilt parallels a trend toward delegitimizing guilt as a political concept altogether. Indeed, some argue that the application of the moral and juridical concept of guilt in politics has given rise to defensive reactions. With guilt in mind, as the story goes, political leaders limit their political actions, especially in Germany (Grunenberg 2001), for fear of being accused of forgetting the country's guilty past. This argument is growing stale, though. Indeed, it has haunted Germany since the debate on Germany's war guilt after World War I (cf. Ethel Matala de Mazza's chapter in this volume).

In this chapter, I will develop my argument in three steps. First, I too doubt the proposition that German memory culture after 1945 can adequately be described as a discourse of guilt. To make this point, I return to the important discussions from 1945 on the issue of collective guilt, since this debate (as I will argue) has shaped the understanding of guilt in post-war Germany in a misleading way, with severe consequences for the official commemoration of Germany's war

Matthias Buschmeier, *The Absence of Productive Guilt in Shame and Disgrace* In: *Guilt*. Edited by: Katharina von Kellenbach and Matthias Buschmeier, Oxford University Press. © Oxford University Press 2022.
DOI: 10.1093/oso/9780197557433.003.0017

crimes. The character of the official national memory culture reveals itself best on the anniversaries of fraught occasions in German history; speeches commemorating such events are ways of rewriting and interpreting their impact on our very present.

Therefore, secondly, I have analyzed all speeches that Germany's federal presidents delivered on occasions such as the commemoration days for the end of World War II and the liberation of concentration camps (mainly Auschwitz, Bergen-Belsen, and Buchenwald), and for anniversaries of massacres committed by the Wehrmacht or the SS. Indeed, every German president has spoken on such occasions. My selection presented in this paper is based on the conspicuousness of the material. The question guiding my research was: "To what extent do the highest representatives of the German state remember and address Germany's crimes against humanity as a matter of guilt? And further, if they address Germany's guilt at all, how do they understand guilt?" As a preview of the main result of this analysis—and hence the main thesis of this article—I can summarize thus: German memory culture (as becomes evident in these sources) is not suffused with a semantic of guilt in the first place. Rather, while the criminal character of the Nazi system and its actions is widely acknowledged, German memory culture drips instead with a rhetoric of shame.

Finally, I argue that such a rhetoric of shame is problematic and that Germany would have done better interpreting its recent history with a concept of productive guilt instead.

The Allied Collective Guilt Campaign and Its Immediate Reception

During the last weeks of the war, in already occupied zones of Germany—and then throughout all of Germany after the defeat—the Allies found a stubborn German population who projected the responsibility for war crimes and for the extinction of European Jews onto a small elite of the Nazi Party—namely, onto the SS or the Gestapo. A bold majority of Germans viewed themselves as free from any concrete guilt. After the capitulation of Hitler-led Germany, the American-British Psychological Warfare Division (PWD) under General Robert A. McClure assumed a primary task of countering such attitudes, in attempts to denazify and demilitarize the population. Even before the war had ended, the Allies decided on a "strategy of truth," expecting the greatest impact on the enemy to be achieved by confronting him with the reality of his inhumane behavior (Matz 1969, 25).

In May 1945, President Truman passed Directive 1067 of the Joint Chiefs of Staff to the supreme commander of the military forces in Europe, Dwight D.

Eisenhower. This directive formulated central elements of the Allied occupation (JCS 1945) and is often interpreted, especially in German research, as a change in US strategy toward a "policy of austerity"—but not one of revenge. Read soberminded, the directive articulates realistic steps and necessary measures for the occupation of a huge, hostile territory. This directive must be understood within the larger European framework; for the Allies, it was essential that the German population not end up better off than its neighbors, who had suffered under German exploitation.[2] Moreover, the directive did not include any form of a collective guilt accusation toward the Germans (Hurwitz 1954, 10–3). On the contrary, the document differentiated carefully between certain categories of perpetrators and how they should be treated juridically.

It is true, however, that General McClure, who headed the propaganda press, did vote for a restrictive policy against the Germans. With his Directive No. 1 for Propaganda Policy of Overt Allied Information Services, he intended to prepare the German population for unconditional surrender; in his view, the Germans should be aware that they had to accept and simply follow the orders of their occupiers. Newspapers were deemed essential to cultivating this attitude: "The first steps of re-education will be limited strictly to the presentation to the Germans of irrefutable facts which will stimulate a sense of Germany's war guilt and of the collective guilt for such crimes as the concentration camps." (Hurwitz 1954, 37)

Although this Directive No.1 was replaced one week later, its intention had clearly not been to confront the Germans with an emotional accusation of guilt. Rather, it was hoped that a mere presentation of facts would stimulate Germans' insight into the collective nature of the Shoah. They should then, in response to learning these facts, acknowledge the reality of the Shoah by themselves (including the fact that far more than a few individuals perpetrated it).

In June 1945, the Jewish-American military sociologist Morris Janowitz, who worked for the Psychological Warfare Branch of the U.S. Army (a subdivision of the PWD), reported on the success of the information campaign on German crimes. He and his team had interviewed hundreds of Germans from different regions and different demographics. In his 1946 article, published in the *American Journal of Sociology*, Janowitz recalled the initial intention of the campaign, which had run only a few weeks: specifically, the Allies wanted to raise awareness among the German population regarding its collective responsibility and involvement, as a basic requirement for their democratic re-education. The Allies differentiated "between those legally guilty of having directly committed atrocities and those morally responsible for having allowed National Socialism to come into being and for having tolerated its crimes" (Janowitz 1946, 141). Here, Janowitz was obviously referring to Directive No. 2, in which General McClure had refined the thesis of collective guilt to avoid further misunderstandings; in a letter from May 28, 1945, McClure spoke of the need to "differentiat(e) clearly between the active guilt of the criminal, which can only be atoned by punishment, and the passive guilt of the people as a whole, which can be atoned for by hard work, national restitution and a change of heart" (Hartenian 1987, 154–5). The campaign's target audience, simply put, was the passively guilty. And in contrast to Directive No. 1, which essentially drew no distinction between the two groups, McClure was now also indicating *how* Germans could atone for such passive guilt: through hard work, economic reparations, and a change in their inner disposition. Nonetheless, in the German reception of the campaign, this pathway to restoration was mostly ignored.

The heart of this campaign were transmissions by Radio Luxembourg and the BBC; newspaper articles about the conditions in the camps; and addresses by important writers in exile, such as Franz Werfel and Thomas Mann, or by the Swiss psychologist Carl Gustav Jung. Furthermore, posters and flyers with images of camp victims and featuring other German war crimes were designed to force the population to acknowledge the character of the Nazi regime of which they had been part (cf. Brink 1998, 21–100). Close to the camps, the Allies went so far as to force local residents to confront the sight of inmates' corpses—in some places, even to participate in the burials. Meanwhile, for those not living near camps, the PWD showed the documentary *Death Mills*—with scenes from Dachau, Auschwitz, Majdanek, Bergen-Belsen, and Buchenwald—in German and Austrian movie theaters until early 1946, to bring the shock to a larger audience. According to Janowitz, almost every German citizen was exposed to this campaign in one way or another (Janowitz 1946, 143).

Janowitz reports that, nonetheless, three-quarters of the Germans interviewed denied the existence of concentration camps, although it was clear that the population had known about them from very early on, as, for instance, popular Nazi-era jokes about Dachau (the first Nazi concentration camp, installed in 1933) prove. Moreover, they claimed to have had no idea of the extent or magnitude of the extermination, even if they admitted to having heard stories about it from soldiers on furlough or reports by escaped inmates. Psychological repression was the Germans' main coping mechanism and could hardly be broken by the Allied information campaign.

In fact, most of those interviewed not only denied any personal guilt but also held that the deeds were justified as acts of war. What is more, the horrifying experience of the extensive Allied bombing had led to a sort of ethical indifference among the bombing victims, resulting in an apathy against any suffering beyond their personal experience (Janowitz 1946, 145).

The only crack in that psychological protective barrier, according to Janowitz, was the acknowledgment of the systematic exclusion of the Jewish population before and during the war. Many Germans saw the painful situation after the war as a retribution for their bad treatment of Jewish people (and a disproportionately harsh retribution at that, believe it or not), in which the Germans did admit to having participated. Thus, Germans considered their victim status similar to that of the Jews. Socialist author Susanne Kerckhoff, in her fictional *Berliner Briefe* (Berlin Letters) *()*, which were published in 1948 and have never been translated, portrayed this widespread attitude: "'We are all victims of Fascism!,' calls out the little man, who once said: 'The Führer knows best'" (Kerckhoff 2020, 22).

The German population's projection of guilt onto a small Nazi elite and their fear of collective punishment after the war both bespeak a primarily juridical understanding of the term "guilt," however. This was a major misunderstanding, since the American-British information campaign did not address juridical guilt among civilians at all. To deal with German guilt in the courtroom, Directive JCS 1067 and the Potsdam Communiqué had outlined explicit guidelines focusing only on individual, not collective guilt. Janowitz pointed to this confusion that Germans received the thesis of collective guilt so affectively with feign emotion already in his 1946 report (Janowitz 1946, 146). Alas, the population's misunderstanding of guilt reveals a major dilemma, not only surrounding the American policy of confrontation, but also the cultural memory discourse after 1945.

It is actually quite astonishing how closely Janowitz's relatively small empirical study matches the results of the large-scale study launched by the Frankfurt Institute for Social Research in 1950–1951. The "group experiment" was the institute's first empirical study after it reopened under the returning directors Max Horkheimer and Theodor Adorno. In "Schuld und Abwehr" (Guilt and Defense) his chapter on the study, Adorno states that Germans received the thesis of collective guilt so affectively because they feared being held legally responsible.[3] In the eyes of a large majority, any German who supported the idea of a collective German guilt was considered a *Nestbeschmutzer* (one who soils the nest). The collective guilt claim is, thus, according to Adorno (1997, 191), the raw nerve around which the collective gathers in order to defend itself against more pain—and, thereby, finds unity again.

Published in 1946, the collection of poems *Dies Irae* (Days of Rage) by the popular author Werner Bergengruen (1946) illustrates just how widespread this fear of collective punishment was among the German public. Bergengruen had remained in Germany under the Nazi regime. His poems underscore a fairly common understanding of the time, particular to German people: because nobody could claim true innocence, everyone feared the apocalyptic tribunal of the Allies—in which, Germans feared, no distinctions were drawn anymore. Moreover, Bergengruen, along with many other Germans, considered a divine judge the only legitimate judge for the failures of the German people, as no earthly juridical sentence could possibly correspond to their wrongdoing; atonement must be made alone and in private, to God. In this conceptualization, the processing of earthly guilt is given over to religious transcendence: a very effective and widely shared strategy to ward off guilt.

Remember, though: the aim of the PWD was not to accuse the German population in a legalistic sense to begin with (Hartenian 1987, 150). The aim of the confrontation policy was to break through the "denial of" and "absence of feelings of guilt" (Janowitz 1946, 143–4) in order to push the Germans to confront their collective participation in the National Socialist system. The Americans attempted this by pursuing a collective understanding of guilt, on the one hand, and criminal prosecution for perpetrators, on the other—but a rational addressing of the facts was unable to inculcate any insight into the inhuman character of National Socialist rule and its crimes. Thus, the Allies then attempted to leverage guilt to elicit an affective-emotional response for this purpose.

In other words, guilt as a collective feeling—not as a concept of punishment—came to be seen by the Allies as a positive instrument for *liberating* the Germans from their ethical indifference and thereby returning them to the circle of "civilized" nations. Indeed, the "Standing Directive for Psychological Warfare against Members of the German Armed Forces," from June 1944, articulates precisely that goal: "Ultimate restoration of Germany to a place in the world family of democratic nations" (Psychological Warfare Division 1945, 142).

On the German side, however, this strategy also failed, and not only Janowitz but also Albert Norman, a member of the Information Control Division (successor organization to the PWD), admitted this as early as 1946: "Collective guilt was never accepted. The efforts on our part 'to arouse a sense of collective responsibility' for the Nazi crimes [. . .] 'never took root'" (Norman 1951, 14). Also, Kerckhoff stated in resignation: "One thing failed in the Beginning of the End: to compel the German public with guilt feelings and the willingness to atone" (Kerckhoff 2020, 15).

Moreover, Germans who had opposed the Nazi regime, according to Janowitz, saw in the induction of collective guilt feelings a disregard for the many Germans who were themselves victims of the Nazis. The Social Democrat Kurt Schumacher, for instance, rigorously argued this position, and the fact that Germans had suffered at the hands of the Nazis eventually developed into a narrative for warding off guilt. Later, this position supported the legend of a widespread German resistance against the regime, which, as the "Bielefeld Memo Study 2019" shows, is now once again a prevalent belief in German families (Rees, Papendick, and Wäschle 2019, 13–7).

By contrast, Kerckhoff's literary alter ego, named Helene, vigorously insists that—even if she *was* a left-wing opponent of the Nazi regime—only by inwardly accepting guilt and responsibility could a human society resurrect itself: "[W]e do have to stand on the grounds of our German guilt and the duty for amends and reparations" (Kerckhoff 2020, 73). However, such voices were rarely raised.

Another variation on the rejection of collective guilt was presented by Eugen Kogon in his much-acclaimed book from 1946, *Der SS-Staat* (1946). In the German version, he reversed the guilt by blaming the Allies' "shock policy" for Germany's defensiveness; that is, it was not the Germans but the *Allies* who were to blame for the fact that "the forces of the German conscience" had not risen (Kogon 1946, 327–8).[4] In a similar vein, others—and this may have been a central reason for the Americans' early departure from their strategy—pointed out that Soviet propaganda was much friendlier toward the Germans by comparison (Janowitz 1946, 146).

Meanwhile, supporters of the bourgeois-national center—in particular, former supporters of the German Centre Party—saw the honor of the German nation tainted by the American-British campaign. Ostensibly, this attitude had led to a refusal to acknowledge German crimes, or to a strategy of projecting them onto a few fanatical SS supporters. Thus, even though the Information Control Division had already dropped collective guilt as a propaganda issue by the end of 1945 (Hartenian 1987, 161), the term (i.e., "collective guilt") still haunted German discussions and was instrumentalized against itself. This, according to my thesis, is also the reason why, in the official German memorial discourse, the focus ultimately shifted away from guilt and onto injured honor, inner shame, and public disgrace.

Thus, I shall summarize here with a brief interim conclusion: Up to this point, it has become clear that the Allies' talk of collective guilt was not meant in a legal sense, nor was it by any means aimed at collective punishment for the population. The occupation directives were *not* primarily conceived to punish the German people; on the contrary, the directives were aimed at permanently ending Nazism, both as a manner of thinking and as a martial expression. Quite simply, to achieve the psychological aspect of their denazification goals, it was necessary for the Allies to make the Germans realize that they themselves—as a result of their ruthless warfare—were indeed responsible for their current situation. The Americans regarded the Germans' strong psychological repression in the face of these matters as the reason why they barely responded to rational arguments. Therefore, the Allies banked on the idea of appealing instead to emotion, specifically by cultivating a sense of collective guilt—and here again, the German population resisted the idea, misunderstanding the concept of collective guilt as a legal category (sometimes intentionally so). Collective guilt, understood in that way, could be more easily rejected and criticized, on the basis that a liberal-individualistic concept of guilt was the grounds for a legal understanding of the same (Lübbe 1997, 687–9).

The German historian Norbert Frei has repeatedly pointed out the exonerative function of the defense against the collective guilt thesis after 1945; Frei's argument is that the defense against the collective guilt thesis in legal understanding is also an attempt to obstruct further legal punishment of Nazi crimes (Frei 1997, 633–4). But this rhetorical move indirectly admits collective involvement; the more you argue against it, the more it seems that you likely admit to its holding some validity—otherwise, why become so invested in refuting it? Nonetheless, by resorting to the broad semantics of "guilt" (in the German language) as a legal category so as to paint the idea of "collective guilt" as absurd and illegitimate—that is, by sticking to a narrow, legal-moral concept of guilt—the population effectively blocked itself from engaging with the psycho-emotional, collective dimensions of the Shoah. In this way, the tremendous presence of the term "collective guilt" in the young Federal Republic reveals the society's repressive tendencies.

From Collective Guilt to Shame

Aleida Assmann has pointed to the Allies' collective guilt campaign as the origin of a German trauma of blocked memory. This trauma did not consist of the admission of guilt; rather, it consisted of shame, which acted "like a lasting shield against moral confrontation and collective memory" (Assmann 1999, 1154, my translation). In drawing this conclusion, she assumes that German citizens actually experienced a shock. This shock, according to Assmann, evoked shame (which the Allies did not intend) instead of guilt (which they *did* intend). Thus, again, Germans become "victims" of their history.

The historians A. Dirk Moses (2007a, 2007b) and Barbara Wolbrink (2009) have proposed describing the debate on collective guilt with the term "stigma" instead. Stigma is, according to Wolbrink (2009, 329), the irreversible sign "of disparagement, degradation, shame or social exclusion." This stigma arises "through the judgment and attribution of the world public" (343, my translations). However, it must be emphasized that this narrative was cultivated primarily by the Germans themselves and—after a very short period, from the spring of 1945 until the year's end—was maintained neither by the Allies *nor* by a broad "world public opinion." In fact, the emerging conflict between the Western Powers and the Soviet Union in 1946, and their associated efforts to win Germans over to their respective political systems, led to a policy of *integration* rather than exclusion. It is precisely for this reason that I, unlike Moses and Wolbrink, hold on to the concept of shame, because it focuses on the reaction of the Germans and not—like the concept of stigma—on "attribution from outside" (361).

In contrast to Assmann, I would like to argue, though, that the shift from guilt to shame is mainly a discursive effect[5] that pervades the memory debate and, regardless of the actual experiences in the post-war period, has long-term consequences for the nature of German memory culture. In this respect, the sociologist Bernhard Giesen (2004) has spoken of the Germans' "perpetrator trauma."

While perpetrators who directly participated in the persecution and murder experienced the trauma of losing their positions (and perhaps also their sense) of omnipotence, bystanders' experienced the trauma of having to face their own moral indifference and proximity to killing—the latter being a sort of trauma that extended to subsequent generations. As Giesen (2004, 116) has noted, immediately after the war, a "tacitly agreed coalition of silence" about the common disgrace became the core of national identity. The legal concept of individualistic guilt played a central role in this process; it allowed most people to distinguish themselves from both the perpetrators and the victims by imagining themselves in a neutral position under the law. The defense against the ascription of collective guilt was crucial in shaping such an understanding.

Assmann pointed out that, ironically, the voices of exiled authors paved the way for this interpretation. As early as March 1945, in a magazine essay titled "The End," the very same Thomas Mann who had cheered Word War I (cf. Ethel Matala de Mazza's chapter) held the German people, as a whole, responsible for the crimes of the Nazis. However, simultaneously—in contrast to the editorial presentation with the subheading "Guilt of the People"—Mann separated this responsibility from the concept of guilt (cf. Mann 1945, 15). His essay by no means absolves the Germans of their guilt, but rather presents their shame as a punishment for their unforgivable fall from grace, specifically invoking a biblical reference: from now on, they are "a people that can never show its face again" (18).

Assmann also observed the introduction of the vocabulary of shame in the famous passage from chapter 46 of Thomas Mann's novel *Doktor Faustus*, published in 1947, in which the narrator reports on the forced visit to Buchenwald made by Weimar's citizens. This scene is directly linked to the collective guilt campaign of the spring of 1945, because Mann inserts almost verbatim a passage from his own radio address of May 8, 1945, published in German on May 18, 1945, in the *Bayerische Landeszeitung*, a newspaper of the Allied Army Group 6, under the title *Thomas Mann über die deutsche Schuld* (Thomas Mann on German Guilt). His text is less about the guilt itself than about the Germans as a dishonored people:

> The thick-walled torture chambers, to which Hitlerism turned Germany, have been broken open, and our shame lies openly before the eyes of the world [...]. Because everything German, everyone who speaks German, writes German, has lived in German, is affected by this dishonorable exposure." (Mann [1945] 1990, 951, my translation)

Not only have the perpetrators brought dishonor on Germany, though, he writes, so have the passive bystanders. Even as an exile, he claims, he "feels deeply ashamed of what had become possible in the land of his fathers and masters." In still more scathing words, Mann asserts that Germany today represents "an abomination of mankind and an example of evil" (951–2). In other words: the Germans have *made* themselves a pariah people.

With this description, in which collective guilt brings shame for the fact that the crimes reflect disgracefully on *all* who still identify with the collective—even those who did not personally perpetrate those crimes—Thomas Mann is in line with the psychological impact of the Allies' collective guilt campaign. However, the conceptual separation of guilt and shame also offers a narrative that henceforth brings the two ideas into mutual opposition. Shame, therefore, opens up the possibility of fending off accusations of guilt. Shame can be felt for the moral misconduct of others, while moral and legal guilt remain a matter for the individual.

Rhetoric of Shame: German Presidents on German Guilt

Such rhetoric of shame is present in many official speeches from the end of the war even until today (cf. Dubiel 1999). An early example of a public figure articulating a rejection of guilt while at the same time emphasizing shame and disgrace is Konrad Adenauer's inaugural speech as Lord Mayor of Cologne on October 1, 1945, before the City Council:

> Guilty of this nameless misery, of this indescribable misery, are those cursed persons who came to power in the disastrous year 1933, those who covered and defiled the German name with disgrace in front of the whole civilized world, who destroyed our Reich, who systematically and deliberately plunged our seduced and paralyzed people, when their own, more than deserved downfall was certain, into the deepest misery. [...] We, you and I, are not to blame for this misery. (Adenauer 1945, 5–6, my translation)

It was probably this passage, among others, that prompted the Allies to dismiss Adenauer from his post only a few days later (Kämper 2005, 277).

On November 25, 1945, Theodor Heuss, not yet federal president, spoke at a memorial service for the victims of National Socialism. In referring to victims, Heuss meant, above all, the Germans who fell victim to the Nazis for being members of the political opposition, mentioning the "counter-movements and counter-enterprises against National Socialist rule" (Heuss 1967a, 52, my translation). While other groups of victims disappear behind ambiguous phrasing—e.g., "millions of strangers tortured to death" (56)—the "inner-German political victims" are explicitly commemorated with biographical memories. (In this phrasing, the approximately 165,000 Germans of Jewish faith who were killed are once again expelled from the community.) With the National Socialists, Heuss laments, Germans had lost "the honor of the German name that sank into the mud." Thus, they were "angry, depressed, ashamed" (56).

This speech by Heuss set a tone that he continues to develop as the first federal president and that has become characteristic of commemorative speeches by German federal presidents since. In his speech "Mut zur Liebe" (Courage to Love) "" before the Society for Christian-Jewish Cooperation on December 7, 1949, he addresses the topic directly again. What at first sounds like a comprehensive admission of guilt for German crimes turns into a lament about a fate that *happened to* non-Jewish Germans. The people's guilt is only articulated in the form of a question; the accusation of collective guilt is branded as Nazi rhetoric; and the discourse of guilt is deflected into a rhetoric of shame, shifting the Germans from perpetrators to victims:

> There's no sense talking around things. The dreadful injustice that has been done to the Jewish people must be brought up in the following sense: are we, am I, are you guilty because we lived in Germany, are we complicit in these crimes? [...] One has spoken of a "collective guilt" of the German people. But the word "collective guilt" and what is behind it is a mere simplification, it is a turnaround, namely the way the Nazis used to look at the Jews: that the fact of being Jewish already included the guilt phenomenon. But something like a collective shame has grown and remained from that time. The worst thing that Hitler did to us [...] was that he forced us into shame [...]. (Heuss 1967b, 114, my translation)

Several years later, in his commemorative speech in Bergen-Belsen on November 30, 1952, when he refers to concentration camps as "a catalogue of horror and shame" and to the National Socialism era as "shameful years" (Heuss 1965, 225, 226, my translation), the demand for remembrance is not tied to guilt, but to shame. The guilt of belonging to a nation of perpetrators is not permanent; rather, what is permanent is the shame of having failed as Germans in the face of their own culture's demands: "And this is our shame, that such a thing took place in the space of a cultural tradition from which Lessing and Kant, Goethe and Schiller entered world consciousness. No one, no one will take this shame from us" (228).

From this point on, the defense against the collective guilt thesis became a fixed component of Federal Republic commemorative speeches. This rejection is combined with the idea that the process of coming to terms with German guilt—construed as the post-war legal proceedings themselves—is now complete. We also, thus, see the return of a line of thought that first appeared in the collective guilt debates of the immediate post-war period: the "generation argument," by which subsequent generations are wholly absolved (cf. Kämper 2015, 201–2). In this way, the commandment to remember fades away, and guilt becomes merely an optional meditation prompt. Richard Weizsäcker spoke of this in his speech to the Bundestag for the 40th anniversary of the end of World War II, on May 8, 1985:

> There is no such thing as the guilt or innocence of an entire nation. Guilt is, like innocence, not collective, but personal. There is discovered or concealed individual guilt. There is guilt that people acknowledge or deny. Everyone who directly experienced that era should today quietly ask himself about his involvement then. The vast majority of today's population were either children then or had not been born. They cannot profess a guilt of their own for crimes that they did not commit. No discerning person can expect them to wear a penitential robe simply because they are Germans. But their forefathers have left them a grave legacy. All of us, whether guilty or not, whether old or young, must accept the past. (Weizsäcker 1985, 4)

Forty years after the war, Weizsäcker sees now an "opportunity to draw a line under a long period of European history" (8). He insinuates that the end of the guilt discourse should lead to the idea of a European reconciliation. Weizsäcker also, by alluding to the symbolic significance of the 40-year mark through a reference to the Bible, indicates that 40 years have brought a change of generations, and thus that a sort of "generational" remembrance is emerging.

Fifteen years later, Roman Herzog (1999, my translations) reflects this manner of thinking in his own January 27, 1999 speech. The speech must be considered in the context of the Walser-Bubis debate on the appropriate remembrance of the Shoah, and the related discussion about the central Holocaust memorial in Berlin. To Herzog, the epoch for any possible "personal guilt" is effectively over. The memorial is not a "demonstration of permanent guilt;" the "large majority of Germans living today are not to blame for Auschwitz [nor] for selection, expulsion, and genocide."

In a speech on November 9 of the following year, Herzog again speaks out against a "blanket assignment of guilt" (Herzog 1998a, my translation), this time against the backdrop of the renewed thesis of the Germans' collective guilt by the American political scholar Daniel Jonah Goldhagen (1996). And on January 19, 1996, Herzog insists that remembrance "is not meant to be a confession of guilt extending into the future," because guilt—as the lawyer Herzog says—is "always (a) very individual matter" (Herzog 1998b, 11, my translation). "Questions of guilt" should therefore no longer be "in the foreground" (11), according to Herzog—only for him then to say explicitly that "we cannot recognize the collective guilt of the German people for the crimes of National Socialism" (17).

Here, once again, the link between the legally focused rejection of the collective guilt thesis and the generational argument becomes apparent, and the concept of responsibility—which all Germans still bear today for the crimes committed—is henceforth put forward repeatedly. For the audience, though—and maybe for the speakers too—it is notoriously unclear precisely what the source of this responsibility is. Is it derived from the character of German crimes, as suggested by Herzog's declaration that "[Germans' responsibility] for seeing that something of this kind never happens again is larger, particularly since many Germans incurred guilt through their actions in the past" (11)?

If so, do groups and nations that were not involved in such crimes bear *less* responsibility for ensuring that such crimes do not happen again? Or does the appeal to the principle of responsibility derive legitimacy from more general ethical premises, which would apply universally—not just to those burdened with historical guilt?

Herzog renounces a rational justification for such responsibility and falls back on the emotion of shame. On April 27, 1995, he takes up Heuss's speech in Bergen-Belsen at the same place, with explicit reference to it. "Shame and outrage" fill Herzog, he claims, when he remembers "that it was Germans who committed these crimes; shame and outrage that they were committed in the land of Lessing, Kant and Goethe" (Herzog 1998c, 5). Switching from guilt to shame, he also formulates the imperative of a culture of remembrance *adapted to* the historical time gap; the "generation of contemporary witnesses" must ensure that they are "presenting the past and recalling the past in ways that make our young people feel it is their own responsibility to struggle against any possible repetition of that past" (7). According to Herzog, the culture of remembrance is thus tasked with obliging an innocent generation to take responsibility for history. The core binding agent for Germans is "collective shame" (as he quotes Heuss's phrase a few days later on May 8)—not guilt—in the context of their historical responsibility (Herzog 1995, my translation). Indeed, the previous year (on August 1, 1994), Herzog in Warsaw referred to the "shame that the name of our country and people will forever be linked to pain and suffering" (Herzog 1994, my translation).

The theme of collective shame repeats again and again. President Johannes Rau says—now in the 21st century—on November 30, 2003:

> I think it is important that we look carefully and distinguish between personal guilt and involvement of individuals and the adaptation and approval of a large majority in the "Third Reich." We must look carefully because there is no collective guilt on the part of all Germans, but there should be collective shame, as Theodor Heuss said. (Rau 2003, my translation)

Similarly, on May 8, 2005, President Horst Köhler (2005) also looks with "horror and shame" at the "the Holocaust, the brutal perversion of all civilized values perpetrated by Germans," at crimes that "brought dishonor upon our country."

With Köhler's successor, Christian Wulff, guilt and responsibility are separated from each other, but shame remains. In 2010, the latter says in Warsaw: "The unimaginable cruelties committed by Germans at that time fill us Germans with shame" (Wulff 2010, my translation). And the year after: "The name 'Auschwitz' stands like no other for the crimes committed by Germans against millions. These fill us Germans with repulsion and shame. They imbue us with a historical responsibility independent of individual guilt" (Wulff 2011, my translation).

When Joachim Gauck assumes the presidency several years later, the assessment of shame slowly changes. At first, he too uses the generational argument to banish the guilt to history, whereas shame is allowed to inhabit the present:

> Today there are most likely only a few individuals left in Germany who bear personal responsibility for the crimes of the National Socialist state. I myself was just five years old when the war ended. However, as the offspring of a generation which either committed or tolerated brutal crimes, and as the offspring of a state which robbed people of their humanity, I feel deep shame and deep sympathy for those who suffered at the hands of the Germans. (Gauck 2014)

Gauck justifies the postulate of responsibility with the historical guilt directly: "[F]or all post-war generations in Germany, yesterday's guilt gives rise to a special responsibility for today and tomorrow" (Gauck 2014).

At the same time, however, Gauck identifies shame as a negative affect. As he explains in 2015, the "remembrance of the genocide of the Jewish people and nascent shame about this simply took precedence over coming to terms with guilt for other crimes in [post-war] Germany" (Gauck 2015b). Already in 2013, he had pointed out in Estonia that guilt and shame both make cultural remembrance more difficult: "It is probably the hardest for the peoples to integrate moments of guilt and shame into the national memory" (Gauck 2013b, my translation). But Gauck did not understand this shame as an emotional response resulting from one's personal reflections on guilt; rather, he understood the shame as an artifact of being *made* to feel humiliated. In construing it thus, he picked up on an early theme of the collective guilt debate, according to which Germany was dishonored not by its own actions but by the victorious powers' disclosure of its crimes.

Meanwhile, Gauck attributes the origins of the collective guilt debate to a different time frame. For him, the generation of the Protests of '68 started the blame game: "A search for culprits, including national guilt, began, and one could say that from then until the end of the 1980s, the country experienced a phase of collective shame." The result was "a very deep split within the West German population" (Gauck 2013a). Gauck thus recognizes collective shame as a destructive emotion, one that dissolves the binding forces of a society, and he invokes the generational argument less with regard to guilt than to shame: "Today's young people often can more openly and fully face a past that is tainted with shame" (Gauck 2015a).

President Frank-Walter Steinmeier, who took office in 2017, however, seems to be returning to the rhetoric of Theodor Heuss. In his speech at the inauguration of a memorial site in Minsk in 2018, he states, "I am filled with shame and grief by the suffering that Germans have brought upon your country" (Steinmeier 2018a). A year later, he practically echoes the sentiment, in very similar wording, while commemorating the massacre in Fivizzano, Italy (Steinmeier 2019b). In his speech of September 1, 2019, though, he uses the rhetoric of shame while simultaneously—like Gauck—now acknowledging that shame might carry negative effects:

> Rest assured that there is not a single German who is not moved when they reflect on this trail of barbarism. This also holds true for those who reject these memories, who feel such strong humiliation that they seek escape through denial and aggression. What German could look at Wieluń, at Warsaw or Palmiry, or at Auschwitz and other places where the Shoah took place, without feeling shame? (Steinmeier 2019a)

Although Steinmeier does not want to subscribe to the collective guilt thesis either, he understands that it was fundamental in dealing with the guilty past: "It has always been controversial to speak of guilt wherever a collective is referred to. And yet this confession of guilt was a necessary step in order to lift the cloak of silence that shrouded these crimes" (Steinmeier 2018a).

Soon after, in fact, in his January 2020 speech at Yad Vashem, Steinmeier dispenses with any shameful vocabulary whatsoever; on the contrary, he now reemphasizes the historical *guilt* of the Germans: "Seventy-five years after the liberation of Auschwitz, [. . .] I stand here laden with the heavy, historical burden of guilt" (Steinmeier 2020). This remarkable exception is, at the same time, linked to the diagnosis that the much-vaunted culture of remembrance was ultimately unable to prevent the resurgence of anti-Semitic attitudes and actions in Germany, as Steinmeier himself admits. The shame-and-disgrace rhetoric led the Germans to believe that they were marked by a collective stigma that could never be erased (Giesen 2004, 135). Such a stigma instilled a sense of futility; Germany could not get rid of the stigma, no matter how fully it came to terms with its guilt or how actively it worked at making amends for the suffering it inflicted—so why bother trying?

My thesis is that Gauck and Steinmeier sensed that guilt is by no means the sentiment that fuels new aggressions or divides a society. Perhaps it is, instead, precisely the coupling of the culture of remembrance with a rhetoric of shame that has not only led to an aversion to acts of remembrance politics (as evident in many statements by the populist right-wing movement), but has also turned into aggression toward the victims themselves—who, in a shortened logic, now became the cause of the shame.

Moral Shame and Productive Guilt

The shame-driven version of Germany's memory of its World War II crimes has consequences for the country's relationship to the victimized groups. What is the nature of such relationships burdened by shame? John Rawls—who defines shame "as the feeling that someone has when he experiences an injury to his self-respect" (Rawls 1972, 442)—distinguishes between what he calls natural shame and moral shame. We feel natural shame when we become aware that we lack a certain quality or ability that we would reasonably consider desirable in ourselves and others; natural shame is thus perceived as a lack intrinsic to our personality, observed either through introspection or through actions that reflect said lack. The decisive factor in whether we perceive or feel this shame, however, is what we expect from ourselves and others. (If you have no wish to become a figure skater, for instance, you do not feel ashamed that you cannot do a pirouette.)

For Rawls, shame is mostly felt when the deficiencies of our personality mean that we cannot build and maintain relationships that seem important to us. Thus, moral shame is to be considered a social emotion.[6] Those who desire certain qualities and virtues in themselves and in others and who, on condition of their own esteem, assume the reciprocity of the moral demands of their relationship, will feel shame if their actions lack these qualities or virtues. The feeling of moral shame, as a result of this self-evaluation, translates into self-abasement and the acceptance of low esteem by one's fellow human beings.

This sentiment simultaneously implies a similarity to and a distinction from guilt: both can be socially regulative emotions, both can be felt in response to the same action (Rawls 1972, 482–3), but each has different implications. According to Maria-Sibylla Lotter, the sense of shame "refers to an egocentric sense of values, a sense of self-worth, in contrast to feelings of guilt, which could be described as allocentric because they are directed at injuries to other people and to norms" (Lotter 2012, 105, my translation). In other words, feelings of guilt primarily address relationships directly, due to the violation of one party's legitimate interests. Therefore, guilt motivates people to compensate for such violations, in order to be able to rebuild the disturbed relationship. Shame, by contrast, impacts relationships by generating "anxiety about the lesser respect that others may have for us and [. . .] our disappointment [over] failing to live up to our ideals" (Rawls 1972, 446). Shame is the feeling of a damaged social identity and of its disturbed relationships; one individual can hardly restore these relationships singlehandedly because shame's obloquy and disgrace are directed at the *identity* of the person, and not at specific *actions*, as in guilt.[7] Shame is directed to the concept of good and value; guilt, to that of right. For shame, this means questioning one's very self as good; guilt questions the legitimacy of a certain behavior (Rawls 1972, 484). In shame, ultimately, "(i)t is the self that is at fault, not the commission of the act" (Stearns 2017, 3).

In theories of shame, therefore, the gaze of others and the experience of having one's failings discovered have repeatedly been thematized as shame's central moment. The result of the violation of self-respect is not limited to a self-directed sense of shame, but can also be construed (in an attempt to defend one's self-concept) as an external violation *by* another—a sense of having *been* shamed, from the outside. Hence, even if one recognizes that one's own deficiencies have indeed brought shame, this depressing realization can easily be twisted into the projection that the cause of the feeling of self-abasement has largely been due to the fact that *others* have judged that deficiency. In shame, therefore, hatred might be directed precisely against the innocent victim, because the victim's role forces us to recognize our own failings—and we experience shame at seeing this aspect of ourselves. "Shame [...] can generate counterproductive anger or aggression [and] produces a desire to lash out against unfair emotional pain" (Stearns 2017, 5).

Another strategy is to deny or trivialize those actions that violate one's self-esteem in order to avoid social exclusion. "The shamed person tends to shrink, characteristically seeks to hide [...]. Often, efforts go into blaming someone or something else for the problem involved, or denying or forgetting it" (Stearns 2017, 4). With regard to the German discourse on collective guilt, these strategies were and are more blatantly obvious than any sort of lasting change to one's own personal identity. "He [the Nazi] picked the Jew as scapegoat for everything. [. . .] He knows that a vast majority of the German people still yearn to free themselves from any responsibility and to shift the burden of blame away from themselves!" (Kerckhoff 2020, 72). At the same time, the effort to build up "a renewed confidence in the excellence of one's person" (Rawls 1974, 484)—the only positive way to lift the sense of shame—is part of the political DNA of the Federal Republic of Germany. Any repeated shaking of this trust then generates new shame.

There is an important distinction, however, between expressing and acknowledging guilt or guilt-driven shame in public speech acts. In contrast to Maria-Sibylla Lotter in this volume, I do argue that in the case of political apologies and guilt confessions, the public does not expect the representative to *feel* the guilt in actuality. What is expected, though, is that the confession bears some sort of *consequences* in the social sphere: morally, to change behavior in the future; legally, to prosecute perpetrators; and/or economically, to make amends financially for the damage and harm that were caused. The sincerity of a political apology, therefore, is not primarily measured by the presumed emotions of the representative offering it. It is assessed by concrete political, social, and financial outcomes.

To express vicarious shame for a wrongdoing, on the contrary, signals something very different. It might also admit guilt, but it does not include an *offer* to the addressed public. What is more, it kicks the ball back into the victims' court—in a way that burdens rather than empowers. It says: "Behold, I am drowned in shame. My feelings are sincere; I can do nothing about the past but feel shame and regret. Please, now it is up to *you* to restore our relationship by accepting me back." Thus, expressing shame in public is a very different sort of apologetic speech act. If we do not believe in the sincerity of the *feelings* a person has expressed, we are not inclined to restore the relationship. This is why the effect of such a speech act depends heavily upon its rhetorical and performative power. My argument in this chapter is that Germany, as a political entity,[8] up to the present day, is afraid to bear the full, particularly financial, consequences of its guilt, and thus prefers playing the shame card rather than the guilt card.

Missed Chances, Shame, and the New Right

After 1945, shame and guilt are two competing interpretive schemes for the emotional treatment of the collective dimension of German crimes. The collective guilt campaign in Germany brought about not only the negation of legal guilt, which the German populace feared would result in retaliation against itself, but also the recoding of guilt into shame, as the populace came to imagine itself as a humiliated pariah people. People who are ashamed fear "derision and contempt" (Rawls 1972, 483).

In reflecting on the German spirit, Thomas Mann spoke of the "disgrace of a disgraceful philosophy" (Mann 1945, 18). Hence, perception of the German character in total was drenched in shame, a shame that not only belongs to the National Socialist era but extends, according to Mann, to the intellectual history of Germany from the early 19th century on. Lost, therefore, was the chance to come to terms with the collective character of the Germans' guilt (a guilt that undoubtedly existed), and to work on their own initiative toward restoring relations—efforts that needed to be perceived as uncoerced. For the collective guilt campaign to work, it would have required an understanding of guilt by the Germans that was not reduced solely to legal concerns.

Karl Jaspers's (1945) entrance into the collective guilt debate, with a newspaper article in October 1945 in the *Neue Zeitung*, and in a series of lectures at the University of Heidelberg in the winter semester of 1945–1946 that were later published as a monograph titled *Die Schuldfrage* [1946] (2012) and translated as *The Question of German Guilt* [1946] (2001), must be understood as an attempt to bring about a differentiated concept of guilt for the Germans. The question that Jaspers, like the Allied victors, asked himself was how the Germans as a collective could ever again join the ranks of civilized nations. For him, the collective guilt campaign is first and foremost the instrument with which the Germans should become aware of their current status as a pariah people whom the world meets "with horror, with hatred, and scorn" (Jaspers [1946] 2001, 21). Jaspers opposed including a collective feeling of shame in the new national identity. He pleaded for making a collective feeling of *guilt* the basis of spiritual renewal instead (as echoed in Kerckhoff's 1948 *Berliner Briefe*). In a certain sense, therefore, Karl Jaspers fell in line with the original intention of the U.S. propaganda.

At the same time, he revealed through his well-known differentiation of the multiple dimensions of guilt that the coupling of guilt, shame, and disgrace in the Allies' campaign threatened to hinder the necessary renewal of the German people. It is certainly true that Jaspers operated (in large parts of his text) on the basis of the tripartite Christian concept of guilt, with *contritio*, *confessio*, and *renovatio/satisfactio*; after all, he addressed a predominantly Christian audience. Such Christian semantics, however, must not lead one to characterize Jaspers' endeavor as an act of Christian penance for human weakness; he firmly opposed attempts to classify German guilt in a "guilt-atonement relationship" or to understand it through the figure of original sin as a historical phenomenon of universal human guilt (Jaspers [1946] 2001, 95).

To Jaspers, it was rather a matter of *feeling* "a co-responsibility for the acts of members of our families," although this form of compassion for guilt "cannot be objectivized" (73) by being translated into the language of the law. As members of a community whose identity we accept as our own, according to Jaspers, we can feel complicit "in the links of tradition" (73) and, through "our feeling of collective guilt [. . .] we [can] (renew) human existence from its origin" (75). Note that, in these sections on collective guilt, Jaspers deliberately chooses the active verbs "to feel" or "to sympathize" several times and insistently. For him, feeling, not reason, represents the link between the collective and the individual—a relationship that was profoundly corrupted under National Socialist rule, where the individual had to merge into the collective and had obviously forgotten how to feel empathy, let alone shame, with respect to the victims. The individual feeling of collective guilt thus becomes an exercise in humanity for the Germans. "Where guilt is not felt, all distress is immediately leveled on the same plane" (113).

Jaspers also rejected any vociferous calls for confessions of guilt, as, in his view, such confessions are themselves essentially aggressive: "[A] confession of guilt wants to force others to confess" (101) and belies a "baneful tendency to feel that confessing guilt makes us better than others" (102). Guilt-defense reactions can only be prevented when a confession of guilt is sincere: "Without guilt consciousness we keep reacting to every attack with a counterattack. [. . .] Where consciousness of guilt has been appropriated, we bear false and unjust accusations with tranquility. For pride and defiance are molten" (115).

The concepts of "consciousness of guilt" and "feeling of guilt" must, in no case, be misunderstood psychoanalytically as pathological, irrational sufferings whose origin has shifted into the subconscious; according to Rawls, these are not actual feelings of guilt at all, but rather are only *similar* to the state of mind that such feelings tend to generate. In one way or another, guilt as a *moral* feeling includes the desire to seek forgiveness (Rawls 1972, 481). In the case of neurotic guilt, by contrast, while one might also feel the pressure to ask for forgiveness, such desire has no counterpart in reality; there is no true guilt to be forgiven. Therefore, it is dangerous to speak of German guilt as if it were some sort of neurotic habit, as some intellectuals do. Doing so, one risks—or even participates in—the possibility of denying the historical guilt altogether.

All said, for Jaspers (113–5) and Kerckhoff (2020), it is only out of a fundamentally true feeling of guilt, and the resulting awareness of such, that guilt can become productive. True feelings of guilt prevent one from denying, trivializing, postponing amends for, or defending against the action that has caused it. When Jaspers, in his 1962 epilogue to the German version, states that the actual intention of his book is "to show the way back to dignity in accepting the guilt ever clearly recognized in its nature" (Jaspers [1946] 2012, 94), it becomes evident how much the book is meant as a counter-argument to the rhetoric of shame, disgrace, and honor.

As famous as the book is, it must nevertheless be conceded that in the representative commemoration of the Federal Republic, it was not guilt but primarily shame that was offered to the Germans as a collective feeling—with perhaps one major exception, articulated in gesture and not in word; namely, Willy Brandt's genuflection in Warsaw, which enabled a new "narrative of national guilt" (Giesen 2004, 131) precisely *because* he was personally innocent (cf. Maria-Sibylla Lotter's chapter). "Brandt's Warsaw Genuflection separated the individual guilt of the ritual actor from the collective guilt of the German nation" (Giesen 2004, 132). Brandt certainly created the prerequisites for remembrance of the victims, for naming the perpetrators, and for reconciliation in this deputy confession of guilt, especially in its perception abroad (132–3).

Nevertheless, the move to reinterpret guilt as shame has clearly remained an attractive means for Germany to free itself from the burden of the past. Evidence for this thesis is documented in the popular reaction to Martin Walser's Peace Prize speech in 1998—a reaction that is still circulating in right-wing populist and some right-wing conservative discourses. Walser deliberately did not speak of German guilt, but instead of the Shoah as a "historical burden, the imperishable shame, [and] not a day [goes by] on which it is not held up to us" (Walser 2008, 18). It is not the *reason* for the disgrace that is unbearable for Walser, but rather the disgrace itself, and the way that this disgrace manifests in the collective accusation from which one would rather hide in shame:

> Easily twenty times I have averted my eyes from the worst filmed sequences of concentration camps. No serious person denies Auschwitz; no person who is still of sound mind quibbles about the horror of Auschwitz; but when this past is held up to me every day in the media, I notice that something in me rebels against this unceasing presentation of our disgrace. Instead of being grateful for this never-ending presentation of our disgrace, I begin to look away. (Walser 2008, 89–90)

Accordingly, Walser calls *the* central Federal Republican act of remembrance of the crime of the Shoah—Berlin's Memorial to the Murdered Jews of Europe—a "monumentalization of disgrace" (91). In this case, the remembrance is not about an externally induced stigma, but rather the *phantasm* of such a stigma, whose emotional consequences of shame and disgrace one would like to escape once and for all.

One could say that by conceptualizing the Shoah as a German disgrace, Martin Walser not only takes the imperative of the Federal Republic of Germany's memory policy to its extremes, but even turns it on its head, by conflating remembrance with the nurturing of a stigma whose emotional consequences (of shame and disgrace) one would rather escape once and for all. Walser's sentiments are echoed by the influential journalist Rudolf Augstein shortly afterward in *Der Spiegel* on November 30, 1998. And even today (almost identically), the right-wing politician Björn Höcke frequently refers to the memorial as the "monument of disgrace." Such rhetoric shows its frightening impact in the negative affect that emerges not only against the implementation of memory policy, but also against the victims and their descendants themselves—when Augstein, for instance, writes in response to anti-Semitic ciphers and stereotypes:

> Now, in the middle of the regained capital Berlin, a memorial is to commemorate our continuing disgrace. [. . .] [T]hey will not dare, in deference to the

> New York press and the sharks in lawyers' clothing, to keep the center of Berlin free of such a monstrosity. (Augstein 1998, 32, my translation)

The official Federal Republican culture of remembrance has by no means reached a "post-guilt" stage yet, as Jaspers and Brandt sought to induce. Instead, it has settled—almost defiantly—into shame. In such a context, Kerckhoff's warning from 1948 seems uncannily relevant:

> Well, if the Germans had changed, if they sighed under the burden of their guilt, it would have been a wrong not to help them rise again. [. . .] I have not met a single guilty Nazi, not one, not a single one! Either they were not Nazis at all or they are [. . .] proud of having been one [. . .]. When a fascist slaps you on the right cheek, do not turn to him your left one! He does not walk away shattered in shame. He hits you: knockout!" (Kerckhoff 2020, 75)

Notes

1. Please note that from 1949 to 1991, I am only referring to West Germany. The case of memory politics in the GDR is very different from that in the Federal Republic of Germany.
2. This was clearly stated in the explanations to the Potsdam Agreement of December 11, 1945. Cf. Warburg 1947, 322–4.
3. Berger (2012, 45–8) also (and erroneously) understands the Allies' collective guilt campaign as a legal campaign, primarily in view of the famous questionnaires that were designed to ascertain involvement with the regime and of the local verdict chambers set up by the Allies, which pronounced sentences for minor offenses. Proceedings were opened for about one million cases, but there were only 23,600 convictions. In the German public's perception, this may have been a collective accusation. However, if one looks even just at the number of members of organizations such as the Gestapo, the SS, the Wehrmacht, police, Nazi Party, and other governmental organizations that were directly involved in crimes, then the number of proceedings opened can be estimated as, in fact, small.
4. It is interesting that Kogon's harsh critique of the Allies' collective guilt policy has not been translated in the English edition (cf. Kogon 1998).
5. Heidrun Kämper (2005, 280) calls the debate on collective guilt a prime example of the linguistic construction of a reality through which argumentation communities have come together.
6. It becomes clear why the transfer of these feelings to cultures is problematic when a hierarchization of social "leading emotions" is associated with it. Stephen Pattison (2000, 131–49), for example, argues that emotions vary in their social dimension and thus also in individual experience, both historically and culturally. Cf. also the contributions in Baucks and Meyer (2011).
7. See also Morris (1976, 61).

8. I am emphasizing the political notion here. There is no doubt that, at least since the 1990s, on the individual, cultural, and social levels, Germans have dealt differently with their individual and collective guilt.

References

Adenauer, Konrad. 2020. "Ansprache des Oberbürgermeisters Adenauer vor der von der britischen Militärregierung ernannten Kölner Stadtverordneten-Versammlung." In *Verhandlungen der Stadtverordneten-Versammlung zu Köln vom Jahre 1945. Köln (o.J.)*. 1. Sitzung vom 1. Oktober 1945, 5–7. Accessed February 12, 2020. https://www.konrad-adenauer.de/quellen/reden/1945-10-01-rede.

Adorno, Theodor W. 1997. "Schuld und Abwehr." In *Soziologische Schriften II*, edited by Rolf Tiedemann, 21–326.Gesammelte Schriften 9. Frankfurt am Main: Suhrkamp.

Assmann, Aleida. 1999. "Ein deutsches Trauma? Die Kollektivschuldthese zwischen Erinnern und Vergessen." *Merkur* 53: 1142–54.

Assmann, Aleida, and Frevert, Ute. 1999. *Geschichtsvergessenheit—Geschichtsversessenheit: Vom Umgang mit deutschen Vergangenheiten nach 1945*. Stuttgart: Deutsche Verlags-Anstalt.

Augstein, Rudolf. 1998. "Wir alle sind verletzbar." *Der Spiegel* 49: 32–33.

Bauks Michaela, and Martin F. Meyer, eds. 2011. *Zur Kulturgeschichte der Scham*. Archiv für Begriffsgeschichte 9. Hamburg: Felix Meiner.

Bergengruen, Werner. 1946. *Dies Irae: Eine Dichtung*. Munich: Zinnen.

Berger, Thomas U. 2012. *War, Guilt, and World Politics after World War II*. New York: Cambridge University Press.

Brink, Cornelia. 1998. *Ikonen der Vernichtung: Zum öffentlichen Gebrauch von Fotografien aus nationalsozialistischen Konzentrationslagern nach 1945*. Berlin: Akademie.

"Directive No. 1 for Propaganda Policy of Overt Allied Information Services, SHAEF/PWD, May 22, 1945." Printed in Hurwitz, Harold Judah. 1954. *Military Government and the German Press: An Experiment in Cross-Cultural Institutional Change*, 37. MA thesis, Columbia University.

Dubiel, Helmut. 1999. *Niemand ist frei von der Geschichte: Die nationalsozialistische Herrschaft in den Debatten des Deutschen Bundestages*. Munich: Hanser.

Frei, Norbert. 1997. "Von deutscher Erfindungskraft oder: Die Kollektivschuldthese in der Nachkriegszeit." *Rechtshistorisches Journal* 16: 621–34.

Gauck, Joachim. 2013a. "'Erinnerung: Fundament für die Zukunft'—Diskussionsveranstaltung im Museum der Okkupationen." Tallinn, Estonia, July 9. Accessed September 1, 2020. http://www.bundespraesident.de/SharedDocs/Reden/DE/Joachim-Gauck/Reden/2013/07/130709-Estland-Diskussion-Museum.html.

Gauck, Joachim. 2013b. "'Erinnerungskultur und Versöhnung in Deutschland'—Rede an der Universidad de los Andes." Bogotá, Colombia, May 10. Accessed September 1, 2020. http://www.bundespraesident.de/SharedDocs/Reden/DE/Joachim-Gauck/Reden/2013/05/130510-Kolumbien-Grundsatzrede.html.

Gauck, Joachim. 2014. "Commemoration of the Second World War." Gdansk, Poland, September 1. Accessed September 1, 2020. http://www.bundespraesident.de/SharedDocs/Reden/EN/JoachimGauck/Reden/2014/140901-Poland-Second-World-War.html.

Gauck, Joachim. 2015b. "Commemoration of the End of the Second World War." Schloß Holte-Stukenbrock, May 6. Accessed September 1, 2020. http://www.bundespraesident.de/SharedDocs/Reden/EN/JoachimGauck/Reden/2015/150506-Holte-Stukenbrock.html.

Gauck, Joachim. 2015b. "Day of Remembrance of the Victims of National Socialism." Berlin, January 27. Accessed September 1, 2020. http://www.bundespraesident.de/SharedDocs/Reden/EN/JoachimGauck/Reden/2015/150127-Gedenken-Holocaust.html.

Giesen, Bernhard. 2004. "The Trauma of Perpetrators: The Holocaust as the Traumatic Reference of German National Identity." In *Cultural Trauma and Collective Identity*, edited by Jeffery C. Alexander et al., 112–54. Berkeley: University of California Press.

Goldhagen, Daniel Jonah. 1996. *Hitler's Willing Executioners: Ordinary Germans and the Holocaust*. New York: Alfred Knopf.

Grunenberg, Antonia. 2001. *Die Lust an der Schuld: Von der Macht der Vergangenheit über die Gegenwart*. Berlin: Rowohlt.

Hartenian, Larry. 1987. "The Role of Media in Democratizing Germany: United States Occupation Policy 1945–1949." *Central European History* 20: 145–90.

Herzog, Roman. 1994. "Ansprache von Bundespräsident Roman Herzog anlässlich des Gedenkens an den 50. Jahrestag des Warschauer Aufstandes in Warschau." Warsaw, Poland, August 1. Accessed September 1, 2020. http://www.bundespraesident.de/SharedDocs/Reden/DE/Roman-Herzog/Reden/1994/08/19940801_Rede.html.

Herzog, Roman. 1995. "Ansprache von Bundespräsident Roman Herzog beim Staatsakt aus Anlaß des 50. Jahrestages des Endes des Zweiten Weltkrieges." Berlin, May 8. Accessed September 1, 2020. http://www.bundespraesident.de/SharedDocs/Reden/DE/Roman-Herzog/Reden/1995/05/19950508_Rede.html.

Herzog, Roman. 1998a. "Rede von Bundespräsident Roman Herzog bei der Gedenkveranstaltung aus Anlaß des 60. Jahrestages der Synagogenzerstörung am 9./10. November ("Reichspogromnacht") in Berlin." Berlin, November 9. Accessed September 1, 2020. http://www.bundespraesident.de/SharedDocs/Reden/DE/Roman-Herzog/Reden/1998/11/19981109_Rede.html.

Herzog, Roman. 1998b. "Speech at Bergen-Belsen on Yom Ha Shoah, April 27, 1995." In *Lessons from the Past Visions for the Future*, 5–9. German Issues 18. Baltimore: American Institute for Contemporary German Studies, The Johns Hopkins University. https://www.aicgs.org/site/wp-content/uploads/1998/06/Roman-Herzog-Speeches-German-Issues-18.pdf.

Herzog, Roman. 1998c. "Speech to the Bundestag in Support of a Day of Commemoration for the Victims of National Socialism, January 19, 1996." In *Lessons from the Past Visions for the Future*, 11–18.German Issues18, Baltimore: American Institute for Contemporary German Studies, The Johns Hopkins University. https://www.aicgs.org/site/wp-content/uploads/1998/06/Roman-Herzog-Speeches-German-Issues-18.pdf.

Herzog, Roman. 1999. "Rede von Bundespräsident Roman Herzog: "Die Zukunft der Erinnerung."" Bundestag, Berlin, January 27. Accessed September 1, 2020. http://www.bundespraesident.de/SharedDocs/Reden/DE/Roman-Herzog/Reden/1999/01/19990127_Rede.html;jsessionid=9308FB367AC915B734B6FEDB524404EF.2_cid362.

Heuss, Theodor. 1965. "Das Mahnmal." In *Die grossen Reden: Der Staatsmann*, 224–30. Tübingen: Wunderlich. https://www.zeit.de/reden/die_historische_rede/heuss_holocaust_200201.

Heuss, Theodor. 1967a. "In memoria." In *Die grossen Reden*, 51–56. Munich: Deutscher Taschenbuch Verlag.

Heuss, Theodor. 1967b. "Mut zur Liebe." In *Die grossen Reden*, 113–118. Munich: Deutscher Taschenbuch Verlag.

Hurwitz, Harold Judah. 1954. *Military Government and the German Press: An Experiment in Cross-Cultural Institutional Change*. MA thesis, Columbia University.

Janowitz, Morris. 1946. "German Reactions to Nazi Atrocities." *American Journal of Sociology* 52, no. 2: 141–46.
Jaspers, Karl. 1945. "Antwort an Sigrid Undset." *Die Neue Zeitung*, October 28.
Jaspers, Karl. (1946) 2001. *The Question of German Guilt*. Translated by E. B. Ashorn. New York: Fordham University Press.
Jaspers, Karl. (1946) 2012. *Die Schuldfrage: Von der politischen Haftung Deutschlands*. Munich: Piper.
JCS (Joint Chiefs of Staff). 1945. *JCS 1067: Directive to Commander-in-Chief of United States Forces of Occupation Regarding the Military Government of Germany*. Printed in Warburg, James Paul, ed. 1947. *Germany—Bridge or Battleground*, 279–302. New York: Harcourt, Brace. https://en.wikisource.org/wiki/JCS_1067.
Kämper, Heidrun. 2005. *Der Schulddiskurs in der frühen Nachkriegszeit: Ein Beitrag zur Geschichte des sprachlichen Umbruchs nach 1945*. Berlin: Walter de Gruyter.
Kämper, Heidrun. 2015. "Der Schulddiskurs als Generationenphänomen." In *Sprache der Generationen*, edited by Eva Neuland, 185–206. Frankfurt am Main: Peter Lang.
Kerckhoff, Susanne. (1948) 2020. *Berliner Briefe: Ein Briefroman*, edited by Peter Graf. Berlin: Das kulturelle Gedächtnis.
Kogon, Eugen. 1946. *Der SS-Staat: Das System deutscher Konzentrationslager*. Munich: Karl Alber.
Kogon, Eugen. (1950) 1998. *The Theory and Practice of Hell: The Classic Account of the Nazi Concentration Camps Used as a Basis for the Nuremberg Investigations*. Translated from the German by Heinz Norden. New York: Berkley Trade.
Köhler, Horst. 2005. "Speech by Federal President Horst Köhler at the Memorial Ceremony in the Plenary Chamber of the German Bundestag Marking the 60th Anniversary of the End of the Second World War in Europe." Berlin, May 8. Accessed September 1, 2020. http://www.bundespraesident.de/SharedDocs/Reden/EN/HorstKoehler/Reden/2005/05/20050508_Rede.html.
Lotter, Maria-Sibylla. 2012. *Scham, Schuld, Verantwortung: Über die kulturellen Grundlagen der Moral*. Frankfurt am Main: Suhrkamp.
Lübbe, Hermann. 1997. "Kollektivschuld: Funktionen eines moralischen und juridischen Unbegriffs." *Rechtshistorisches Journal* 16: 687–95.
Mann, Thomas. 1945. "The End." *The Free World* 9, no. 3: 15–18.
Mann, Thomas. (1945) 1990. "Die Lager." In *Reden und Aufsätze*, edited by Hans Bürgin and Peter de Mendelssohn, 951–53. Gesammelte Werke 12. Frankfurt am Main: Fischer.
Matz, Elisabeth. 1969. *Die Zeitungen der US-Armee für die deutsche Bevölkerung (1944–1946)*. Münster: C. J. Fahle.
Morris, Herbert. 1976. *On Guilt and Innocence: Essays in Legal Philosophy and Moral Psychology*. Berkeley: University of California Press.
Moses, A. Dirk. 2007a. *German Intellectuals and the Nazi Past*. Cambridge: Cambridge University Press.
Moses, A. Dirk. 2007b. "Stigma and Sacrifice in the Federal Republic of Germany." *History and Memory* 19, no. 2: 139–80.
Norman, Albert. 1951. *Our German Policy: Propaganda and Culture*. New York: Vintage Press.
Pattison, Stephen. 2000. *Shame: Theory, Therapy, Theology*. Cambridge: Cambridge University Press.
Psychological Warfare Division. 1945. *The Psychological Warfare Division Supreme Headquarters Allied Expeditionary Force: An Account of its Operations in the Western European Campaign, 1944–1945*. Bad Homburg: Psychological Warfare Division.

Rau, Johannes. 2003. "Gedächtnisvorlesung von Bundespräsident Johannes Rau aus Anlass des sechzigsten Jahrestags der Hinrichtung der Mitglieder der 'Weißen Rose.'" Ludwig-Maximilians-Universität Munich, January 30. http://www.bundespraesident.de/SharedDocs/Reden/DE/Johannes-Rau/Reden/2003/01/20030130_Rede.html.

Rawls, John. 1972. *A Theory of Justice*. Oxford: Oxford University Press.

Rees, Jonas, Andreas Zick, Michael Papendick, and Franziska Wäschle. 2019. "MEMO Multidimensionaler Erinnerungsmonitor STUDIE II/2019." Forschungsbericht IKG. https://pub.uni-bielefeld.de/record/2934984.

Salzborn, Samuel. 2020. *Kollektive Unschuld: Die Abwehr der Shoah im deutschen Erinnern*. Berlin: Hentrich und Hentrich.

Stearns, Peter N. 2017. *Shame: A Brief History*. Urbana: University of Illinois Press.

Steinmeier, Frank-Walter. 2018a. "Federal President Frank-Walter Steinmeier at the Ceremony Marking the 60th Anniversary of Action Reconciliation—Service for Peace." Berlin, May 27. Accessed September 11, 2020. https://www.bundespraesident.de/SharedDocs/Downloads/DE/Reden/2018/05/180527-Aktion-Suehnezeichen-Englisch.pdf?__blob=publicationFile.

Steinmeier, Frank-Walter. 2018b. "Opening of the Maly Trostenets Memorial Site." Minsk, Republic of Belarus, June 29. Accessed September 1, 2020. http://www.bundespraesident.de/SharedDocs/Reden/EN/Frank-Walter-Steinmeier/Reden/2018/06/180629-Maly-Trostenets-Memorial-Belarus.html.

Steinmeier, Frank-Walter. 2019a. "Commemoration of Second World War at Warsaw, September 1, 2019." Warsaw, Poland, September 1. Accessed September 1, 2020. https://www.bundespraesident.de/SharedDocs/Reden/EN/Frank-Walter-Steinmeier/Reden/2019/09/190901-Poland-Commemoration-Warsaw.html.

Steinmeier, Frank-Walter. 2019b. "75th Anniversary of the Massacres in Fivizzano." Fivizzano, Italy, August 25. Accessed September 1, 2020. http://www.bundespraesident.de/SharedDocs/Reden/EN/Frank-Walter-Steinmeier/Reden/2019/08/190825-Fivizzano.html.

Steinmeier, Frank-Walter. 2020. "Fifth World Holocaust Forum at Yad Vashem." Jerusalem, January 23. Accessed September 1, 2020. http://www.bundespraesident.de/SharedDocs/Reden/EN/Frank-Walter-Steinmeier/Reden/2020/01/200123-World-Holocaust-Forum-Yad-Vashem.html.

Walser, Martin. 2008. "Experiences While Composing a Sunday Speech (1998)." In *The Burden of the Past: Martin Walser on Modern German Identity: Texts, Contexts, Commentary*, edited by Thomas A. Kovach and Martin Walser, 85–94. Rochester, NY: Camden House.

Warburg, James Paul, ed. 1947. *Germany—Bridge or Battleground*. New York: Harcourt, Brace.

Weizsäcker, Richard von. 1985. "Speech by Federal President Richard von Weizsäcker during the Ceremony Commemorating the 40th Anniversary of the End of War in Europe and of National-Socialist Tyranny." Bundestag, Bonn, May 8. Accessed September 1, 2020. http://www.bundespraesident.de/SharedDocs/Downloads/DE/Reden/2015/02/150202-RvW-Rede-8-Mai-1985-englisch.pdf?__blob=publicationFile.

Wolbrink, Barbara. 2009. "Nationales Stigma und persönliche Schuld: Die Debatte über Kollektivschuld in der Nachkriegszeit." *Historische Zeitschrift* 289: 325–64.

Wulff, Christian. 2010. "Bundespräsident Christian Wulff auf der Konferenz 'Europa—Kontinent der Versöhnung?'" Warsaw, Poland, December 7. Accessed September 1, 2020 http://www.bundespraesident.de/SharedDocs/Reden/DE/Christian-Wulff/Reden/2010/12/20101207_Rede.html.

Wulff, Christian. 2011. "Bundespräsident Christian Wulff bei der offiziellen Gedenkveranstaltung anlässlich des 66. Jahrestags der Befreiung des Konzentrationslagers Auschwitz." Auschwitz-Birkenau, Poland, January 27. Accessed September 1, 2020. http://www.bundespraesident.de/SharedDocs/Reden/DE/Christian-Wulff/Reden/2011/01/20110127_Rede.html.

Index

For the benefit of digital users, indexed terms that span two pages (e.g., 52–53) may, on occasion, appear on only one of those pages.

The manufacturer's authorised representative in the EU for product safety is Oxford University Press España S.A. of El Parque Empresarial San Fernando de Henares, Avenida de Castilla, 2 – 28830 Madrid (www.oup.es/en or product.safety@oup.com). OUP España S.A. also acts as importer into Spain of products made by the manufacturer.

Printed in the USA/Agawam, MA
May 21, 2026

905357.019